Lecture Notes in Bioinformatics 7632

Edited by S. Istrail, P. Pevzner, and M. Waterman

Subseries of Lecture Notes in Computer Science

T0280332

Tetsuo Shibuya   Hisashi Kashima
Jun Sese   Shandar Ahmad (Eds.)

Pattern Recognition in Bioinformatics

7th IAPR International Conference, PRIB 2012
Tokyo, Japan, November 8-10, 2012
Proceedings

 Springer

Series Editors

Sorin Istrail, Brown University, Providence, RI, USA
Pavel Pevzner, University of California, San Diego, CA, USA
Michael Waterman, University of Southern California, Los Angeles, CA, USA

Volume Editors

Tetsuo Shibuya
The University of Tokyo, Institute of Medical Science
4-6-1 Shirokanedai, Minato-ku, 108-8639 Tokyo, Japan
E-mail: tshibuya@hgc.jp

Hisashi Kashima
The University of Tokyo, Department of Mathematical Informatics
7-3-1 Hongo, Bunkyo-ku, 113-8654 Tokyo, Japan
E-mail: kashima@mist.i.u-tokyo.ac.jp

Jun Sese
Tokyo Institute of Technology, Department of Computer Science
2-12-1 Ookayamama, Meguro-ku, 152-8550 Tokyo, Japan
E-mail: sesejun@cs.titech.ac.jp

Shandar Ahmad
National Institute of Biomedical Innovation
7-6-8 Saito-Asagi, Suita, 567-0085 Osaka, Japan
E-mail: shandar@nibio.go.jp

ISSN 0302-9743   e-ISSN 1611-3349
ISBN 978-3-642-34122-9   e-ISBN 978-3-642-34123-6
DOI 10.1007/978-3-642-34123-6
Springer Heidelberg Dordrecht London New York

Library of Congress Control Number: 2012948870

CR Subject Classification (1998): J.3, I.5, F.2.2, I.2, I.4, H.3.3, H.2.8

LNCS Sublibrary: SL 8 – Bioinformatics

Typesetting: Camera-ready by author, data conversion by Scientific Publishing Services, Chennai, India

Printed on acid-free paper

Springer is part of Springer Science+Business Media (www.springer.com)

Preface

The Pattern Recognition in Bioinformatics (PRIB) meeting was established in 2006 under the auspices of the International Association for Pattern Recognition (IAPR) to create a focus for the development and application of pattern recognition techniques in the biological domain. Since its establishment, PRIB has brought together top researchers, practitioners, and students from around the world to discuss the applications of pattern recognition methods in the field of bioinformatics so as to solve problems in the life sciences.

The seventh PRIB conference was held in Tokyo, Japan, during November 8–10, 2012. This year we received 33 high-quality submissions from various countries around the world, and 24 of them were finally included in these conference proceedings. Their topics range widely from fundamental techniques, sequence analysis to biological network analysis. We were fortunate to have three leading researchers for invited speakers; Hwanjo Yu from Postech, Kwong-Sak Leung from The Chinese University of Hong Kong, and Takayuki Aoki from Tokyo Institute of Technology.

We would like to thank all authors for submitting the high-quality papers, the reviewers for their efforts to keeping the high quality of this conference, and the sponsors for generously providing financial support. Finally, we are grateful to Springer for their professional support in preparing these proceedings and for the continued support of PRIB.

November 2012

Tetsuo Shibuya
Hisashi Kashima
Jun Sese
Shandar Ahmad

Organization

Organizing Committee

General Chairs

Jun Sese	Tokyo Institute of Technology
Shandar Ahmad	National Institute of Biomedical Innovation

Program Chairs

Tetsuo Shibuya	The University of Tokyo
Hisashi Kashima	The University of Tokyo

Publicity Chairs

Tsuyoshi Kato	Gunma University
Koji Tsuda	National Institute of Advanced Industrial Science and Technology

Invited Speaker Chairs

Masakazu Sekijima	Tokyo Institute of Technology
Tsuyoshi Hachiya	Keio University

Program Committee

Jesus S. Aguilar-Ruiz
Tatsuya Akutsu
Kiyoko Aoki-Kinoshita
Sebastian Böcker
Jaume Bacardit
Hideo Bannai
Rainer Breitling
Frederic Cazals
Florence D'Alché-Buc
Dick De Ridder
Tjeerd Dijkstra
Federico Divina
Richard Edwards
Maurizio Filippone
Rosalba Giugno
Jin-Kao Hao
Morihiro Hayashida
Tom Heskes
Seiya Imoto

Zhenyu Jia
Yoshimoto Junichiro
Giuseppe Jurman
R. Krishna Murthy
    Karuturi
Tsuyoshi Kato
Tetsuji Kuboyama
Xuejun Liu
Marco Loog
Elena Marchiori
Francesco Masulli
Alison Motsinger-Reif
Vadim Mottl
Sach Mukherjee
Allioune Ngom
Josselin Noirel
Carlotta Orsenigo
Andrea Passerini
Thang Pham

Esa Pitkänen
Clara Pizzuti
Beatriz Pontes
Miguel Rocha
Juho Rousu
Yvan Saeys
Hiroto Saigo
Taro Saito
Guido Sanguinetti
Masakazu Sekijima
Kim Seyoung
Evangelos Simeonidis
Johan Suykens
Roberto Tagliaferri
Koji Tsuda
Alexey Tsymbal
Hong Yan
Haixuan Yang

Sponsors

International Association for Pattern Recognition (IAPR)
Tokyo Institute of Technology
Japanese Society for Bioinformatics (JSBi)

Table of Contents

Applications of Pattern Recognition Techniques

Protein Structure and Docking

Complex Data Analysis

Sequence Analysis

Robust Community Detection Methods with Resolution Parameter for Complex Detection in Protein Protein Interaction Networks

Twan van Laarhoven and Elena Marchiori

Institute for Computing and Information Sciences,
Radboud University Nijmegen, The Netherlands
{tvanlaarhoven,elenam}@cs.ru.nl

Abstract. Unraveling the community structure of real-world networks is an important and challenging problem. Recently, it has been shown that methods based on optimizing a clustering measure, in particular modularity, have a resolution bias, e.g. communities with sizes below some threshold remain unresolved. This problem has been tackled by incorporating a parameter in the method which influences the size of the communities. Methods incorporating this type of parameter are also called multi-resolution methods. In this paper we consider fast greedy local search optimization of a clustering objective function with two different objective functions incorporating a resolution parameter: modularity and a function we introduced in a recent work, called w-log-v. We analyze experimentally the performance of the resulting algorithms when applied to protein-protein interaction (PPI) networks. Specifically, publicly available yeast protein networks from past studies, as well as the present BioGRID database, are considered. Furthermore, to test robustness of the methods, various types of randomly perturbed networks obtained from the BioGRID data are also considered. Results of extensive experiments show improved or competitive performance over MCL, a state-of-the-art algorithm for complex detection in PPI networks, in particular on BioGRID data, where w-log-v obtains excellent accuracy and robustness performance.

1 Introduction

The development of advanced high-throughput technologies and mass spectrometry has boosted the generation of experimental data on protein-protein interaction and shifted the study of protein interaction to a global, network level. In particular, it has been shown that groups of proteins interacting more with each other than with other proteins, often participate in similar biological processes and often form protein complexes performing specific tasks in the cell. Detecting protein complexes, consisting of proteins sharing a common function, is important, for instance for predicting a biological function of uncharacterized proteins. To this aim protein-protein interaction (PPI) networks have been used as a convenient graph-based representation for the comparative analysis and detection

T. Shibuya et al. (Eds.): PRIB 2012, LNBI 7632, pp. 1–13, 2012.

of (putative) protein complexes [18]. A PPI network is a graph where nodes are proteins and edges represent interactions between proteins.

Detecting protein complexes in a PPI network can be formalized as a graph-clustering problem. Clustering amounts to divide data objects into groups (clusters) in such a way that objects in the same cluster are more similar to each other than to objects in the other clusters. Since clustering is an ill-posed and computationally intractable problem, many methods have been introduced, in particular for graph-clustering (see e.g. the recent review by Fortunato [7]). Effective methods for graph-clustering contain a parameter whose tuning affects the community structure at multiple resolution scales. These methods are also called multi-resolution methods (see e.g. [15]). The resolution parameter(s) can be used in two main ways: as a parameter to be tuned; or as a way to generate clusterings at multiple resolution scales, which can then be used to analyze the clustering behavior of objects across multiple resolutions [17], or to ensemble the results to produce a consensus clustering [22].

In [27], the resolution bias of state-of-the-art community detection methods has been analyzed, and a simple yet effective objective function was introduced. Results indicated that methods based on greedy local search optimization are robust to the choice of the clustering objective function, when a multi-resolution parameter is added to the objective function.

The goal of this paper is to investigate experimentally the performance of such multi-resolution methods when applied to PPI networks, with respect to data generated from different laboratory technologies as well as with respect to random removal or shuffling of edges in the network. This latter investigation is motivated by the fact that PPI data are still not fully reliable, with the potential inclusion of both false positive and false negative interactions (see e.g. the discussion in [18]). Specifically, we consider fast greedy local search optimization of a clustering objective function with two different objective functions incorporating a resolution parameter: modularity [11] and a function we introduced in a recent work, called w-log-v [27].

To analyze their performance we consider the yeast Saccharomyces cerevisiae, which is a well studied model organism for higher eukaryotes with several protein interaction data generated from diverse laboratory technologies. Specifically, we consider six PPI networks from past studies and the present BioGRID curated database of protein interactions [24]. In order to assess robustness with respect to random perturbations of the graph, we generate a large collection of networks using the BioGRID data, by either removing or by adding a percentage of randomly selected edges, or by randomly shuffling edges while keeping the original degree of each node.

Results of the experiments indicate improved performance of modularity and w-log-v over MCL (the Markov Cluster Algorithm) [26], a state-of-the art method for community detection in PPI networks based on stochastic flow in graphs. MCL was found to achieve best overall performance in yeast PPI networks [2] and competitive performance with methods for overlapping community detection in PPI networks [20].

In particular best performance is achieved by w-log-v on the BioGRID data, and excellent robustness on randomly perturbed versions of this network. Since PPI networks are known to be noisy with respect to the presence of both false positive and false negative interactions, the high robustness shown by the proposed algorithm substantiates its effectiveness on this type of data.

1.1 Related Work

A vast literature on protein complex detection with PPI networks exists (see e.g. the review [18]). Previous related works on multi resolution algorithms for clustering PPI networks either apply an algorithm multiple times with different values of the resolution parameter in order to investigate how proteins cluster at different resolution scales, e.g. [17], or tune the resolution parameter in order to choose a best setting for the considered type of networks, e.g. [2, 20]. Here we aim at investigating thoroughly effectiveness and robustness of two such algorithms by means of an extensive experimental analysis.

In [2] a comparative assessment of clustering algorithms for PPI networks was conducted. In particular, robustness was analyzed, with respect to alterations (addition and/or removal of randomly selected edges) of a test graph which was constructed using a number of yeast complexes annotated in the MIPS database, by linking each pair of proteins belonging to the same complex. The considered algorithms with parameters tuned on the test graph, were then applied to various yeast datasets. Results showed that MCL with inflation (resolution) parameter value equal to 1.8 was performing best on the considered datasets. Robustness of MCL when applied to yeast PPI networks has previously also been analyzed in [21]. According to their results, MCL is rather robust across different networks and with respect to missing or noisy information on protein-protein associations. Here we show that greedy local search optimization of a clustering objective function (e.g. w-log-v) incorporating a resolution parameter achieves improved robustness (and accuracy) on the BioGRID data.

2 Methods

2.1 Greedy Local Search Optimization

A recent experimental study by Lancichinetti and Fortunato [16] showed that the best methods for graph community detection are those of Blondel et al. [1] and Rosvall and Bergstrom [23]. Both of these methods use a similar greedy local search optimization procedure (LSO), which is based on moving nodes between clusters, and constructing a clustering bottom-up. The only difference between these methods is the objective that is optimized. We briefly outline this LSO method here.

LSO is a discrete optimization method that finds a clustering without overlapping clusters. Initially, each node is assigned to a singleton cluster. Then, iteratively, nodes are moved between neighboring clusters as long as the objective improves.

Eventually a local optimum will be reached, but the clusters in this local optimum will often be too small. The next step is to repeat the optimization procedure, but this time moving these small clusters instead of single nodes. Effectively, we are then clustering a condensed graph, where each node in the condensed graph is a small cluster. When another local optimum is reached, the condensed graph is further condensed and clustered, and so on.

Because the condensed graphs are much smaller than the original graph, most of the time is spend clustering the original graph. Since this is done in an local and greedy fashion, the overall algorithm is very fast. For instance, it takes less than a second to cluster a graph with 6000 nodes.

2.2 Objectives

The considered optimization method is independent of the objective function that is optimized. Hence we are essentially free to choose the objective to best fit the application. In this paper we will limit ourselves to two objectives. The first is the popular modularity [12],

$$\text{modularity}(Cl) = \sum_{A \in Cl} - \text{nwithin}(A) + \text{nvol}(A)^2.$$

Here Cl denotes a clustering, i.e. a set of clusters. For a particular cluster $A \in Cl$, its volume is $\text{vol}(A) = \sum_{i \in A, j \in V} w_{ij}$, i.e. the sum of the weight of edges incident to nodes in A, which is equivalent to the sum of degrees. Based on the volume we define the normalized volume as $\text{nvol}(A) = \text{vol}(A)/\text{vol}(V)$, where V is the set of all nodes. Finally $\text{within}(A) = \sum_{i,j \in A} w_{ij}$ is the within cluster volume, and $\text{nwithin}(A) = \text{within}(A)/\text{vol}(V)$ is its normalized variant.

The second objective we consider is w-log-v, which was introduced in van Laarhoven and Marchiori [27]. An advantage of this objective over modularity is that it allows more diverse cluster sizes. Because the sizes of protein complexes can differ widely, we believe that this is a useful property. The w-log-v objective is defined as

$$\text{w-log-v}(Cl) = \sum_{A \in Cl} \text{nwithin}(A) \log(\text{nvol}(A)).$$

Using either of the above objectives directly for the task of clustering a PPI network is not advisable. Both objectives were designed for community detection; and communities are usually relatively large, much larger than protein complexes. Therefore, optimizing these objectives will lead to a clustering with a small number of large clusters. This inability to find small clusters is termed the resolution limit of the objective [8].

To overcome the resolution limit, we add a parameter to the objectives as follows,

$$\text{modularity}_\alpha(Cl) = \text{modularity}(Cl) + \alpha \sum_{A \in Cl} \text{nwithin}(A),$$

$$\text{w-log-v}_\alpha(Cl) = \text{w-log-v}(Cl) + \alpha \sum_{A \in Cl} \text{nwithin}(A).$$

Table 1. Sizes of the different datasets. The last column lists the number of MIPS complexes that are (partially) contained in each dataset.

Dataset	Nodes	Edges	Complexes
Uetz *et al.* (2000) [25]	927	823	20
Ho *et al.* (2002) [13]	1563	3596	43
Gavin *et al.* (2002) [9]	1352	3210	50
Gavin *et al.* (2006) [10]	1430	6531	53
Krogan *et al.* (2006) [14]	2674	7075	75
Collins *et al.* (2007) [5]	1620	9064	63
BioGRID, all physical	5967	68486	97

By increasing the parameter α, the clustering is punished for within cluster edges, and hence the optimal clustering will have smaller clusters. Alternatively, by decreasing the parameter α, the clustering is rewarded for within cluster edges, so the optimal clustering will then have larger clusters.

It can be shown that, because the overall scale of the objective is irrelevant for optimization, the modification is equivalent to assuming that the overall volume of the graph is different. For modularity, the adjustment corresponds to assuming that the graph has volume $(1 - \alpha) \text{vol}(V)$, while for w-log-v it corresponds to assuming the volume is $e^{-\alpha} \text{vol}(V)$. This equivalent interpretation provides some intuition for the resolution parameter: when we tune α to find smaller clusters, the objective is equivalent to that for finding the clusters in a smaller graph.

3 Experiments

3.1 PPI Networks

We downloaded a set of protein interactions from version $3.1.88^1$ of the BioGRID database [24]. This database contains a collection of protein interactions from different sources, and discovered with different methods. In this work we only consider interactions found by physical experiments, not those based on genetics.

The BioGRID also contains in full several datasets from high throughput experimental studies, including Uetz *et al.* [25], Ho *et al.* [13], Gavin *et al.* [9, 10], Krogan *et al.* [14], Collins *et al.* [5]. We consider these subnetworks as separate datasets in our experiments. These datasets are generated with different experimental techniques: the Collins [5], Krogan [14] and Gavin [10] datasets include the results of TAP tagging experiments only, while the BioGRID dataset contains a mixture of TAP tagging, Y2H and low-throughput experimental results [20]. Table 1 lists the sizes of these datasets.

---

[1] This version was released on April 25th, 2012.

3.2   Complex Validation

For validation, we compare clusters with the complexes from the MIPS database [19][2], which we take as the gold standard. MIPS specified a hierarchy of complexes and subcomplexes. Since we deal only with non-overlapping clustering, we only include (sub)complexes at the bottom of this hierarchy. And to avoid degenerate cases, we include only complexes with at least 3 proteins. We also exclude the complexes in category 550, since these are unconfirmed. They were found with computational clustering methods, using as input the same high throughput datasets that we consider.

In addition to the complexes from MIPS, we also use a set of complexes derived from the Gene Ontology annotations of the Saccharomyces Genome Database [3]. This dataset was created and also used by [20].

To compare clusters found by a method to either of these sets of gold standard complexes, we use the overlap score [2],

$$\omega(A, B) = \frac{|A \cap B|^2}{|A||B|}.$$

We consider a cluster to *match* a complex if their overlap score is at least 0.25. This threshold is also used in other works, e.g. [20, 4]. When the cluster and complex have the same size, a match then corresponds to the intersection containing at least half of the nodes in the complex and cluster.

Based on this matching we define *precision* as the fraction of clusters that are matched to any complex. Conversely, we define *recall* as the fraction of complexes that are matched to any cluster. Note that we use the terminology from other works such as [4]. These notions differ from the more standard definitions of precision and recall, because a cluster can match more than one complex and vice versa.

It is clearly possible to achieve a high precision or a high recall with a degenerate clustering. For example, by returning just a single easy to find cluster that matches a complex, the precision will be 1 at the cost of a low recall. And by returning all possible (overlapping) clusters, the recall will be 1 at the cost of a low precision. We therefore use the F_1 score, which is the harmonic mean of precision and recall, as a trade-off between the two scores.

For each of the methods, we include only clusters that contain at least 3 proteins. As a result, not all proteins will be in a cluster. We call the fraction of proteins that are in a cluster the *coverage* of a clustering.

The precision and recall as defined above depend heavily on the chosen threshold; and when few complexes are matched, the scores are very sensitive to noise. Therefore, we also look at the positive predictive value (PPV) and cluster-wise sensitivity scores [2], which are based directly on the size of the intersection between complexes and clusters,

---

[2] We used the latest version at the time of writing, which was released on May 18th, 2006.

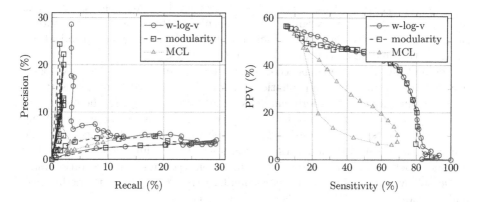

Fig. 1. Precision vs. Recall (left) and Sensitivity vs. PPV (right) on the BioGRID dataset

$$PPV = \frac{\sum_{A \in Cl} \max_{B \in Co} |A \cap B|}{\sum_{A \in Cl} \sum_{B \in Co} |A \cap B|} \qquad Sensitivity = \frac{\sum_{B \in Co} \max_{A \in Cl} |A \cap B|}{\sum_{B \in Co} |B|},$$

where Cl is the set of predicted clusters and Co is the set of gold standard complexes. Note that the asymmetry between the denominators is to account for the case of overlapping clusters.

3.3 Precision vs. Recall

We took the BioGRID all physical dataset, and computed the precision and recall for a wide range of settings of the resolution control parameter α. These results are shown in figure 1 (left). For comparison we also include results with the MCL algorithm for different settings of the inflation parameter. For readability we have applied smoothing in the form of merging points that are very close together.

The first thing that we observe is that despite smoothing, the figure is very noisy in some places. This is not very surprising considering how precision and recall are calculated. Consider a small change in the clustering, such as removing a protein from a cluster. This change might cause the cluster to no longer match a particular complex. If there are no other clusters that matched that complex, then the recall goes down, otherwise it stays the same. Similarly, if this are no other complexes matching the cluster, then the precision goes down. While obviously the change in the two scores is related, the relation is not monotonic, one can change while the other does not.

As the resolution control parameter α goes up, the methods find more clusters; and as a result the recall goes up while the precision goes down. However, after a certain point many of the clusters will become too small, and they will be removed before matching. This decreased coverage causes the recall to go down again.

To get a less noisy picture, we have also plotted the PPV and sensitivity scores, in figure 1 (right). The overall trend in this plot is the same as for the precision and recall: the w-log-v method slightly dominates modularity optimization, which in turn has significantly better results than MCL.

The best parameter settings according to the F_1 score are $\alpha = 2.8$ for w-log-v, $\alpha = 0.97$ for modularity, and inflation 2.7 for MCL. We will use these settings for the remainder of the experiments. As discussed in section 2.2, the parameter α corresponds to assuming a different volume of the graph. The optimal setting for w-log-v corresponds to considering a graph with 16 times fewer edges, while the optimal setting for modularity corresponds to 33 times fewer edges. The difference between the two objectives comes from their inherent resolution bias, by default w-log-v has a bias towards smaller clusters compared to modularity, and therefore the objective needs less adjustment.

3.4   Networks from Single Studies

We next compare the scores on the subnetworks from single studies. The results of this experiment are shown in table 2. The "MIPS method" is based on the gold standard complexes, but including only proteins that occur in the dataset under investigation. It represents the best possible scores.

On most datasets w-log-v has the best recall and F_1 score, except on the datasets from Gavin et al. Gavin *et al.* [9, 10], where MCL performs significantly better. The precision of modularity optimization is often slightly better than that for w-log-v optimization. This is due to the fact that with the settings chosen in the previous paragraph, we find more clusters with w-log-v optimization. Hence, in general recall will be higher at the cost of lower precision.

3.5   Randomly Perturbed Graphs

To further test the robustness of the methods, we applied them to randomly perturbed networks. We performed three different experiments, all starting from the BioGRID network.

1. Removing a randomly chosen subset of the interactions.
2. Randomly adding new spurious interactions between pairs of proteins.
3. Randomly rewire a subset of the edges, while maintaining the degree of each node. Note that such a move both removes an observed interaction and adds a new spurious one.

We varied the amount of edges affected by each type of perturbation. Each experiment was repeated 10 times with different seeds for the random number generator, we calculated the mean and standard deviation of the F_1 score across these repetitions. The results are shown in figure 2. When edges are removed, the performance of all methods degrades similarly. On the other hand, the LSO methods are much more robust to the addition of extra edges than MCL. Also

Table 2. Results of applying the different methods to subnetworks for single studies. The best result for each dataset is highlighted in bold.

Dataset	Method	Clusters	Coverage	Precision	Recall	Sens.	PPV	F_1
Uetz *et al.* (2000)	MIPS	20	2.5%	85.0%	11.6%	9.3%	69.1%	20.5%
	w-log-v	173	21.1%	4.6%	**6.2%**	6.7%	73.6%	**5.3%**
	modularity	160	**21.8%**	**5.0%**	5.5%	**7.0%**	70.1%	5.2%
	MCL	143	17.2%	4.2%	4.1%	5.3%	**75.0%**	4.2%
Ho *et al.* (2002)	MIPS	43	5.2%	88.4%	26.0%	17.5%	67.8%	40.2%
	w-log-v	278	46.0%	2.5%	**6.8%**	11.8%	**69.3%**	3.7%
	modularity	257	**46.2%**	**2.7%**	6.2%	**12.0%**	67.4%	**3.8%**
	MCL	227	35.5%	1.8%	2.7%	10.7%	64.4%	2.1%
Gavin *et al.* (2002)	MIPS	50	7.9%	92.0%	32.9%	27.9%	62.5%	48.4%
	w-log-v	202	**39.7%**	12.4%	**21.2%**	**21.6%**	**72.0%**	15.6%
	modularity	199	39.0%	11.6%	19.9%	19.9%	69.6%	14.6%
	MCL	177	34.5%	**14.7%**	**21.2%**	21.4%	71.1%	**17.4%**
Gavin *et al.* (2006)	MIPS	53	7.8%	92.5%	34.9%	28.3%	63.6%	50.7%
	w-log-v	193	**40.9%**	13.5%	21.2%	23.3%	**72.3%**	16.5%
	modularity	188	37.9%	13.3%	19.9%	22.6%	70.6%	15.9%
	MCL	164	33.4%	**16.5%**	**21.9%**	**24.0%**	71.6%	**18.8%**
Krogan *et al.* (2006)	MIPS	75	8.6%	97.3%	50.7%	38.3%	66.1%	66.7%
	w-log-v	401	58.9%	8.0%	**25.3%**	26.6%	72.6%	12.1%
	modularity	314	**60.4%**	9.6%	23.3%	**26.8%**	70.3%	**13.5%**
	MCL	380	43.9%	7.4%	21.2%	22.4%	**73.5%**	10.9%
Collins *et al.* (2007)	MIPS	63	8.9%	98.4%	43.8%	32.9%	65.6%	60.7%
	w-log-v	194	**38.8%**	20.1%	**32.2%**	27.3%	**75.1%**	**24.8%**
	modularity	177	34.3%	20.3%	29.5%	26.7%	72.8%	24.1%
	MCL	172	36.8%	**20.9%**	30.1%	**29.1%**	69.9%	24.7%
BioGRID, all physical	MIPS	97	7.9%	96.9%	64.4%	55.3%	70.2%	77.4%
	w-log-v	505	**84.5%**	**5.9%**	**22.6%**	**38.6%**	69.4%	**9.4%**
	modularity	599	83.6%	4.2%	19.2%	35.6%	**72.1%**	6.9%
	MCL	283	69.0%	5.3%	11.6%	28.7%	40.6%	7.3%

note that the standard deviation is much larger with the MCL method. That means that for some rewired graphs the method gives reasonably good results, while for others the result is very bad. Unsurprisingly, the experiment with rewired edges sits somewhere in between the two other experiments.

4 Discussion

Because of the incompleteness of both the PPI data and of knowledge on true complexes, care must be taken in the interpretation of the results. The "MIPS method", that is, the best possible method based on the MIPS complexes, covers only a small part of the proteins present in each of the datasets. Conversely, not all MIPS complexes are covered by the datasets, so the recall is always smaller than the precision. In general, results show that for each dataset, the majority of clusters induced by the intersection of that dataset with the complexes in MIPS, match a complex; with a percentage varying between 85% and 98%. These values

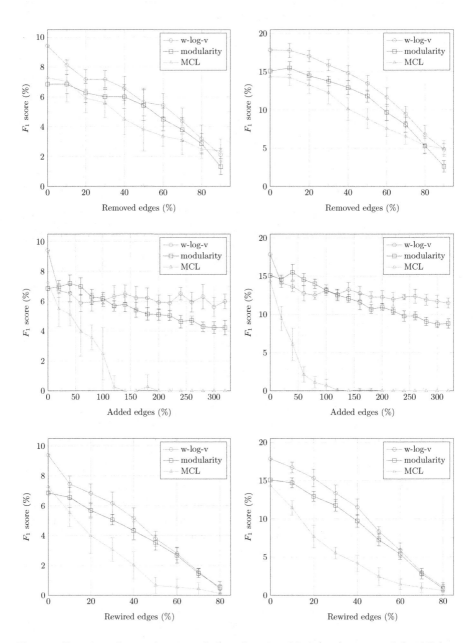

Fig. 2. F_1 score when a fraction of the edges is added (top), removed (middle) or rewired (bottom) at random in the BioGRID dataset. Error bars indicate standard deviation, measured over 10 runs. The left plots use MIPS complexes as the gold standard, the right plots use SGD complexes.

provide upper bounds on the maximum precision and recall achievable on the considered dataset.

On all datasets the algorithms obtain precision smaller than recall: the difference of these values provides information on the fraction of clusters matching more than one complex. For instance on the Uetz dataset, there is almost a one-to-one correspondence between clusters and matched complexes (e.g. 4.6% precision and 6.2% recall for w-log-v), while on the BioGRID this relation is clearly one-to-many (e.g. 5.9% precision and 22.6% recall for w-log-v).

There are complexes that are matched by only one method: specifically, 6 complexes are matched only by w-log-v, 5 only by modularity, and 4 only by MCL. Comparing w-log-v and MCL, there are 18 complexes found only by w-log-v and 4 found only by MCL. This is not too surprising, since MCL has a rather low recall. An example of a complex detected by w-log-v and not by MCL is the Signal recognition particle (SRP) complex, consisting of six proteins, one of the complexes involved in Transcription and/or in the Nucleus[3].

The improved performance of w-log-v on the BioGRID data appears mainly due to its capability to generate a large number of clusters matching multiple complexes (high recall). Nevertheless, figure 1 shows that the precision vs. recall curve of w-log-v dominates the curve of the other two methods.

Robustness of a community detection method is and important issue also in the context of PPI networks, since they are known to contain a high amount of false negative and false positive interactions. Indeed, limitations of experimental techniques as well as the dynamic nature of protein interaction are responsible for the high rate of false-positives and false-negatives generated by high-throughput methods. For instance, Y2H screens have false negative rates in the range from 43% to 71% and TAP has false negative rates of 15%-50%, and false positive rates for Y2H could be as high as 64% and for TAP experiments they could be as high as 77% [6]. Results show that w-log-v achieves best robustness the under random addition, removal and rewiring of a percentage of edges in the BioGRID network. Such high robustness substantiates the effectiveness of w-log-v on this type of data.

5   Conclusions

This paper analyzed the performance of two fast algorithms on PPI networks that optimize in a greedy way a clustering objective function with resolution parameter. An extensive experimental analysis was conducted on PPI data from previous studies as well as on the present BioGRID database. Results indicated improved performance of the considered algorithms over a state-of-the-art method for complex detection in PPI networks, in particular with respect to robustness. These results indicate that the considered algorithms provide an efficient, robust and effective approach for protein complex discovery with PPI

---

[3] See   e.g.   http://pin.mskcc.org/web/align.SubtreeServlet?dbms=mysql&db= interaction&species=SC

networks. Interesting issues for future work include the assessment of the algorithms' robustness with respect to tailored models of false positive and false negative interactions which are present in data generated by specific technologies, as well as the extension of the considered methods to detect overlapping clusters of high quality.

Our implementation of the LSO method in Octave/C++ is available from http://cs.ru.nl/~T.vanLaarhoven/prib2012/.

Acknowledgments. This work has been partially funded by the Netherlands Organization for Scientific Research (NWO) within the NWO project 612.066.927.

References

[1] Blondel, V.D., et al.: Fast unfolding of communities in large networks. Journal of Statistical Mechanics: Theory and Experiment 2008(10), P10008+ (2008)
[2] Brohée, S., van Helden, J.: Evaluation of clustering algorithms for protein-protein interaction networks. BMC Bioinformatics 7(1), 488+ (2006)
[3] Cherry, J.M., et al.: Saccharomyces Genome Database: the genomics resource of budding yeast. Nucleic Acids Research (2011)
[4] Chua, H.N., et al.: Using indirect protein-protein interactions for protein complex prediction. Journal of Bioinformatics and Computational Biology 6(3), 435–466 (2008)
[5] Collins, S., et al.: Towards a comprehensive atlas of the physical interactome of saccharomyces cerevisiae. In: Molecular Cellular Proteomics, pp. 600200–600381 (2007)
[6] Edwards, A.M., et al.: Bridging structural biology and genomics: assessing protein interaction data with known complexes. Trends in Genetics 18(10), 529–536 (2002)
[7] Fortunato, S.: Community detection in graphs. Physics Reports 486, 75–174 (2010)
[8] Fortunato, S., Barthélemy, M.: Resolution limit in community detection. Proceedings of the National Academy of Sciences 104(1), 36–41 (2007)
[9] Gavin, A.C., et al.: Functional organization of the yeast proteome by systematic analysis of protein complexes. Nature 415(6868), 141–147 (2002)
[10] Gavin, A.-C., Aloy, P., Grandi, P., Krause, R., Boesche, M., Marzioch, M., Rau, C., Jensen, L.J., Bastuck, S., Dumpelfeld, B., Edelmann, A., Heurtier, M.-A., Hoffman, V., Hoefert, C., Klein, K., Hudak, M., Michon, A.-M., Schelder, M., Schirle, M., Remor, M., Rudi, T., Hooper, S., Bauer, A., Bouwmeester, T., Casari, G., Drewes, G., Neubauer, G., Rick, J.M., Kuster, B., Bork, P., Russell, R.B., Superti-Furga, G.: Proteome survey reveals modularity of the yeast cell machinery. Nature 440(7084), 631–636 (2006)
[11] Girvan, M., Newman, M.E.J.: Community structure in social and biological networks. Proceedings of the National Academy of Sciences of the United States of America 99(12), 7821–7826 (2002a)
[12] Girvan, M., Newman, M.E.J.: Community structure in social and biological networks. Proc. National. Academy of Science 99, 7821–7826 (2002b)
[13] Ho, Y., et al.: Systematic identification of protein complexes in saccharomyces cerevisiae by mass spectrometry. Nature 415(6868), 180–183 (2002)
[14] Krogan, N.J., et al.: Global landscape of protein complexes in the yeast saccharomyces cerevisiae. Nature 440(7084), 637–643 (2006)

[15] Lambiotte, R.: Multi-scale Modularity in Complex Networks. In: Modeling and Optimization in Mobile, Ad Hoc and Wireless Networks, pp. 546–553 (2010)

[16] Lancichinetti, A., Fortunato, S.: Community detection algorithms: A comparative analysis. Phys. Rev. E 80, 56117 (2009)

[17] Lewis, A., et al.: The function of communities in protein interaction networks at multiple scales. BMC Syst Biol. 4, 100 (2010)

[18] Li, X., et al.: Computational approaches for detecting protein complexes from protein interaction networks: a survey. BMC Genomics, 11(suppl. 1), S3+ (2010)

[19] Mewes, H.W., et al.: MIPS: a database for genomes and protein sequences. Nucleic Acids Res. 30, 31–34 (2002)

[20] Nepusz, T., et al.: Detecting overlapping protein complexes in protein-protein interaction networks. Nature Methods (2012)

[21] Pu, S., et al.: Identifying functional modules in the physical interactome of Saccharomyces cerevisiae. Proteomics 7(6), 944–960 (2007)

[22] Ronhovde, P., Nussinov, Z.: Multiresolution community detection for megascale networks by information-based replica correlations. Physical Review E, 80(1), 016109+ (2009)

[23] Rosvall, M., Bergstrom, C.T.: Maps of random walks on complex networks reveal community structure. Proceedings of the National Academy of Sciences of the United States of America 105(4), 1118–1123 (2008)

[24] Stark, C., et al.: Biogrid: a general repository for interaction datasets. Nucleic Acids Research 34(Database-Issue), 535–539 (2006)

[25] Uetz, P., et al.: A comprehensive analysis of protein-protein interactions in saccharomyces cerevisiae. Nature 403(6770), 623–627 (2000)

[26] Van Dongen, S.: Graph Clustering Via a Discrete Uncoupling Process. SIAM Journal on Matrix Analysis and Applications 30(1), 121–141 (2008)

[27] van Laarhoven, T., Marchiori, E.: Graph clustering with local search optimization: does the objective function matter? (submitted, 2012)

Machine Learning Scoring Functions
Based on Random Forest and Support Vector Regression

Pedro J. Ballester

European Bioinformatics Institute, Cambridge, UK
pedro.ballester@ebi.ac.uk

Abstract. Accurately predicting the binding affinities of large sets of diverse molecules against a range of macromolecular targets is an extremely challenging task. The scoring functions that attempt such computational prediction exploiting structural data are essential for analysing the outputs of Molecular Docking, which is in turn an important technique for drug discovery, chemical biology and structural biology. Conventional scoring functions assume a predetermined theory-inspired functional form for the relationship between the variables that characterise the complex and its predicted binding affinity. The inherent problem of this approach is in the difficulty of explicitly modelling the various contributions of intermolecular interactions to binding affinity.

Recently, a new family of 3D structure-based regression models for binding affinity prediction has been introduced which circumvent the need for modelling assumptions. These machine learning scoring functions have been shown to widely outperform conventional scoring functions. However, to date no direct comparison among machine learning scoring functions has been made. Here the performance of the two most popular machine learning scoring functions for this task is analysed under exactly the same experimental conditions.

Keywords: molecular docking, scoring functions, machine learning, chemical informatics, structural bioinformatics.

1 Introduction

Docking has two stages: predicting the position, orientation and conformation of a molecule when docked to the target's binding site (pose generation), and predicting how strongly the docked pose of such putative ligand binds to the target (scoring). Whereas there are many relatively robust and accurate algorithms for pose generation, the inaccuracies of current scoring functions continue to be the major limiting factor for the reliability of docking [1]. Indeed, despite extensive research, accurately predicting the binding affinities of large sets of diverse protein-ligand complexes remains one of the most important and difficult active problems in computational chemistry.

Scoring functions are traditionally classified into three groups: force field, knowledge-based and empirical. Force-field scoring functions parameterise the potential energy of a complex as a sum of energy terms arising from bonded and non-bonded interactions [2]. The functional form of each of these terms is characteristic of the

T. Shibuya et al. (Eds.): PRIB 2012, LNBI 7632, pp. 14–25, 2012.

particular force field, which in turn contains a number of parameters that are estimated from experimental data and computer simulations. Knowledge-based scoring functions use the three dimensional co-ordinates of a large set of protein-ligand complexes as a knowledge base. In this way, a putative protein-ligand complex can be assessed on the basis of how similar its features are to those in the knowledge base. The features used are often the distributions of atom-atom distances between protein and ligand in the complex. Features commonly observed in the knowledge base score favourably, whereas less frequently observed features score unfavourably. When these contributions are summed over all pairs of atoms in the complex, the resulting score is converted into a pseudo-energy function, typically through a reverse Boltzmann procedure, in order to provide an estimate of the binding affinity (e.g. [3]). Lastly, empirical scoring functions calculate the free energy of binding as a sum of contributing terms, each identified with a physicochemically distinct contribution to the binding free energy such as: hydrogen bonding, hydrophobic interactions, van der Waals interactions and the ligand's conformational entropy. Each of these terms is multiplied by a weight and the resulting parameters estimated from binding affinities. In addition to scoring functions, there are other computational techniques, such as those based on molecular dynamics simulations, that provide a more accurate prediction of binding affinity. However, these expensive calculations remain impractical for the evaluation of large numbers of protein–ligand complexes and are generally limited to series of congeneric molecules binding to a single target [2,4,5].

For the sake of efficiency, scoring functions do not fully account for some physical processes that are important for molecular recognition, which in turn limits their ability to select and rank-order small molecules by computed binding affinities. It is generally believed [4] that the two major sources of error in scoring functions are their limited description of protein flexibility and the implicit treatment of solvent. In addition to these enabling simplifications, there is an important computational issue that has received little attention until recently [6]. Each scoring function assumes a predetermined theory-inspired functional form for the relationship between the variables that characterise the complex, which also include a set of parameters that are fitted to experimental or simulation data, and its predicted binding affinity. Such relationship takes the form of a sum of weighted physicochemical contributions to binding in the case of empirical scoring functions or a reverse Boltzmann methodology in the case of knowledge-based scoring functions. The inherent problem of this rigid approach is that it leads to poor predictivity in those complexes that do not conform to the modelling assumptions (see [7] for an insightful discussion of this issue). As an alternative to these conventional scoring functions, nonparametric machine learning can be used to implicitly capture binding interactions that are hard to model explicitly. By not imposing a particular functional form for the scoring function, intermolecular interactions can be directly inferred from experimental data, which should lead to scoring functions with greater generality and prediction accuracy. This unconstrained approach was likely to result in performance improvement, as it is well-known that the strong assumption of a predetermined functional form for a scoring function constitutes an additional source of error (e.g. imposing an additive form for the considered energetic contributions [8]). Incidentally, recent experimental results have resulted in a redefinition of molecular interactions such as the hydrogen bond [9] or the hydrophobic interaction [10] which means that previously proposed functional forms may need to be revised accordingly.

While there have been a number of machine learning classifiers exploiting x-ray structural data for discriminating between binders and non-binders (e.g. [11,12]), it is only recently that machine learning for nonlinear regression has been shown [6] to be a powerful approach to build generic scoring functions. This trend has been highlighted [13-15] as a particularly promising approach. Indeed, a growing number of studies showing the benefits of these techniques are being presented [6,15-18]. However, these new scoring functions have all been using different benchmarks to evaluate their performance. This prevents us from being able to compare them to each other, as the performance of a scoring function can vary dramatically depending not only on the selection of test set, but also that of the training set and interaction features. In this paper, the performance of the two most popular machine learning approaches to scoring, Random Forest (RF) [19] and SVM epsilon-regression (SVR) [20], is investigated. The focus will be on generic, rather than family-specific (e.g. [21]), scoring functions, which constitute a harder regression problem due to the higher nonlinearity introduced by diverse protein-ligand complexes.

2   Machine Learning Scoring Functions

2.1   RF-Score [6]

RF-Score uses RF as the regression model. A RF is an ensemble of many different decision trees randomly generated from the same training data. RF trains its constituent trees using the CART algorithm [22]. As the learning ability of an ensemble of trees improves with the diversity of the trees [19], RF promotes diverse trees by introducing the following modifications in tree training. First, instead of using the same data, RF grows each tree without pruning from a bootstrap sample of the training data (i.e. a new set of N complexes is randomly selected with replacement from the N training complexes, so that each tree grows to learn a closely related but slightly different version of the training data). Second, instead of using all features, RF selects the best split at each node of the tree from a typically small number (m_{try}) of randomly chosen features. This subset changes at each node, but the same value of m_{try} is used for every node of each of the P trees in the ensemble. RF performance does not vary significantly with P beyond a certain threshold and thus P=500 was set as a sufficiently large number of trees. In contrast, m_{try} has some influence on performance and thus constitutes the only tuning parameter of the RF algorithm. In regression problems, the RF prediction is given by arithmetic mean of all the individual tree predictions in the forest. RF also has a built-in tool to measure the importance of individual features across the training set based on the process of "noising up".

RF-Score outperformed [6] 16 state-of-the-art scoring functions on the same independent test set (2007 PDBbind core set [23]). To investigate the impact of chance correlation [24], the relationship between features and binding affinity in the training set was destroyed by performing a random permutation of binding affinities, while leaving the interaction features untouched (a process known as Y-randomisation). After training, the resulting RF model was used to predict the test set. Over ten independent trials, performance on the test set was on average R=−0.018 with standard deviation S_R=0.095, which demonstrated the negligible contribution of chance correlation to RF-Score's prediction ability. Additional methodological considerations are discussed in [13].

2.2   Breneman and Co-workers [16]

The next three scoring functions used SVR as the regression model. SVR searches for the hyperplane that best discriminates between two classes of feature vectors: those for which the error in the value predicted by the regression model is below a sufficiently small value ε and those with a higher error. Vectors with higher error are used to guide this search. As in SVM classifiers, nonlinear kernels may be used to map input features onto a higher dimensional feature space where better discriminating hyperplanes are possible.

The 2005 release of the PDBbind benchmark was used in this study. Five different non-overlapping training/test data partitions were made: (refined-core)/core with 977/278 complexes, core/(refined-core) with 278/977 complexes and three random partitions with 278/977 complexes each. Each complex was represented by a set of Property-Encoded Shape Distributions (PESD) features encoding geometry, electrostatic potential and polarity for both the protein and the ligand interaction surfaces [16]. Several scoring functions based on SVM regression as implemented in the e1071 SVM R package [25] were presented. No feature selection was employed except for the removal of invariant columns prior to training. SVM was trained with two control parameters: the gamma parameter of the default radial kernel and cost of contraints violation parameter. A range of models was defined by considering a number of values for both parameters and for each of these five-fold cross-validation over the training set was carried out (this cross-validation process was repeated 10 times with different random seeds). The selected SVM model was that with the highest average correlation coefficient with measured binding affinity over the validation sets.

The performance of PESD-SVM scoring functions were compared against SFCScore, as the latter family of multivariate linear regression scoring functions was shown to perform better than 14 other scoring functions on a common test set [26]. The comparison between PESD-SVM and SFCScore could only be semiquantitative on comparably sized training/test partitions, as there was only an overlap of 700 complexes between the data sets used in each study. The performance was comparable in general and slightly improved in some cases. These results are particularly valuable taking into account that, unlike PESD-SVM, SFCScore also included nonsurface-based features, its training set had complexes in common with the test set and it was enriched with industrial data through the Scoring Function Consortium, a collaborative effort with various pharmaceutical companies and the Cambridge Crystallographic Data Center.

2.3   Xie, Bourne and Co-workers [15]

This study presents a SVM regression model to predict IC50 values. While this is not a generic scoring function, the study is relevant to our analysis in that it builds upon the idea that performance improvement can be achieved by circumventing error-prone modelling assumptions with nonparametric machine learning. In particular, the authors focus on the fact that noncovalent interactions often depend on one another in a nonlinear manner and hence a nonlinear function of energy terms should lead to more accurate scoring functions that the linear combinations widely used in standard scoring functions. The docking program eHiTS [27] was selected for this study because it calculates a large number of individual energy terms, which contribute to the overall energy score also known as the eHiTS-Energy scoring function.

SVM-light [28] was used to train a regression model with 80 InhA experimental IC50 values in negative log units. 67 of these 80 molecules were not co-crystallised with the target and hence had to be docked into a InhA structure (PDB code: 1BVR). The eHiTS output provided a total of 20 different energy terms contributing to the overall energy score. In order to determine the optimal combination of energy terms for regression, 128 different combinations of these features and four SVM kernel functions (linear, polynomial, radial basis function and sigmoid tanh) were considered. Five-fold cross-validation was applied to select the final model. The model with the highest mean Spearman's correlation coefficient over all five partitions was selected, which corresponded to the linear kernel and one of the considered combinations of features. Feature importance was measured by re-training the selected model using all but a given feature. The left-out feature that resulted in the largest decrease in performance was deemed as the most important feature for regression.

The selected SVM model obtained in a large improvement in the Spearman's correlation coefficient (0.607), when compared with that achieved by the eHiTS-Energy scoring function (0.117). The mean correlation coefficient in 100 Y-randomisation trials of this SVM model was 0.079, which means that chance correlation makes a very minor contribution to performance. These results demonstrate that assuming an additive form for empirical scoring functions is a suboptimal setting.

2.4   Meroueh and Co-workers [17]

Two generic scoring functions based on SVR were presented: SVR-KB and SVR-EP. SVR-KB employs the same representation as RF-Score, as it 1-tier encoding counts atomic contacts within the same distant cutoff. Additional features were considered through several binning strategies, different atom types and scaling the pairwise counts as pair knowledge-based potentials [17]. To build the SVR-EP model, a feature selection protocol using Simulated Annealing [29] was applied leading to the use of four of the 14 physicochemical properties considered. LibSVM v3.0 [30] was used for model training and prediction using the Gaussian radial basis function (RBF) kernel. Grid search was conducted on some of the most important learning control parameters, such as ε in the loss function, gamma in the RBF kernel as well as the trade-off between training error and margin, to give the best performance in a five-fold cross-validation. In each cross-validation, 20 runs were performed on a random split bases and the quantity of average was recorded.

Two new test data sets from CSAR [31] were used. SVR-KB trained with the 2010 release of the PDBbind refined set (2292 complexes) resulted in the best performance as measured by several measures such as the square of Pearson's correlation coefficient ($R^2=0.67$). Compared to seven widely used scoring functions on these test sets, SVR-KB outperformed the best of these by nearly 0.2 in R^2. The SVR-EP also resulted in superior performance, although at a lower level than SVR-KB. In contrast, conventional scoring functions tested on the same test set obtained an R^2 in the range 0.44 to 0.00.

3   Experimental Setup

These machine learning techniques for regression are used here to learn the nonlinear relationship between the atomic-level description of the protein-ligand complex as provided by a X-ray crystal structure and its binding affinity. This approach requires the characterisation of each structure as a set of features relevant for binding affinity (Figure 1 illustrates such characterisation for a particular protein-ligand complex).

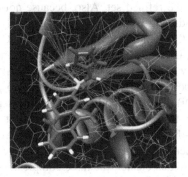

Fig. 1. Visualisation of the GAJ ligand molecule complexed with *Helicobacter Pylori* Type II Dehydroquinase (PDB code 2C4W). Protein-ligand atomic contacts are pictured as blue lines (only a fraction of these contacts are shown to avoid cluttering the figure).

Usually, each feature will comprise the number of occurrences of a particular protein-ligand atom type pair interacting within a certain distance range. This representation can have a significant impact on performance, as a number of conflicting objectives have to be balanced such as selecting atom types that result in dense features while allowing a direct interpretation in terms of which intermolecular interactions contribute the most to binding in a particular complex. On the other hand, the independent variable of this regression is the binding affinity of the ligand for this target. Binding affinities uniformly span many orders of magnitude and hence are typically log-transformed. It is also a common practice to merge dissociation constant (K_d) and inhibition constant (K_i) measurements in a single binding constant K, as this increments the amount of data that can be used to train the machine learning algorithm and has been seen elsewhere that distinguishing between both data types does not lead to significant performance improvement.

The PDBbind benchmark [23] is an excellent choice for validating generic scoring functions. It is based on the 2007 version of the PDBbind database [32], which contains a particularly diverse collection of protein-ligand complexes, assembled through a systematic mining of the entire Protein Data Bank [33]. The first construction step was to identify all the crystal structures formed exclusively by protein and ligand molecules. This excluded protein-protein and protein-nucleic acid complexes, but not oligopeptide ligands as they do not normally form stable secondary structures by themselves and therefore may be considered as common organic molecules. Secondly, Wang et al. collected binding affinity data for these complexes from the literature. Emphasis was placed on reliability, as the PDBbind curators manually reviewed all binding affinities from the corresponding primary journal reference in the PDB.

In order to generate a refined set suitable for validating scoring functions, the following data requirements were additionally imposed. First, only complete and binary complex structures with a resolution of 2.5Å or better were considered. Second, complexes were required to be non-covalently bound and without serious steric clashes. Third, only high quality binding data were included. In particular, only complexes with known K_d or K_i were considered, leaving those complexes with assay-dependent IC_{50} measurements out of the refined set. Also, because not all molecular modelling software can handle ligands with uncommon elements, only complexes with ligand molecules containing just the common heavy atoms (C, N, O, F, P, S, Cl, Br, I) were considered. In the 2007 PDBbind release, this process led to a refined set of 1300 protein-ligand complexes with their corresponding binding affinities. Still, the refined set contains a higher proportion of complexes belonging to protein families that are overrepresented in the PDB. This was considered detrimental to the goal of identifying those generic scoring functions that will perform best over all known protein families. To minimise this bias, a core set was generated by clustering the refined set according to BLAST sequence similarity (a total of 65 clusters were obtained using a 90% similarity cutoff). For each cluster, the three complexes with the highest, median and lowest binding affinity were selected, so that the resulting set had a broad and fairly uniform binding affinity coverage. By construction, this core set is a large, diverse, reliable and high quality set of protein-ligand complexes suitable for validating scoring functions. The PDBbind benchmark essentially consists of testing the predictions of scoring functions on the 2007 core set, which comprises 195 diverse complexes with measured binding affinities spanning more than 12 orders of magnitude.

Regarding representation, atom types are selected so as to generate features that are as dense as possible, while considering all the heavy atoms commonly observed in PDB complexes (C, N, O, F, P, S, Cl, Br, I). As the number of protein-ligand contacts is constant for a particular complex, the more atom types are considered the sparser the resulting features will be. Therefore, a minimal set of atom types is selected by considering atomic number only. Furthermore, a smaller set of interaction features has the additional advantage of leading to computationally faster scoring functions. In this way, the features are defined as the occurrence count of intermolecular contacts between elemental atom types i and j:

$$x_{j,i} \equiv \sum_{k=1}^{K_j} \sum_{l=1}^{L_i} \Theta(d_{cutoff} - d_{kl})$$

where d_{kl} is the Euclidean distance between k^{th} protein atom of type j and the lth ligand atom of type i calculated from the PDBbind structure; K_j is the total number of protein atoms of type j and L_i is the total number of ligand atoms of type i in the considered complex; Θ is the Heaviside step function that counts contacts within a d_{cutoff} neighbourhood of the given ligand atom. For example, $x_{7,8}$ is the number of occurrences of protein nitrogen hypothetically interacting with a ligand oxygen within a chosen neighbourhood. This representation led to a total of 81 features, of which 45 are necessarily zero across PDBbind complexes due to the lack of proteinogenic amino acids with F, P, Cl, Br and I atoms. Therefore, each complex was characterised by a vector with 36 integer-valued features.

Lastly, just as in Cheng et al. [23], the 1105 complexes in the PDBbind 2007 refined set that are not in the core set will be used as the training set, whereas the core set of 195 complexes will be used as the independent test set. In this way, a set of protein-ligand complexes with measured binding affinity can be processed to give two non-overlapping data sets, where each complex is represented by its feature vector $\vec{x}^{(n)}$ and its binding affinity $y^{(n)}$:

$$D_{train} = \left\{ \left(y^{(n)}, \vec{x}^{(n)} \right) \right\}_{n=1}^{1105}; D_{test} = \left\{ \left(y^{(n)}, \vec{x}^{(n)} \right) \right\}_{n=1106}^{1300}; y \equiv -\log_{10} K$$

4   Results and Discussion

The SVR RBF kernel implementation in the caret package [34] of the statistical software suite R was used. As with previous studies [16], grid search was conducted on the gamma parameter in the RBF kernel (γ) and the cost of constraint violation parameter (C) to give the best performance in a five-fold cross-validation of the training set. In each cross-validation, SVR was trained using the 36 combinations of parameter values arising from $\gamma \in \{0.01, 0.1, 1, 10, 100, 1000\}$ and $C \in \{0.25, 0.5, 1, 2, 4, 8\}$. Thereafter, the average root mean square error between predicted and measured binding affinity across the five cross-validation sets (i.e. those not used to train the SVR) was calculated for each (γ,C) combination and that with the lowest value was selected to train on the entire training set to give SVR-Score$\equiv$SVR(γ=0.1,C=1). This model selection procedure is intended to find the model that is most likely to generalize to independent test data sets. When ran on the independent test set, SVR-Score achieved a Pearson's correlation of R=0.726, Spearman's correlation Rs=0.739 and standard deviation SD=1.70 as illustrated in Figure 2 (left).

The same R package was employed to build and run this version of RF-Score. Model selection was carried out by five-fold cross-validation. In each cross-validation, RF was trained using the 35 mtry values that cover all the feature subset sizes up to the number of interaction features, i.e. mtry $\in \{2, 3, \ldots, 36\}$. Thereafter, the average root mean square error between predicted and measured binding affinity across the five cross-validation sets was calculated for each mtry value and that with the lowest value was selected to train on the entire training set to give RF-Score≡RF(mtry=5). In the independent test set, RF-Score achieved a Pearson's correlation of R=0.774, Spearman's correlation Rs=0.762 and standard deviation of 1.59 as illustrated in Figure 2 (right).

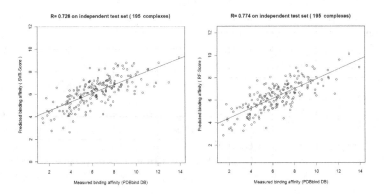

Fig. 2. SVR-Score predicted versus measured binding affinity (left) and RF-Score predicted versus measured binding affinity (right) on the independent test set (195 complexes)

Table 1. Performance of scoring functions on the PDBbind benchmark

scoring function	R	R_s	SD
RF-Score	0.774	0.762	1.59
SVR-Score	0.726	0.739	1.70
X-Score::HMScore	0.644	0.705	1.83
DrugScore^{CSD}	0.569	0.627	1.96
SYBYL::ChemScore	0.555	0.585	1.98
DS::PLP1	0.545	0.588	2.00
GOLD::ASP	0.534	0.577	2.02
SYBYL::G-Score	0.492	0.536	2.08
DS::LUDI3	0.487	0.478	2.09
DS::LigScore2	0.464	0.507	2.12
GlideScore-XP	0.457	0.435	2.14
DS::PMF	0.445	0.448	2.14
GOLD::ChemScore	0.441	0.452	2.15
SYBYL::D-Score	0.392	0.447	2.19
DS::Jain	0.316	0.346	2.24
GOLD::GoldScore	0.295	0.322	2.29
SYBYL::PMF-Score	0.268	0.273	2.29
SYBYL::F-Score	0.216	0.243	2.35

Next, the performance of RF-Score and SVR-Score is compared against that of a broad range of scoring functions on the PDBbind benchmark [23]. Using a pre-existing benchmark, where other scoring functions had previously been tested, ensures the optimal application of such functions by their authors and avoids the danger of constructing a benchmark complementary to the presented scoring function. Table 1 reports the performance of all scoring functions on the independent test set, with RF-Score obtaining the best performance followed by SVR-Score. In contrast, conventional scoring functions tested on the same test set obtained a lower correlation spanning from 0.216 to 0.644.

Given the secrecy of proprietary scoring functions, it is not possible to obtain full implementation details of these, often including training set composition. Consequently, in the context of this benchmark, it could only be reported [23] that, unlike RF-Score and SVR-Score, top scoring functions such as X-Score::HMScore, DrugScoreCSD, SYBYL::ChemScore and DS::PLP1 have an undetermined number of training complexes in common with this test set, which constitutes an advantage for the latter set of functions. On the other hand, calibration sets for conventional scoring functions typically contain around 100-300 selected complexes and hence training these functions with the 1105 complexes from this study could in principle lead to some improvement (note however that the latter strongly depends on whether the adopted regression model is sufficiently flexible to assimilate larger amounts of data and still keep overfitting under control). This issue was investigated in [23], where the third best performing function in Table 1 (best performing in that study), X-Score::HMScore, was recalibrated by its authors using exactly the same 1105 training complexes as RF-Score and SVR-Score (i.e. ensuring that training and test sets have no complexes in common). This gave rise to X-Score::HMScore v1.3, which obtained practically the same performance as v1.2 (R=0.649 versus R=0.644). Since RF-Score, SVR-Score and X-Score::HMScore v1.3 used exactly the same training set and were tested on exactly the same test set, this result also means that all the performance gain (R=0.774 and R=0.726 versus R=0.649) is guaranteed to come from the scoring function characteristics, ruling out any influence of using different training sets on performance. While this recalibration remains to be investigated for the remaining scoring functions (this can only be done by their developers), the fact that these perform much worse than RF-Score/SVR-Score along with the very small improvement obtained by recalibrating X-Score::HMScore strongly suggests that the top part of the ranking in Table 1 would remain exactly the same.

5   Conclusions and Future Prospects

Machine learning for nonlinear regression is a largely unexplored approach to develop generic scoring functions. Here, a comparison between RF and SVR as the regression models has been carried out. Using the same training set, test set, interaction features and model selection strategy, it was observed that RF-Score performs better than SVR-Score at predicting binding affinity. In turn, both machine learning scoring functions outperformed a set of 16 established scoring functions on the same independent test set, which demonstrate the benefits of circumventing problematic modeling assumptions via nonparametric machine learning.

Future prospects for this new class of scoring functions are exciting, as there is a number of promising research avenues which are likely to lead to further performance improvements. First, only three nonparametric machine learning techniques have been used to date (RF, SVR and Multi-Layer Perceptron) and hence alternative techniques might be more suitable for this problem. Second, only two model selection strategies have been applied so far (OOB and five-fold cross-validation) and therefore it remains to be seen whether other strategies could lead to reduced overfitting. Third, unlike models with fixed structure, nonparametric machine learning techniques are sufficiently flexible to effectively assimilate large volumes of training data. Indeed, it has been observed [6,17] that performance on the test set improves dramatically with increasing training set size. This means that ongoing efforts to compile and curate additional experimental data should eventually lead to more accurate and general scoring functions. Finally, in order to facilitate the use, analysis and future development of machine learning-based scoring functions, RF-Score code is made available at http://www.ebi.ac.uk/~pedrob/software.html.

Acknowledgements. The author thanks the Medical Research Council for a Methodology Research Fellowship.

References

1. Moitessier, N., et al.: Towards the development of universal, fast and highly accurate docking/scoring methods: a long way to go. Br. J. Pharmacol. 153, S7–S26 (2008)
2. Huang, N., et al.: Molecular mechanics methods for predicting protein-ligand binding. Phys. Chem. Chem. Phys. 8, 5166–5177 (2006)
3. Mitchell, J.B.O., et al.: BLEEP - potential of mean force describing protein-ligand interactions: I. Generating potential. J. Comput. Chem. 20, 1165–1176 (1999)
4. Guvench, O., MacKerell Jr., A.D.: Computational evaluation of protein-small molecule binding. Curr. Opin. Struct. Biol. 19, 56–61 (2009)
5. Michel, J., Essex, J.W.: Prediction of protein–ligand binding affinity by free energy simulations: assumptions, pitfalls and expectations. J. Comput. Aided Mol. Des. 24, 639–658 (2010)
6. Ballester, P.J., Mitchell, J.B.O.: A machine learning approach to predicting protein-ligand binding affinity with applications to molecular docking. Bioinformatics 26, 1169–1175 (2010)
7. Marshall, G.R.: Limiting assumptions in structure-based design: binding entropy. J. Comput. Aided Mol. Des. 26(1), 3–8 (2012)
8. Baum, B., Muley, L., Smolinski, M., Heine, A., Hangauer, D., Klebe, G.: Non-additivity of functional group contributions in protein-ligand binding: a comprehensive study by crystallography and isothermal titration calorimetry. J. Mol. Biol. 397, 1042–1054 (2010)
9. Arunan, E., et al.: Definition of the hydrogen bond (IUPAC Recommendations 2011). Pure and Applied Chemistry 83, 1637–1641 (2011)
10. Snyder, P.W., et al.: Mechanism of the hydrophobic effect in the biomolecular recognition of arylsulfonamides by carbonic anhydrase. Proceedings of the National Academy of Sciences 108, 17889–17894 (2011)
11. Li, L., Li, J., Khanna, M., Jo, I., Baird, J.P., Meroueh, S.O.: Docking to Erlotinib Off-Targets Leads to Inhibitors of Lung Cancer Cell Proliferation with Suitable in Vitro Pharmacokinetics. ACS Med. Chem. Lett. 1(5), 229–233 (2010)
12. Durrant, J.D., McCammon, J.A.: NNScore: A Neural-Network-Based Scoring Function for the Characterization of Protein–Ligand Complexes. J. Chem. Inf. Model. 50(10), 1865–1871 (2010)

13. Ballester, P.J., Mitchell, J.B.O.: Comments on 'Leave-Cluster-Out Cross-Validation is appropriate for scoring functions derived from diverse protein data sets': Significance for the validation of scoring functions. J. Chem. Inf. Model. 51, 1739–1741 (2011)
14. Cheng, T., Li, Q., Zhou, Z., Wang, Y., Bryant, S.H.: Structure-Based Virtual Screening for Drug Discovery: a Problem-Centric Review. The AAPS Journal 14(1), 133–141 (2012)
15. Kinnings, S.L., Liu, N., Tonge, P.J., Jackson, R.M., Xie, L., Bourne, P.E.: A Machine Learning-Based Method to Improve Docking Scoring Functions and its Application to Drug Repurposing. J. Chem. Inf. Model. 51, 408–419 (2011)
16. Das, S., Krein, M.P., Breneman, C.M.: Binding Affinity Prediction with Property-Encoded Shape Distribution Signatures. J. Chem. Inf. Model. 50, 298–308 (2010)
17. Li, L., Wang, B., Meroueh, S.O.: Support Vector Regression Scoring of Receptor-Ligand Complexes for Rank-Ordering and Virtual Screening of Chemical Libraries. J. Chem. Inf. Model. 51, 2132–2138 (2011)
18. Durrant, J.D., McCammon, J.A.: NNScore 2.0: A Neural-Network Receptor–Ligand Scoring Function. J. Chem. Inf. Model. 51(11), 2897–2903 (2011)
19. Breiman, L.: Random Forests. Mach. Learn. 45, 5–32 (2001)
20. Vapnik, V.: The nature of statistical learning theory. Springer, New York (1995)
21. Amini, A., et al.: A general approach for developing system-specific functions to score protein-ligand docked complexes using support vector inductive logic programming. Proteins 69, 823–831 (2007)
22. Breiman, L., et al.: Classification and regression trees. Chapman & Hall/CRC (1984)
23. Cheng, T., Li, X., Li, Y., Liu, Z., Wang, R.: Comparative Assessment of Scoring Functions on a Diverse Test Set. J. Chem. Inf. Model. 49, 1079–1093 (2009)
24. Rucker, C., Rucker, G., Meringer, M.: y-Randomization and its variants in QSPR/QSAR. J. Chem. Inf. Model. 47, 2345–2357 (2007)
25. The Comprehensive R Archive Network (CRAN) Package e1071, http://cran.r-project.org/web/packages/e1071/index.html (last accessed November 2, 2011).
26. Sotriffer, C.A., Sanschagrin, P., Matter, H., Klebe, G.: SFCscore: scoring functions for affinity prediction of protein-ligand complexes. Proteins 73, 395–419 (2008)
27. Zsoldos, Z., Reid, D., Simon, A., Sadjad, S.B., Johnson, A.P.: eHiTS: a new fast, exhaustive flexible ligand docking system. J. Mol. Graph. Model. 26, 198–212 (2007)
28. Joachims, T.: Making large-Scale SVM Learning Practical. In: Schölkopf, B., Burges, C., Smola, A. (eds.) Advances in Kernel Methods - Support Vector Learning. MIT Press (1999)
29. Kirkpatrick, S.C., Gelatt, D., Vecchi, M.P.: Optimization by simulated annealing. Science 220, 671–680 (1983)
30. LIBSVM - A Library for Support Vector Machines, http://www.csie.ntu.edu.tw/~cjlin/libsvm/ (last accessed November 2, 2011).
31. CSAR, http://www.csardock.org (last accessed November 2, 2011).
32. The PDBbind database, http://www.pdbbind-cn.org/ (last accessed November 2, 2011).
33. Berman, H.M., et al.: The Protein Data Bank. Nucleic Acids Res. 28, 235–242 (2000)
34. The Comprehensive R Archive Network (CRAN) Package caret, http://cran.r-project.org/web/packages/caret/index.html (last accessed November 2, 2011).

A Genetic Algorithm
for Scale-Based Translocon Simulation

Sami Laroum[1], Béatrice Duval[1], Dominique Tessier[2], and Jin-Kao Hao[1]

[1] LERIA, 2 Boulevard Lavoisier, 49045 Angers, France
[2] UR 1268 Biopolymères Interactions Assemblages,
INRA, 44300 Nantes, France
{laroum,bd,hao}@info.univ-angers.fr,
tessier@nantes.inra.fr

Abstract. Discriminating between secreted and membrane proteins is a challenging task. A recent and important discovery to understand the machinery responsible of the insertion of membrane proteins was the results of Hessa experiments [9]. The authors developed a model system for measuring the ability of insertion of engineered hydrophobic amino acid segments in the membrane. The main results of these experiments are summarized in a new "biological hydrophobicity scale". In this scale, each amino acid is represented by a curve that indicates its contribution to the process of protein insertion according to its position inside the membrane. We follow the same hypothesis as Hessa but we propose to determine "in silico" the hydrophobicity scale. This goal is formalized as an optimization problem, where we try to define a set of curves that gives the best discrimination between signal peptide and protein segments which cross the membrane. This paper describes the genetic algorithm that we developed to solve this problem and the experiments that we conducted to assess its performance.

Keywords: Membrane Proteins, Classification, Optimization, Genetic Algorithm.

1 Introduction

Membrane proteins play an important role in many processes in living cells, and they are the targets of many pharmaceutical developments. In fact, 50% of these proteins are used in human and veterinarian medicine [4]. Despite their number and their importance, membrane proteins with known three-dimensional structures represent only 2% of the protein data bank (PDB) [2]. It is difficult to determine their structure because they are difficult to express and crystallize [13]. The great importance of these proteins promoted their study and particularly the search of the machinery responsible of addressing these proteins towards the membrane [23].

The proteins transported across the endoplasmic reticulum (ER) membrane include soluble proteins and membrane proteins. Soluble proteins completely

T. Shibuya et al. (Eds.): PRIB 2012, LNBI 7632, pp. 26–37, 2012.

cross the membrane and usually include a short N-terminal segment[1], called Signal Peptide (SP), that will be cleaved after transport. Membrane proteins have one or several segments that get inserted into the membrane, called trans-membrane (TM) segments. Both types of proteins use the same machinery for transport across the ER membrane, a protein complex located in the ER membrane called the translocon. The translocon channel allows the soluble proteins to cross the membrane and permits the hydrophobic TM segment of membrane proteins to fit in the membrane. SP and TM segments have very close biochemical properties, and particularly they both contain a hydrophobic region. Nevertheless, a TM segment possesses the "key" to open sideways the translocon, which permits the insertion of the protein in the membrane.

Bioinformatics methods suggest some solutions to deal with the recognition of membrane proteins. One of the first prediction method was proposed by Kyte and Doolittle [16]. This method was based on an experimentally determined hydrophobicity index where each amino acid was given a score based on its preference to water or lipid. A hydrophobicity plot was performed by summing the hydrophobicity index over a window of a fixed length and values superior to a cutoff threshold indicates possible TM segments. However, this method performed poorly and is outperformed by machine learning algorithms. More recent works propose methods based on hidden Markov models such as TMHMM [15] and HMMTOP [21], on artificial neural networks such as PHDhtm [20] and Memsat [10]. Other methods combine the prediction of TM segments and SP such as Phobius [14], Philius [19], and SPOCTOPUS [22]. These methods give good discrimination performances, but it is difficult to link their results to a biological interpretation of the translocon machinery. Furthermore, they still sometimes confuse a SP and a TM segment. This is particulary true for the first TM segment of a membrane protein, that is located in the N-terminal region.

In 2005, Hessa et al. carried out a series of in vitro experiments with the aim to compute the energy required for the insertion of a designed TM segment in the membrane [8]. Unlike Kyte and Doolittle which attribute to each amino acid a single hydrophobicity index, Hessa et al. determine for each amino acid a contribution profile - the potential of insertion - according to its position in the segment. As a result, Hessa et al. suggested a 'biological hydrophobicity scale' [9] where each amino acid is represented by a curve. The experiments leading to these curves are very complex to realize, and the predictive system issued from this work such as SCAMPI [3] was only designed to predict TM segments. In fact, SCAMPI does not offer a good distinction between SP and TM segments.

In our work, we follow the same hypothesis as Hessa et al. and we assume that we can elaborate "in silico" a new scale for the amino acids, by studying two sets of protein segments which cross the translocon and share the same chemical hydrophobic profile: SP and TM segments. This scale could benefit from a large quantity of data stored in the protein databases and consequently could be more

---

[1] This paper only considers the primary structure of a protein, represented by a sequence of amino acids. One extremity of this sequence (the first synthesized by the ribosome) is called N-terminal.

precise than the scales derived by in vitro experiments. This paper introduces a genetic algorithm to optimize this scale. As suggested by Hessa *et al.*, an amino acid may have different hydrophobic indexes according to its position inside the translocon channel, and consequently we represent its hydrophobic profile by a symmetric curve. However, we shall see that for some amino acids the profile can be represented by a straight line.

The remainder of this paper is organized as follows. In section 2, we describe our formalization of this problem and the genetic algorithm designed to optimize the curves. Section 3 presents the learning dataset that we have built, and the validation protocol while Section 4 reports the experimental results. Finally conclusions are provided in section 5.

2   Optimization of a Hydrophobicity Scale

2.1   Overview of Our Approach

Figure 1 summarizes the approach that we propose to determine the amino acid insertion profiles (curves of amino acids). Our study relies on a learning dataset composed of two types of sequences: SP (class SP) and TM segments (class TM). We define a simple classifier inspired from the work of Kyte and Doolittle. This classifier slides a window to compute an insertion score for each sequence in order to correctly recognize the SP and the TM segments of the learning dataset. The classifier is determined by 20 curves that represent the insertion profiles of the 20 amino acids. To obtain the best discrimination between SP and TM, a genetic algorithm is executed in order to optimize the set of 20 curves. In the following of this section, we describe more precisely each component of this approach.

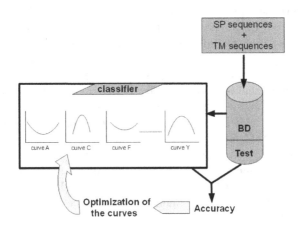

Fig. 1. In silico determination of a hydrophobicity scale

2.2 Discrimination Function: A Sliding Window Classifier

If a denotes one of the 20 amino acids, we note $C[a]$ the curve associated to a in the scale. The curve $C[a]$ gives the value of the hydrophobicity index of the amino acid a depending on its position during the process of protein insertion in the membrane. As the membrane thickness is about 20 amino acids, the curves are defined on a window of length l with $l \simeq 20$. An appropriate window length l will be determined experimentally as explained in section 3.3.

For $j \in [1, l]$, $C[a, j] = C[a](j)$ denotes the index of the amino acid a when it is in position j in the window. For a sequence Seq of amino acids of length l, we use the notation $Seq = < a_1 a_2 \ldots a_l >$ and we define the insertion average of this sequence as the average of its indexes :

$$E(Seq) = \frac{\sum_{j=1}^{l} C[a^j, j]}{l} \qquad (1)$$

In the case of a longer sequence $Seq = < a_1 a_2 \ldots a_n >$ of length $n > l$, a sliding window of fixed length l is scanned on the sequence and we define the insertion index of this sequence as the maximum average calculated on a sub-sequence of length l:

$$E_{max}(Seq) = \max_{1 \leq k \leq n-l+1} \{E(Seq_k)\} \qquad (2)$$

where $Seq_k = < a_k a_{k+1} \ldots a_{k+l-1} >$.

Classifier SP/TM: The distinction between the class SP and the class TM is given by the insertion index $E_{max}(Seq)$ and a threshold τ. $E_{max}(Seq)$ corresponds to the maximum value of hydrophobicity of the sequence, whereas the threshold τ represents a value separating the two classes SP and TM.

A segment TM is generally hydrophobic [16] and therefore our classification rule is:

$$\begin{cases} Seq \in class \; SP & \text{if } \; E_{max}(Seq) < \tau \\ Seq \in class \; TM & \text{otherwise} \end{cases}$$

The set of curves and the threshold τ determine our classifier. Its quality is evaluated by the accuracy and the area under the ROC curve (AUC), which measure the ability of the curves to discriminate between SP sequences and TM sequences.

2.3 Curves Encoding

According to the results of [9] that suggest symmetric insertion profiles, we represent each curve by a parabola defined by an equation $H = \alpha(x - X_0)^2 + \beta$, and we determine a curve by the pair of parameters $(H_{extremity}, H_{middle})$ (figure 2 (A)). $H_{extremity}$ is the value of the curve at the extremities of the window of length l ($X_{min} = 1$ and $X_{max} = l$), whereas H_{middle} is the value of the curve at the middle of the window.

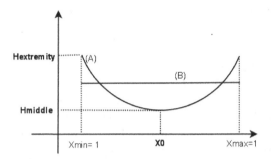

Fig. 2. Representation of an insertion curve

To modify the curve we change $(H_{extremity}, H_{middle})$. The modification of $H_{extremity}$ means that we can act on the behavior of the amino acid at the interfaces of the membrane, while changing H_{middle} allows us to act on the behavior of the amino acid in the middle of the membrane. Note that the manipulated curves can be straight lines (figure 2 (B)) defined by $H_{extremity} = H_{middle}$ ($\alpha = 0$ for the equation of the parabola).

2.4   A Genetic Algorithm for Optimizing the Curves

To optimize the amino acid curves, we use a genetic algorithm (GA). Our algorithm follows the classic schema of a genetic algorithm by evolving a population of individuals. Each individual in our population is a set of 20 curves and each curve is coded by a couple $(H_{extremity}, H_{middle})$. The algorithm begins with an initial population generated by random modifications on a known hydrophobicity scale. The population evolves through the generations by the application of crossover and mutation operators that act directly on the curves of the individuals. An elitism mechanism is used to keep the best individuals of a population. This process is repeated until a predefined number of generations is reached.

Initial Population: The litterature proposes several hydrophobicity scales where each amino acid is assigned a constant index. In our GA algorithm, we decide to generate an initial population by modifications applied on the Eisenberg hydrophobicity scale [5].

Each individual in our initial population is a set of 20 straight lines. To build such an individual, we first initiate a random number k between 1 and 20, which represents the number of amino acid indexes that will be modified. We randomly choose these k amino acids, and we change the value H_{middle} of these k amino acids by the addition of a real value Δ_{mid} randomly selected between $[-3, 3]$ (interval determined experimentally). This process is repeated to generate the required number n of individuals in the initial population. The population size is fixed at $n = 100$ in this work.

Fitness Function: The purpose of the GA algorithm is to optimize a set of 20 curves, in order to correctly discriminate SP sequences from TM sequences. Therefore, the fitness function is given by the classification accuracy measured on the learning dataset.

Evolution: The individuals of the current population P are sorted according to the fitness function. The 10% best individuals of P are directly copied to the next population P' and removed from P. The remaining 90% individuals are then generated by using crossover between two parents selected from the current population by following the principle of wheel selection [7].

Crossover Operator: Our crossover operator considers two parents and generates two children, by exchanging a certain number of curves between two parents (figure 3 (A)).

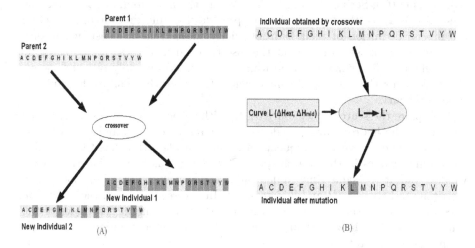

Fig. 3. Figure A shows the crossover operator between two parents where the curves of the amino acids aspartic (D), histidine (H), méthionine (M), proline (P), tyrosine (Y) are exchanged. Figure B shows the mutation operator modifying the curve of the amino acid leucine (L).

To generate the new individuals, we perform the following steps :

1. Randomly select in [1,15] (interval determined experimentally) the number M of amino acids that will be modified between the two parents.
2. Randomly choose the M amino acids which are going to be exchanged.
3. Exchange the curves of the amino acids of the first parent with the curves of the same amino acids of the second parent.

Mutation Operator: The mutation operator is designed to enrich the diversity of the population by manipulating the structure of individual. In our mutation operator we apply a local modification (figure 3 (B)) to a given individual by performing the following steps:

1. Randomly select in [1,12] (interval determined experimentally) the number N of amino acids that will be modified.
2. Randomly choose the N amino acids which are going to be modified.
3. Modify the parameters $(H_{extremity}, H_{middle})$ of the curve by adding a couple $(\Delta_{Hext}, \Delta_{Hmid})$ of real values randomly selected between $[-2,2]$. For a straight line, we modify only the parameter H_{middle} by adding a Δ_{Hmid}.

The mutation operator is applied every 10 generations on 90% of the population.

Strategies of Optimization of the Curves: In a previous work [17], we presented statistics of the amino acid frequencies in our datasets. We have observed that the amino acids Alanine (A), Phenylalanine (F), Isoleucine (I), Leucine (L), Glycine (G), Serine (S) and Valine (V) are the most frequent ones. For the other amino acids, the learning dataset provides few information and therefore the precise definition of their hydrophobicity curves relies on insufficient support. So, we propose to represent the insertion index of these amino acids by a straight line, for which the algorithm has to determine only one parameter. Conversely, for the 7 frequent amino acids, the algorithm has to determine a symmetric curve.

To explore the search space, our GA algorithm operates in two stages. In the first stage, the hydrophobicity indexes are represented by straight lines for all the amino acids, even the frequent ones and the algorithm has to determine the optimal values in this search space. The purpose is to position the sliding window and at the same time to determine the best individual (solution S_1) to discriminate between SP and TM segment. In the second stage, the algorithm fixes the values of the amino acids which are not frequent to the values of the solution S_1 and optimizes symmetric curves for the frequent amino acids by applying specific operators.

Specific Operators: The specific operators are very similar to the precedent operators but they only modify the curves of the frequent amino acids. Thus, the specific crossover operator exchanges the curves of the frequent amino acids between two parents. It chooses randomly M amino acids ($M \leq 7$) among the frequent amino acids and exchanges their curves. As well, the specific mutation modifies the curves of the frequent amino acids. It chooses randomly N amino acids ($N \leq 7$) among the frequent amino acids and modifies their curves by adding a real value Δ_{mid} .

These specific operators allow us to limit the search space by limiting the number of parameters that must be optimized, only the curves of seven amino acids are optimized. The learning dataset provides more information for these amino acids and for this reason, it seems natural to concentrate our search on the frequent amino acids. The more information we have about the amino acids, the more accurately the algorithm can optimize their curves.

3 Experiments and Results

3.1 Learning Dataset: SWP

Our approach requires a dataset containing SP and first TM segments. So, we built a data set, called "SWP" by extracting from the UniprotKB/Swiss-Prot database [12] 684 proteins with TM segments and the same number of proteins with SP. In the case of a soluble protein, the sequence stored in SWP corresponds to the signal peptide (SP). We represent it by the first 35 amino acids of the protein because the length of SP for eucaryotic proteins ranges from 22 to 32 amino acids [1]. In the case of a membrane protein, the sequence stored in SWP corresponds to the first transmembrane segment as annotated in the database. As the annotation of the proteins in SwissProt is the result of TM prediction programs such as TMHMM [15] and MEMSAT [11], we consider that the TM segments in SwissProt are not precisely located on the sequence. So, we add to the TM segment representation the 10 adjacent amino acids before and after the annotated position. To summarize, in our dataset SWP, a secreted protein is represented by a SP which corresponds to the first 35 amino acids, while a membrane protein is represented by its first TM segment with the 10 adjacent amino acids before and after the annotated position.

Note that the constructed dataset relies on the lastest version of the Uniprot-KB/Swiss-Prot database and overlaps the datasets used by other methods. It seems unfair to learn on our dataset and test on other sets of proteins. So, we test all the methods on our dataset using the validation protocol described below.

3.2 Validation Protocol

To assess the performance of our method, we perform a 10-fold cross-validation on the dataset SWP. The initial dataset SWP is split into $K = 10$ subsets of the same size. The method builds a classifier with $(K-1)$ subsets as training set and estimates the error on the remaining subset (test set). We repeat K times the same process by varying the subset that plays the role of test set. The accuracy estimated by K-fold cross-validation is then the average of the accuracies of these K experiments.

3.3 Experimental Results

The purpose of this experiment is 1) to evaluate the influence of the length of the sliding window of the classifier and 2) to optimize the values of the curves for each amino acid. As explained before, the membrane length is about 20 amino acid positions. Hessa *et al.* optimize a profile contribution for each amino acid on a window of 19 amino acids which means that the length of the curves is 19 amino acids. However, the statistical distribution of TM segments in proteins with known 3D structure shows that most TM segments have a length ranging between 21 and 30 amino acids [18]. So, in this experiment we assess a window with 19, 21, and 23 amino acids.

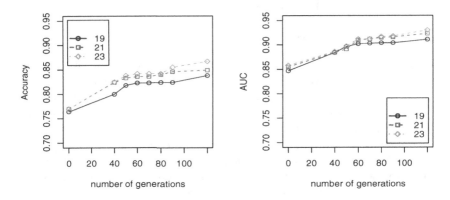

Fig. 4. Performance comparison with different window lengths. The figure shows the average accuracy and average AUC of the best individual of the population.

We run our GA according to the protocol described in section 2.4. In figure 4, the X axis represents the number of generations, while Y axis represents the average accuracy (left figure) and the AUC (right figure) of the best individual of a population. The best results of the accuracy and the AUC are obtained when we use a window with 23 amino acids (green line). This means that the curves with 23 amino acids are better suited to discriminate between signal peptides and TM segments. The figure also displays the performance evolution through the generations. From the generation 1 to the generation 80, our algorithm is in its first stage and it optimizes a straight line for each amino acid. The second stage starts from the optimum solution obtained in the first step to optimize the curves of the frequent amino acids. We observe that this second stage, which only modifies the curves of 7 frequent amino acids, still improves the performance. We can also notice that several runs of our GA give similar results.

This evolution process ends when a predefined number of generations is reached. We do not use any system to avoid overfitting in the training phase.

Table 1 summarizes a comparison with other algorithms. We compare our algorithm with Kyte & Doolittle (KD), Eisenberg (EIS), Engelman (GES) [6] scales and two of the best methods for the discrimination between signal peptides and TM segments: Phobius and Philius. The first method, Phobius, is based on Hidden Markov model and the second method, Philius uses Dynamic Bayesian Networks for its prediction. We also compare the GA algorithm with a method that we developed previously, MN-LS [17] which uses a local search approach for determining the curves of the amino acids. Each prediction method is applied on each fold and then we calculate the average accuracy evaluated according to the same process of 10-fold cross-validation, and table 1 reports the average accuracy with the standard deviation.

The prediction methods based on the hydrophobicity scales slide a fixed length window along the sequence and use a cutoff value to decide if the sequence is a possible TM segment or SP. These methods, as well as MN-LS and GA method,

Table 1. Comparison of our GA and other methods. The table gives the average accuracy and the standard deviation obtained on SWP dataset.

Method	KD	EIS	GES	Phobius	Philius	MN-LS	GA algorithm
average accuracy	0.765	0.755	0.768	0.855	0.842	0.853	0.867
Standard deviation	0.031	0.029	0.0035	0.035	0.032	0.022	0.031

only require as inputs SP and TM segments. For Phobius and Philius, both methods only accept the complete sequence in their web server. These methods are trained to predict SP sequences and TM segments using the complete protein sequence which allow them to take into account additional information like the different composition between cytoplasmic or reticulum exposed loops. MN-LS and GA are developed to optimize curves representing the potential contribution of each amino acid during the insertion of segments in the membrane and use only the SP or the TM segment.

A performance evaluation is presented table 1. We can observe that the hydrophobicity scales perform poorly on SWP dataset, while GA gives the best result. Our GA improved the values of the Eisenberg scale which we used to generate the initial population. However, Phobius and Philius obtain also good predictive performances, but it is easier to drive intuitively-simple reason related to the translocon mechanism for each prediction produced by MN-LS or GA.

Our previous MN-LS method uses a local search approach to determine the curves. As a result, the curves depend on the values of the initial solution, while the GA algorithm optimizes the curves by exploring a large search space in a diversified way and with a good discrimination. The standard deviation shows that the performances of the method are stable on the different runs of the 10-fold cross-validation.

4   Conclusion

In this work, we have presented a genetic algorithm to optimize the curves that represent the contribution of the 20 amino acids to the mechanism of insertion into the membrane. By using a simple sliding window classifier which computes an insertion score of sequences, we demonstrated that the GA algorithm is able to optimize a set of curves that discriminate between two classes of close sequences: signal peptides and TM segments. Despite the simplicity of the classifier, our approach provides classification performances that are equal to two of the best methods of the domain. Furthermore, our approach provides a clear biological interpretation of the insertion phenomenon, which is not the case of sophisticated machine learning methods. Indeed, the curves which we optimize provide an explanation of the contribution of the amino acids during the insertion of the proteins in the membrane.

For future work, we want to introduce more knowledge about the phenomena of membrane proteins insertion to provide more effective guidance of the genetic algorithm.

Acknowledgments. This research was partially supported by the region Pays de la Loire (France) with its "Bioinformatics Program" (2007-2011) and Radapop Project (2009-2013). The authors are grateful to the reviewers for their useful comments.

References

1. Bendtsen, J.D., Nielsen, H., von Heijne, G., Brunak, S.: Improved prediction of signal peptides: SignalP 3.0. Journal of Molecular Biology 340(4), 783–795 (2004)
2. Berman, H.M., Westbrook, J., Feng, Z., Gilliland, G., Bhat, T.N., Weissig, H., Shindyalov, I.N., Bourne, P.E.: The Protein Data Bank. Nucleic Acids Research 28(1), 235–242 (2000)
3. Bernsel, A., Viklund, H., Falk, J., Lindahl, E., von Heijne, G., Elofsson, A.: Prediction of membrane-protein topology from first principles. Proceedings of the National Academy of Sciences of the Unites States of America 105(20), 7177–7181 (2008)
4. Cuthbertson, J.M., Doyle, D.A., Sansom, M.S.P.: Transmembrane helix prediction: a comparative evaluation and analysis. Protein Engineering Design and Selection 18(6), 295–308 (2005)
5. Eisenberg, D., Weiss, R.M., Terwilliger, T.C.: The helical hydrophobic moment: a measure of the amphiphilicity of a helix. Nature 299(5881), 371–374 (1982)
6. Engelman, D.M., Steitz, T.A., Goldman, A.: Identifying nonpolar transbilayer helices in amino acid sequences of membrane proteins. Annual Review of Biophysics and Biophysical Chemistry 15, 321–353 (1986)
7. Goldberg, D.E.: Genetic Algorithms in Search, Optimization, and Machine Learning, 1st edn. Addison-Wesley (January 1989)
8. Hessa, T., Kim, H., Bihlmaier, K., Lundin, C., Boekel, J., Andersson, H., Nilsson, I., White, S.H., von Heijne, G.: Recognition of transmembrane helices by the endoplasmic reticulum translocon. Nature 433(7024), 377–381 (2005)
9. Hessa, T., Meindl-Beinker, N.M., Bernsel, A., Kim, H., Sato, Y., Lerch-Bader, M., Nilsson, I., White, S.H., von Heijne, G.: Molecular code for transmembrane-helix recognition by the Sec61 translocon. Nature 450(7172), 1026–U2 (2007)
10. Jones, D.T.: Improving the accuracy of transmembrane protein topology prediction using evolutionary information. Bioinformatics 23(5), 538–544 (2007)
11. Jones, D.T., Taylor, W.R., Thorton, J.M.: A model recognition approach to the prediction of all-helical membrane protein structure and topology. Biochemistry 33(10), 3038–3049 (1994)
12. Junker, V.L., Apweiler, R., Bairoch, A.: Representation of functional information in the SWISS-PROT data bank. Bioinformatics 15(12), 1066–1067 (1999)
13. Kall, L.: Prediction of transmembrane topology and signal peptide given a protein's amino acid sequence. Method. In: Molecular Biology, vol. 673, pp. 53–62 (2010)
14. Kall, L., Krogh, A., Sonnhammer, E.L.L.: A combined transmembrane topology and signal peptide prediction method. Journal of Molecular Biology 338(5), 1027–1036 (2004)
15. Krogh, A., Larsson, B., von Heijne, G., Sonnhammer, E.L.L.: Predicting transmembrane protein topology with a hidden markov model: application to complete genomes. Journal of Molecular Biology 305(3), 567–580 (2001)
16. Kyte, J., Doolittle, R.F.: A simple method for displaying the hydropathic character of a protein. Journal of Molecular Biology 157(1), 105–132 (1982)

17. Laroum, S., Duval, B., Tessier, D., Hao, J.-K.: Multi-Neighborhood Search for Discrimination of Signal Peptides and Transmembrane Segments. In: Pizzuti, C., Ritchie, M.D., Giacobini, M. (eds.) EvoBIO 2011. LNCS, vol. 6623, pp. 111–122. Springer, Heidelberg (2011)
18. Pasquier, C., Promponas, V.J., Palaios, G.A., Hamodrakas, J.S., Hamodrakas, S.J.: A novel method for predicting transmembrane segments in proteins based on a statistical analysis of the SwissProt database: the PRED-TMR algorithm. Protein Engineering 12(5), 381–385 (1999)
19. Reynolds, S.M., Kaell, L., Riffle, M.E., Bilmes, J.A., Noble, W.S.: Transmembrane Topology and Signal Peptide Prediction Using Dynamic Bayesian Networks. Plos Computational Biology 4(11) (2008)
20. Rost, B., Fariselli, P., Casadio, R.: Topology prediction for helical transmembrane proteins at 86% accuracy. Protein Science 5(8), 1704–1718 (1996)
21. Tusnady, G.E., Simon, I.: The HMMTOP transmembrane topology prediction server. Bioinformatics 17(9), 849–850 (2001)
22. Viklund, H., Bernsel, A., Skwark, M., Elofsson, A.: SPOCTOPUS: a combined predictor of signal peptides and membrane protein topology. Bioinformatics 24(24), 2928–2929 (2008)
23. White, S.H., von Heijne, G.: How translocons select transmembrane helices. Annual Review of Biophysics 37, 23–42 (2008)

A Framework of Gene Subset Selection
Using Multiobjective Evolutionary Algorithm

Yifeng Li, Alioune Ngom, and Luis Rueda

School of Computer Sciences, 5115 Lambton Tower, University of Windsor,
401 Sunset Avenue, Windsor, Ontario, N9B 3P4, Canada
{li11112c,angom,lrueda}@uwindsor.ca
http://cs.uwindsor.ca/uwinbio

Abstract. Microarray gene expression technique can provide snap shots of gene expression levels of samples. This technique is promising to be used in clinical diagnosis and genomic pathology. However, the curse of dimensionality and other problems have been challenging researchers for a decade. Selecting a few discriminative genes is an important choice. But gene subset selection is a NP hard problem. This paper proposes an effective gene selection framework. This framework integrates gene filtering, sample selection, and multiobjective evolutionary algorithm (MOEA). We use MOEA to optimize four objective functions taking into account of class relevance, feature redundancy, classification performance, and the number of selected genes. Experimental comparison shows that the proposed approach is better than a well-known recursive feature elimination method in terms of classification performance and time complexity.

Keywords: gene selection, sample selection, non-negative matrix factorization, multiobjective evolutionary algorithm.

1 Introduction

Microarray gene expression data are obtained through monitoring the intensities of mR-NAs corresponding to tens of thousands of genes [1]. There are two types of microarray data: gene-sample data, which compile the expression levels of various genes over a set of biological samples; and gene-time data, which record the expression levels of various genes over a series of time-points. Both types of data can be represented by a two-dimensional (2D) gene expression matrix. This technique provides a huge amount of data to develop decision systems for cancer diagnosis and prognosis, and to find co-regulated genes, functions of genes, and genetic networks. In this study, we focus on this first application through devising efficient and effective gene selection and classification approaches for gene-sample data. The gene-sample data includes data from two classes, for example, healthy samples and tumorous samples. However, noise, curse of dimensionality, and other problems substantially affect the performance of analysis algorithms devised for microarray data. There are two computational solutions for this problem: feature extraction or feature selection. Feature extraction methods are devised to generate new features, for example the research in [2] extracted non-negative new features/metagenes [3] using *non-negative matrix factorization* (NMF) [4]. And feature

T. Shibuya et al. (Eds.): PRIB 2012, LNBI 7632, pp. 38–48, 2012.
© Springer-Verlag Berlin Heidelberg 2012

selection aims to select a few number of features/genes, which is termed *gene selection*. Gene selection is based on the assumption that only few number of genes contribute to a specific biological phenotype, while most of genes are irrelevant with this. The advantage of gene selection is that it provides directly a small gene subset for biological interpretation and exploration. Any gene selection method needs a gene (or a subset of genes) evaluation criterion to score a gene (or a subset of genes). A search strategy is required for any gene subset selection method, while few gene ranking methods need this strategy.

In the past decade, most feature selection methods were employed for selecting genes, and some feature selection methods are invented specifically for microarray data [5]. *minimum redundancy - maximum relevance* (mRMR) [6] [7] is reported as an efficient and effective method and enjoying much attention. In this method, the mutual information based criteria are proposed to measure the class relevance and feature redundancy. The size of gene subset is fixed by mRMR, and a linear/gready search strategy is proposed. However, it is difficult to decide the weights when combine the two measures into one criterion. *Support vector machine recursive feature elimination* (SVM-RFE) is another successful method for gene selection [8] [9] [10]. SVM-RFE only uses support vectors to rank genes, which is an idea of combining sample selection in gene selection because SVM-RFE selects the boundary samples. There are also some other ideas that prototypic samples are selected to avoid using outliers. Interested reader are referred to [9] for a concise review of sample selection. SVM-RFE can be viewed as both gene ranking method and gene subset selection method. Take Algorithm 2 for example, if the first step (backward search) is only used to sort genes, it is a ranking method; whereas if it involves forward search after backward search to include the sorted genes one by one until the classification performance degenerates, then it is a gene subset selection method. SVM-RFE does not fix the size of gene subset. Mundra and Rajapakse combined the mRMR measure with SVM-RFE (SVM-RFE-mRMR) [11] and reported better accuracy than the original mRMR and SVM-RFE methods. Even a linear search is used to decide the weight of the combination, this is not practically efficient, and the weighting issue between the relevance and redundancy measures is not solved either. Another issue is that SVM-RFE-mRMR may includes unnecessary genes in the gene subset in two cases. Firstly, if the current best validation accuracy in the validation step meets 1, SVM-RFE-mRMR may continue adding genes in the subset until the current validation accuracy is less than 1. For instance, the sequence of the best validation accuracy is $[0.6, 0.8, 1, 1, 1, 0.9]$ and the sorted genes in ascent order is $[\cdots, g_8, g_3, g_{10}, g_2, g_9, g_6]$, SVM-RFE-mRMR may return $[g_6, g_9, g_2, g_{10}, g_3]$, but the algorithm should terminate at the third iterations and return $[g_6, g_9, g_2]$. Secondly, if the current best validation accuracy is less than 1, and this is unchanged until the current validation accuracy is less than it. SVM-RFE-mRMR may keep adding all genes before this. Let us use the above example. If we change 1 to 0.95, similarly SVM-RFE-mRMR may return $[g_6, g_9, g_2, g_{10}, g_3]$. Moreover, since SVM-RFE-mRMR uses a variant of backward search and the number of genes is usually very large, it is too computationally expensive to apply in practice. Computational intelligence approaches, for example evolutionary algorithm, have been used for searching gene subsets. The most crucial part of these approaches is the fitness functions. Good performance has been reported in [12] [13]. This encourages us to design new fitness functions for better result.

In order to apply all the advantages and overcome the disadvantages discussed above, we propose a comprehensive framework to select gene subsets. This framework includes a NMF based gene filtering method, a SVM based sample selection method, and a search strategy use *multiobjective evolutionary algorithm* (MOEA). This MOEA optimizes four fitness functions. Let us call this framework: the MOEA based method for notational simplicity. We also revise the SVM-RFE-mRMR algorithm to solve all its problems, except the weighting issue. In this study, we compared both of the MOEA based method and SVM-RFE-mRMR.

2   Methods

2.1   MOEA Based Gene Subset Selection

In this section, the MOEA based gene subset selection is described in Algorithm 1, and is detailed as below.

Algorithm 1. *MOEA Based Gene Subset Selection*

Input: D, of size m(genes) $\times$ n(samples), and the class labels c

Output: the selected gene subsets: G, the best validation accuracy av and its corresponding gene subsets G_b ($G_b \subseteq G$), and the list of survived genes f

1. split D into training set D_{tr} and validation subset D_{val}. Partition D_{tr} into training subset D_{tr}^{tr} and test subset D_{tr}^{te}
2. NMF based gene filtering (**input**: D_{tr}^{tr} and the number of survived genes K; **output**: K survived genes f)
3. SVM based sample selection (**input**: $D_{tr}^{tr} = D_{tr}^{tr}(f,:)$ and c_{tr}^{tr}; **output**: $D_{tr}^{tr} = D_{tr}^{tr}(:,s)$, where s is the selected samples)
4. search gene subsets by MOEA (**input**: D_{tr}^{tr}, c_{tr}^{tr}, D_{tr}^{te}, and c_{tr}^{te}; **output**: p gene subsets $G = \{g_1, \cdots, g_p\}$)
5. obtain the best validation accuracy and its corresponding gene subsets(**input**: $D_{tr}(f,:)$, c_{tr}, $D_{val}(f,:)$, c_{val}, and G; **output**: the best validation accuracy av and its corresponding gene subsets G_b)

NMF Based Gene Filtering. Gene filtering methods aim to remove some genes which have low ranking scores. This idea is based on the assumption that the the genes with low variations across classes do not contribute to classification. Many gene filtering criteria based on t-test, variance, entropy, range, and absolute values. has been widely used [14]. In this study, we use a novel non-negative matrix factorization (NMF) [4] based criteria, because microarray gene expression intensities are non-negative, and it has been experimentally proved that this criterion works well on microarray data [15] [2]. Suppose D_{tr}^{tr} contains m genes and l samples, it can be decomposed as follows

$$D_{tr}^{tr} \approx AY, \quad D_{tr}^{tr}, A, Y \geq 0, \tag{1}$$

where D_{tr}^{tr}, A, and Y are of size $m \times l$, $m \times r$, and $r \times l$, respectively. $r < \min(m, l)$. A and Y are the basis matrix and the coefficient matrix, respectively. In the application of clustering and feature extraction, columns of A are called metagenes [3] [2] which spans the *feature space*. Each sample is a non-negative linear combination of metagenes. Metagenes are hidden patterns extracted from the original intensity data. Instead of analyzing the original data, we use a criterion on A, as below

$$Gene\_score(i) = 1 + \frac{1}{\log_2(r)} \sum_{j=1}^{r} p(i,j) \log_2 p(i,j), \qquad (2)$$

where $p(i,q) = \frac{A[i,q]}{\sum_{j=1}^{r} A[i,j]}$. This criterion is based on entropy in information theory. the assumption that if the ith row, corresponding to the ith gene, exhibits discriminability across the metagenes, we say this gene contribute to classification. We select K genes with the top K scores. The differences between this and the above mentioned feature ranking methods are that this criterion is unsupervised and operates on the extracted features, instead of directly on the original data.

SVM Based Sample Selection. Since we use a MOEA as search strategy, we hope the fitness values are calculated as fast as possible. Meanwhile, we also expect the gene selection can use essential samples. In this study, we therefore use a simple sample selection to select bounder samples. A linear SVM [16] [17] is trained over D_{tr}^{tr}, and the support vectors are used as input of the MOEA gene selection module to calculate the fitness values.

Multiobjective Evolutionary Algorithm. The following four points should be considered when a high-quality gene subset method is being designed. 1) All the genes in a subset should be relevant to classify the samples as correct as possible. 2) The genes in a subset should be as diverse as possible rather than most of selected genes have the similar profiles. 3) The prediction accuracy and generalization of the selected subsets should be as good as possible. 4) At the same time, the gene subsets should be as small as possible. However, 1) and 2) conflict to some extent. 3) is also conflict with 4). MOEA can optimize more than one (conflicting) objectives and return the Pareto front which are a collection of the non-inferior solutions [18]. Since we have the above four criteria, MOEA should naturally be used to solve the weighting problem instead of using the classical methods to combine them into a single objective using weights as [6] and [11] did. NSGA-II [19], a well-known MOEA algorithm, is used in this study. We customize this algorithm for our application as below.

An individual in the population should be a gene subset. Suppose the length of the survived gene list f in Algorithm 1 is h, we encode a gene subset into a 0-1 binary array of length h. For an individual b, $b[i] = 1$ indicates the ith gene is selected in the subset.

Four fitness functions, considering to class relevance, gene redundancy, prediction accuracy, and gene size, are used. They are formulated as below:

$$f_1(b) = \frac{1}{\frac{1}{\text{sum}(b)} \sum_{b[i]=1} I(i, c_{tr}^{tr})}, \qquad (3)$$

where $I(i, c_{tr}^{tr})$ is the mutual information of the ith discretized gene profile and the class labels on data $\boldsymbol{D}_{tr}^{tr}$;

$$f_2(\boldsymbol{b}) = \frac{1}{\text{sum}(\boldsymbol{b})} \sum_{\boldsymbol{b}[i]=1, \boldsymbol{b}[i']=1} I(i, i'), \tag{4}$$

where $I(i, i')$ is the mutual information of the ith and i'th discretized gene profiles;

$$f_3(\boldsymbol{b}) = linearSVM(\boldsymbol{D}_{tr}^{tr}, \boldsymbol{D}_{tr}^{te}), \tag{5}$$

where $linearSVM$ is a linear SVM classifier trained on $\boldsymbol{D}_{tr}^{tr}$, and returns the prediction accuracy of $\boldsymbol{D}_{tr}^{te}$; and

$$f_4(\boldsymbol{b}) = \frac{\text{sum}(\boldsymbol{b})}{\text{length}(\boldsymbol{b})}. \tag{6}$$

Scattered crossover operation is used in our implementation. For two parents from the mating pool, each parent has equal chance to pass its gene to its child at each position. In the mutation step, for a parent selected for mutation, each position has the probability of p_m to be chosen to have 0-1 flip. Suppose the portion of 0s and 1s in this parent are p_0 and p_1, respectively. And suppose a position is chosen to mutate. If the value at this position is 1(0), it has the probability of $p_0(p_1)$ to be 0 (1). In this way, we can keep the child has the similar 0-1 portions as its parent.

Classification. If the gene subsets G, found by MOEA, are used to predict the class labels of new samples, different prediction accuracies may be obtained. We need to select some gene subsets with the best generalization from G. In order to do this, we use $\{\boldsymbol{D}_{tr}^{tr}, \boldsymbol{D}_{tr}^{te}\}$ to train a linear SVM classifier for any gene subset from G, respectively, and use $\boldsymbol{D}_{tr}^{val}$ to test the classifier. The validation accuracy is used to decide the generalization of a gene subset. The best gene subsets with respect to generalization form a *gene subset committee*. If we use the gene subset committee to train respective linear SVM classifiers over D, we can obtain a *classifier committee*, the class label of a new sample (independent with D) is voted by the committee.

2.2   Revised SVM-RFE-mRMR

For the purpose of application and comparison, we revised the SVM-RFE-mRMR method to solve the weaknesses (except the weighting problem) as discussed in Section 1. [8] and [11] only described the gene ranking step, which is actually incomplete, we therefore append the validation step to find the best gene subset. See Algorithm 2 for details.

3   Experiments

We use three well-cited gene-sample datasets in our experiment. See Table 1 for details. We did two experiments.

First, we used 10-fold *cross-validation* (CV) to evaluate the performance of the MOEA based framework, and compared it with the revised SVM-RFE-mRMR. The experiment procedures are shown in Fig. 1 and 2. During each fold of CV of the MOEA

Algorithm 2. *Revised SVM-RFE-mRMR Gene Subset Selection*

Input: D, of size m(genes) $\times$ n(samples), and the class labels c

Output: selected gene subset g, the best validation accuracy av, and list of survived genes f

    split D into training set D_{tr} and validation set D_{val}

    filter out the genes over D_{tr}, and get gene list f left

    $D_{tr} = D_{tr}(f, :)$

    $D_{val} = D_{val}(f, :)$

    ————————gene ranking step————————

    set β

    given set of genes s initially including by all genes

    ranked set of genes, $r = \{\}$

    repeat

        train linear SVM over D_{tr} with gene set s

        calculate the weight of each gene w_i

        for each gene $i \in s$ **do**

            compute class relevance $R_{s,i}$ and feature redundancy $Q_{s,i}$ over D_{tr}

            compute $r_i = \beta|w_i| + (1 - \beta)\frac{R_{s,i}}{Q_{s,i}}$

        end for

        select the gene with smallest ranking score, $i^* = \arg\min\{r_i\}$

        update $r = r \cup \{i^*\}$; $s = s \setminus \{i^*\}$

    until all gene are ranked

    ————————validation step————————

    $g = \{\}$

    set the best validation accuracy $av = 0$

    for i=length(r) **to** 1 **do**

        $s = s \cup \{r_i\}$

        train linear SVM classifier over D_{tr}

        obtain the validation accuracy a through validating the classifier over D_{val}

        if $av \leq a$ **then**

            if $av < a$ **then**

                $g = s$

            end if

            if $av == 1$ **then**

                break

            end if

        else

            break

        end if

    end for

Table 1. Gene-Sample Datasets

Dataset	#Classes	#Genes	#Samples
Leukemia [3, 20]	2	5000	27+11=38
CNC [3, 21]	2	5893	25+9=34
Colon [22]	2	2000	40+22=62

based method, the whole data O is partitioned into training set O_{tr} and test set O_{te}. Algorithm 1 is employed to find the selected gene subsets G, the best validation accuracy av and its corresponding gene subsets G_b, and the list of survived genes f. After that the best prediction accuracy and that using voting strategy, depicted in Section 2.1, are obtained. The linear SVM classifier is used in the classification step. After 10-fold CV, these two measures are averaged. We designed the same experiment procedure for SVM-RFE-mRMR method. Since this method only returns a gene subset in each fold, the prediction accuracies of the 10 gene subsets are averaged at the end of CV.

The experiment results are shown in Table 2. The "Pred. Acc." column shows the prediction accuracies. For the MOEA based approach, the values outside the parenthesis are the average of the best prediction accuracies. We trained linear SVM classifiers over O_{tr} with different gene subsets, and used O_{te} to test these classifiers. The best prediction accuracy among them are reported at each fold. The values in the parenthesis are the prediction accuracies obtained by the voting method. From this column, we can see that our proposed MOEA based method works well and outperforms SVM-RFE-mRMR in terms of the prediction accuracy. The next column tells us that both our MOEA based method and SVM-RFE-mRMR can obtain a small number of genes. The last column shows the execution time of the whole procedure. We can find that, though using four fitness functions, the MOEA based method is much faster than the revised SVM-RFE-mRMR. Our experimental procedure can avoid *false high prediction accuracy problem* (FHPAP) which is the case that the reported accuracy is higher than the actual one. FHPAP occurs when the whole dataset is used to select features, after that the performance of the selection method is evaluated through dividing the whole dataset into training set and test set, for example FHPAP occurs in [6].

Table 2. Prediction Accuracy

Data	Method	Pred. Acc.	#Genes	Time
Leukemia	MOEA	1(0.9050)	18.1	1.7×10^4
	SVM-RFE-mRMR	0.9083	30.9	6.2×10^4
CNC	MOEA	0.9167(0.7417)	29.9	1.5×10^4
	SVM-RFE-mRMR	0.7417	39.8	6.2×10^4
Colon	The Proposed	0.9357(0.8429)	36.7	1.7505×10^4
	SVM-RFE-mRMR	0.7881	4.8	6.4×10^4

Second, we used the whole datasets as input of Algorithm 1 and 2, respectively to find gene subsets for devising decision system and future biological exploration. The result is shown in Table 3. From this table, we can see that the MOEA based approach can obtain better validation accuracy than the revised SVM-RFE-mRMR approach.

4   Discussion

It is still an open problem of how to choose the most promising one or more gene subsets, G_b, from the gene subsets, G, returned by MOEA. We propose to find G_b according to the validation accuracy. Since G_b may contains more than one gene subsets,

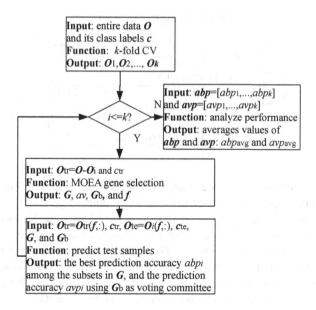

Fig. 1. Procedure of Evaluating the MOEA Gene Selection

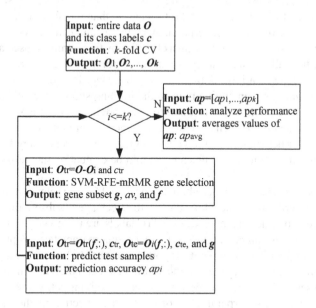

Fig. 2. Procedure of Evaluating the Revised SVM-RFE-mRMR Gene Selection

a voting strategy can be used to determine the class labels of the new coming samples. [18] has a general discussion on this issue. Domain knowledge should be consider

Table 3. Validation Accuracy

Data	Method	Valid. Acc.	#Genes
Leukemia	MOEA	1	14.2
	SVM-RFE-mRMR	1	1
CNC	MOEA	1	24.9
	SVM-RFE-mRMR	0.8182	1
Colon	MOEA	1	19.5
	SVM-RFE-mRMR	0.8571	2

to select the best point on the Pareto front for specific application. Therefore, more thought should be inspired to discover the most discriminative gene subsets from G.

After designing a feature selection method, two steps have to be followed. The first step aims to computationally evaluate the performance of the designed method, and to compare with other existing methods. This requires two substeps: training substep and test substep. Note that the generalized definition of training substep should include both feature selection and training a classifier. The working data should be split into two exclusive parts: the training set and test set (perhaps by cross-validation). If we need to decide some superparameters of the feature selection model, we need to further split the training set into training subset and validation subset. The superparameters could be, for example, the parameter of a scoring function, the size of the feature subset, or the best feature subset if the feature method returns more than one feature subsets. During training, we need to estimate the superparameters. For example, if we need to decide the best size of gene subsets, we need to train a classifier by the training subset, and validate its accuracy. If the validation accuracy is not satisfactory, we need to adjust the size of gene subsets, and repeat until we find the proper size. Once the proper superparameters are found, the training set, including both of the training subset and validation subset, is used to train a classifier. In the test substep, the prediction accuracy is obtained to measure the classification performance of the designed feature selection method, and to compare with other benchmark methods. It is unnecessary to report any feature subset selected, because the main task of this step should be evaluating the performance of a method.

After the first step, the confidence about the designed feature selection method is obtained. The next step is to use the whole dataset to select a gene subset, train a classifier, and wait for predicting new samples whose class labels are unknown. At this step, only the validation accuracy can be obtained if there are superparameters to optimize. However, there is no prediction accuracy to report, because the class labels are unknown. When optimizing the superparameters, the whole data can be divided into training set and validation set. The feature selection runs over the training set, while the validation set is used to adjust the superparameters of the feature selection method according to its output. After the promising superparameters are obtained, the whole dataset with the selected feature subset is used to learn a classifier. If the feature subset needs to be reported, the feature selection method should take the whole data as input. There is no need to worry about the quality of the reported feature subset, because the confidence of its quality comes from the first step. Furthermore, the prediction accuracy of the reported feature subset is expected higher than the prediction accuracy at the first step. The reason is that the reported feature subset uses larger number of samples at the second step.

Some researchers may mixed up the above two steps. For example, the whole dataset is firstly preprocessed and used to select the feature subset (this is actually the task of the second step), and then k-fold CV is employed to split the whole dataset into training sets and test sets. And the training set with the selected feature subset is used to learn a classifier; after that, predicted accuracy is reported through testing the classifier by the test set. Unfortunately, the prediction accuracy is overestimated because the test set has already been used during feature selection. If a sensitive feature selection method is subject to overfitting easily, then the prediction accuracy would be overestimated significantly. Also, some researchers may be wondering how to report the feature subset because they have k feature subsets from k-fold CV, respectively. The issue here is that they try to report the feature subset right after the first step. If the feature subset is reported at the second step, this issue can be avoided.

5   Conclusion and Future Works

This paper proposes a MOEA based framework to select gene subsets. This approach mainly includes a NMF based gene filtering method, a SVM based sample selection method, and a MOEA search strategy. We revise the SVM-RFE-mRMR method for comparison. Our approach overcomes the drawback of the mRMR and the revised SVM-RFE-mRMR methods. Experimental results show that the MOEA based approach outperforms the revised SVM-RFE-mRMR method. We also clarify some experimental issues when estimating designed feature selection methods. Since MOEA outputs more than one gene subsets, our future research will focus on finding better methods to identify the best gene subset after running MOEA. The biological relevance of the genes selected will be investigated as well.

Acknowledgments. This research has been supported by IEEE CIS Walter Karplus Summer Research Grant 2010, Ontario Graduate Scholarship 2011-2012, and Canadian NSERC Grants #RGPIN228117-2011.

References

1. Zhang, A.: Advanced Analysis of Gene Expression Microarray Data. World Scientific, Singapore (2009)
2. Li, Y., Ngom, A.: Non-Negative Matrix and Tensor Factorization Based Classification of Clinical Microarray Gene Expression Data. In: BIBM, pp. 438–443. IEEE Press, New York (2010)
3. Brunet, J.P., Tamayo, P., Golub, T.R., Mesirov, J.P.: Metagenes and Molecular Pattern Discovery Using Matrix Factorization. PNAS 101(12), 4164–4169 (2004)
4. Lee, D.D., Seung, S.: Learning the Parts of Objects by Non-Negative Matrix Factorization. Nature 401, 788–791 (1999)

5. Saeys, Y., Inza, I., Larrañaga, P.: A Review of Feature Selection Techniques in Bioinformatics. Bioinformatics 23(19), 2507–2517 (2007)
6. Ding, C., Peng, H.: Munimun Redundancy Feature Selection from Microarray Gene Expression Data. Journal of Bioinformatics and Computational Biology 3(2), 185–205 (2005)
7. Peng, H., Long, F., Ding, C.: Feature Selection Based on Mutual Information: Criteria of Max-Dependency, Max-Relevance, and Min-Redundancy. IEEE Transactions on Pattern Analysis and Machine Intelligence 27(8), 1226–1238 (2005)
8. Guyon, I., Weston, J., Barnhill, S.: Gene Selection for Cancer Classification Using Support Vector Machines. Machine Learning 46, 389–422 (2002)
9. Mundra, P.A., Rajapakse, J.C.: Gene and Sample Selection for Cancer Classification with Support Vectors Based t-statistic. Neurocomputing 73(13-15), 2353–2362 (2010)
10. Mundra, P.A., Rajapakse, J.C.: Support Vectors Based Correlation Coefficient for Gene and Sample Selection in Ccancer Classification. In: CIBCB, pp. 88–94. IEEE Press, New York (2010)
11. Mundra, P.A., Rajapakse, J.C.: SVM-RFE with MRMR Filter for Gene Selection. IEEE Transactions on Nanobioscience 9(1), 31–37 (2010)
12. Liu, J., Iba, H.: Selecting Informative Genes Using A Multiobjective Evolutionary Algorithm. In: CEC, vol. 1, pp. 297–302. IEEE Press, New York (2002)
13. Paul, T.K., Iba, H.: Selection of The Most Useful Subset of Genes for Gene Expression-Based Classification. In: CEC, vol. 2, pp. 2076 - 2083. IEEE Press, New York (2004)
14. Kohane, I.S., Kho, A.T., Butte, A.J.: Microarrays for An Integrative Genomics. MIT Press, Cambridge (2003)
15. Kim, H., Park, H.: Sparse Non-Negatice Matrix Factorization via Alternating Non-Negative-Constrained Least Squares for Microarray Data Analysis. Bioinformatics 23(12), 1495–1502 (2007)
16. Vapnik, V.: The Nature of Statistical Learning Theory. Springer, Berlin (1995)
17. Chang, C., Lin, C.: LIBSVM : A Library for Support Vector Machines. ACM Transactions on Intelligent Systems and Technology 2(2), 27:1–27:27 (2001), http://www.csie.ntu.edu.tw/~cjlin/libsvm
18. Deb, K.: Multi-Objective Optimization Using Evolutionary Algorithm. Wiley, West Sussex (2001)
19. Deb, K., Pratap, A., Agarwal, S., Meyarivan, T.: A Fast and Elitist Multiobjective Genetic Algorithm: NSGA-II. IEEE Transactions on Evolutionary Computation 6(2), 182–197 (2002)
20. Golub, T.R., Slonim, D.K., Tamayo, P., et al.: Molecular Classification of Cancer: Class Discovery and Class Prediction by Gene Expression Monitoring. Science 286(15), 531–537 (1999), http://www.broadinstitute.org/cgi-bin/cancer/datasets.cgi
21. Pomeroy, S.L., Tamayo, P., Gaasenbeek, M., et al.: Prediction of Central Nervous System Embryonal Tumour Outcome Based on Gene Expression. Nature 415, 436–442 (2002), Data Available at http://www.broadinstitute.org/cgi-bin/cancer/datasets.cgi
22. Alon, U., Barkai, N., Notterman, D.A., et al.: Broad Patterns of Gene Expression Revealed by Clustering of Tumor and Normal Colon Tissues Probed by Oligonucleotide Arrays. PNAS 96(12), 6745–6750 (1999), Data Available at http://genomics-pubs.princeton.edu/oncology

Multiple Tree Alignment with Weights Applied to Carbohydrates to Extract Binding Recognition Patterns

Masae Hosoda, Yukie Akune, and Kiyoko F. Aoki-Kinoshita*

Dept. of Bioinformatics, Faculty of Engineering, Soka University,
1-236 Tangi-machi, Hachioji, Tokyo, Japan 192-8577
{e12d5605,e10d5601,kkiyoko}@soka.ac.jp

Abstract. The purpose of our research is the elucidation of glycan recognition patterns. Glycans are composed of monosaccharides and have complex structures with branches due to the fact that monosaccharides have multiple potential binding positions compared to amino acids. Each monosaccharide can potentially be bound by up to five other monosaccharides, compared to two for any amino acid. Glycans are often bound to proteins and lipids on the cell surface and play important roles in biological processes. Lectins in particular are proteins that recognize and bind to glycans. In general, lectins bind to the terminal monosaccharides of glycans on glycoconjugates. However, it is suggested that some lectins recognize not only terminal monosaccharides, but also internal monosaccharides, possibly influencing the binding affinity. Such analyses are difficult without novel bioinformatics techniques. Thus, in order to better understand the glycan recognition mechanism of such biomolecules, we have implemented a novel algorithm for aligning glycan tree structures, which we provide as a web tool called MCAW (Multiple Carbohydrate Alignment with Weights). From our web tool, we have analyzed several different lectins, and our results could confirm the existence of well-known glycan motifs. Our work can now be used in several other analyses of glycan structures, such as in the development of glycan score matrices as well as in state model determination of probabilistic tree models. Therefore, this work is a fundamental step in glycan pattern analysis to progress glycobiology research.

Keywords: glycomics, glycans, bioinformatics, multiple tree alignment algorithm.

1 Introduction

The purpose of our research is the elucidation of glycan recognition patterns. Glycans are composed of monosaccharides and have complex structures with branches because glycans have more than one binding site compared with the amino acid sequences. Many glycans are bound to proteins and lipids on the

---

* Corresponding author.

T. Shibuya et al. (Eds.): PRIB 2012, LNBI 7632, pp. 49–58, 2012.

cell surface and play important roles in biological processes such as determination of blood type, cellular adhesion, antigen-antibody reactions, and virus infections [15].

Moreover, a family of proteins called lectins are known to recognize and bind to glycans. There are many lectin binding and steric mechanisms involving glycan structures in the control of protein-protein interactions. Many signaling events are also known to be regulated by lectin binding [11].

In general, lectins bind to the terminal monosaccharides of glycans on glycoconjugates. However, it is suggested that some lectins recognize not only terminal monosaccharides, but also internal monosaccharides, possibly influencing the binding affinity [15]. The same may be surmized for other glycan-binding biomolecules such as viruses and bacteria as well. Thus, in order to better understand the glycan recognition mechanism of such biomolecules, glycan arrays were developed [9,2]. Glycan arrays consist of a variety of immobilized glycans on a chip, and are used to assess the binding reaction with fluorescently-labeled proteins, viral glycan binding proteins, antibodies and cells. The Consortium for Functional Glycomics (CFG) [12] has furthermore made their glycan array experimental data available on the web [19]. Therefore, it is now possible to obtain many glycan structures that bind with high affinity to a particular lectin, virus, or bacteria that has been analyzed by the CFG.

With the increasing availability of such glycan binding data, we developed Profile PSTMM [14,4] (probabilistic sibling-dependent tree Markov model) to probabilistically extract recognition patterns of glycans using a probabilistic model similar to HMM [7]. However, the complexity of the algorithm brought forth several challenges that needed to be solved. First, a simplified model with similar probabilistic predictive performance was developed called Ordered Tree Markov Model (OTMM) [10]. However, the development of a "Profile OTMM" model first required the determination of an appropriate state model to learn, which was one of the original challenges of Profile PSTMM.

Therefore, we decided to focus on tree alignment of glycans to obtain glycan profiles that may be recognized by a particular glycan-binding biomolecule. In order to do this, we decided to base our algorithm on ClustalW [13] to progressively build a multiple tree alignment. This was possible due to the existence of a pairwise glycan algorithm which we had previously implemented. Moreover, we also developed a web-based tool on RINGS [1,18] to visualize the resulting glycan profiles on the web such that they could be easily analyzed. We will describe our new multiple glycan alignment algorithm called MCAW (multiple carbohydrate alignment with weights) and briefly introduce some preliminary analytical results.

2   Background

In order for readers to understand our algorithm, we first describe notations that will be used throughout this paper. Glycans are usually classified based on their core structure, which is a particular subtree pattern of monosaccharides

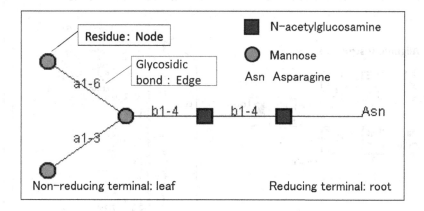

Fig. 1. Example of the N-glycan core structure, and description of related terminology

including the root. The glycan in Figure 1 is an N-glycan, or N-linked glycan structure, which is usually found on asparagine residues on the outer surface of proteins. In mammalian organisms, these glycans on average contain from 10-15 monosaccharides each. As shown in this figure, glycan structures are represented as unordered tree structures, where residues such as monosaccharides and amino acids are nodes, and glycosidic bonds are edges, and the root is usually placed on the right side, branching out towards the left. The right side of the figure is called the reducing terminal, and the left side is the non-reducing terminal end.

2.1 Representation of Glycan Profiles

In order to begin implementing our algorithm, we first needed to define a new text format for representing glycan profiles by expanding the KCF format [5]. PKCF (Profile KCF) contains information indicating alignment order, alignment position, and state (gap, missing, or residue) of each node. The left side of Figure 2 is an example of a glycan alignment of two structures, which is depicted in PKCF format on the right. The ordering of the nodes corresponds to the ordering of the glycans whose names appear in the ENTRY field. Edge information is ordered similarly. PKCF can also represent gaps and missing portions of alignments. In the NODE section, residues are listed by their names, gaps are represented as "-", and missing portions of trees are represented as "0" (the number zero). 'End' in the figure corresponds to "0" in PKCF which indicates that the aligned position does not exist in the indicated glycan structure; that is, it is beyond the terminal residue of the glycan structure.

2.2 KCaM

KCaM [6] is a pairwise glycan alignment algorithm that combines the maximum common subtree and Smith-Waterman local protein sequence alignment

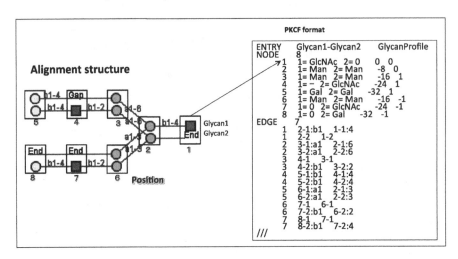

Fig. 2. PKCF format of multiple glycan alignment

algorithms. This algorithm has been preceded by a number of related algorithms, such as the tree edit distance [16] and multiple protein sequence alignment [17]. However, a description of these algorithms are beyond the scope of this manuscript, and the interested reader may refer to the original literature. The dynamic programming algorithm of KCaM is described below.

$$
Q[u,v] = \max \left\{
\begin{array}{l}
0, \\
\max_{v_i \epsilon sons(v)} \{Q[u,v_i] + d(v)\}, \\
\max_{u_i \epsilon sons(u)} \{Q[u_i,v] + d(u)\}, \\
w(u,v) + \max_{\psi \epsilon M(u,v)} \left\{ \sum_{u_i \epsilon sons(u)} Q[u_i, \psi(u_i)] \right\}
\end{array}
\right\}
$$

Here, u and v refer to a particular node u in one tree and node v in the other, and $Q[u,v]$ computes the alignment score of the subtrees rooted at u and v. $sons(x)$ refer to the children of node x, $d(x)$ refers to the gap penalty of deleting node x, $M(u,v)$ refers to the mapping of $sons(u)$ with $sons(v)$, and $w(u,v)$ refers to the score of matching nodes u and v. Thus by computing the scores of all pairs of nodes in the two input glycan structures in breadth-first order, the final score of matching the two glycans can be obtained by finding the pair of nodes with the highest score, and the alignment can be found by backtracking down to the leaves. In most cases, the best score involves the root node of at least one of the input glycans.

With this algorithm, it is possible to align most of the monosaccharides in two glycan structures. However, one must also be careful about the terminal ends. For example, let us assume that the highest scoring node pair are nodes x and y, and that node x is not the root node. Then the parent and further ancestors of node x are not aligned to any other nodes. In this case, we add "missing" nodes to the parent (and possibly grandparent, grand-grandparent, etc.) of node

y to match with the ancestors of x. The same approach is used for nodes at the non-reducing end where the leaf of one glycan is matched to an internal node in the other.

3   Methods

3.1   MCAW Algorithm

Multiple glycan alignment is based on comparing the nodes and edges of glycan profiles, similar to KCaM. In order to distinguish between single glycans and glycan profiles, we use the term *position* to indicate a node of a profile.

The procedure of the MCAW algorithm is as follows.

1. Calculate a distance matrix from the pairwise alignments (using KCaM) of all vs. all of the input glycans.
2. Create a guide tree based on the distance matrix.
3. Calculate weights of each glycan based on distance as indicated from the guide tree.
4. Add glycans to the alignment in the order of the guide tree, adding the most similar glycans first.

We generated the guide tree using the Fitch-Margoliash method [8]. The weights of each glycan structure is computed from the distance to the root of the guide tree. This is performed in order to avoid the inclusion of too many gaps in the alignment by first aligning the most similar glycan structures. Glycans are aligned according to the guidetree. Aligned glycan structures become a single profile structure. However, an alignment may also be performed on a glycan and a profile. Therefore, for the MCAW algorithm, we consider even a single glycan as a profile (containing one glycan), and thus progressively align two profiles with one another with this algorithm. The dynamic programming algorithm of MCAW is described below.

$$
Q[u,v] = \max \left\{
\begin{array}{l}
0, \\
\max_{v_i \epsilon sons(v)} \left\{ Q[u, v_i] + d(v) \right\}, \\
\max_{u_i \epsilon sons(u)} \left\{ Q[u_i, v] + d(u) \right\}, \\
\frac{1}{|A||B|} \left\{ \sum_{n=1}^{|A|} \sum_{m=1}^{|B|} w(u_n, v_m) a_n b_m \right\} + \\
\qquad \max_{\psi \epsilon M(u,v)} \left\{ \sum_{u_i \epsilon sons(u)} Q[u_i, \psi(u_i)] \right\}
\end{array}
\right\}
$$

$Q[u,v]$ is the glycan alignment score for positions u and v in profiles A and B, respectively. $|A|$ (resp. $|B|$) is the number of glycans in profile A (resp. B). a_n (resp. b_m) signifies the weight of the nth glycan in profile A (resp. mth glycan in profile B). $w(u_n, v_m)$ is the score between the nodes in positions u and v of the nth and mth glycans of profiles A and B, respectively. $sons(u)$ (resp. $sons(v)$) are the child positions of u (resp. v), and $M(u,v)$ is the mapping between the children of position u and those of position v.

3.2 MCAW Tool

We implemented steps 1 through 3 of the MCAW procedure in Perl, and step 4 was implemented in Java. The Perl program stores the resulting guide tree as a text file including the weights computed for each glycan structure. The Java program then reads in this file to progressively build up the multiple alignment. The resulting alignment is output in PKCF format. CGI-Perl was used to implement the web interface for reading in the input glycan structures and alignment parameters and also to display the results, which is a Java applet that takes the PKCF results from the MCAW program and draws the profile graphically.

4 Results

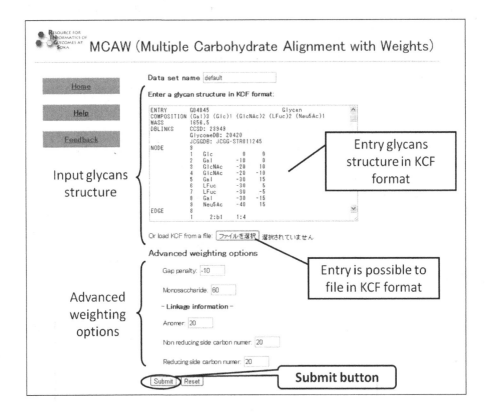

Fig. 3. A snapshot of the input screen for the MCAW tool, where glycan structures are specified in KCF format. Input glycans can also be specified as a file. There are also options to weight the gaps, residues (monosaccharides), and glycosidic bond information (anomers, non-reducing side carbon number and reducing side carbon number) in the "Advanced weighting options" which are provided with default values.

4.1   MCAW Tool

We implemented MCAW as a web tool in RINGS to output a multiple glycan alignment of an input data set of glycans on the web. The URL of the MCAW Tool is http://www.rings.t.soka.ac.jp/cgi-bin/tools/MCAW/mcaw_index.pl. Figure 3 is a snapshot of the input screen, where glycan structures are specified in KCF format.

Input glycans can also be specified as a file. Additionally, it is possible to add weighting options when calculating the alignment score. There are options to weight the gaps, residues (monosaccharides), and glycosidic bond information (anomers, non-reducing side carbon number and reducing side carbon number) in the "Advanced weighting options" in which default values are provided.

4.2   CFG Array Experiment

We have analyzed several data sets of glycan structures from binding affinity data which we obtained from the CFG. Here we present one example of our analyses. We performed an alignment of high-affinity glycan structures (illustrated in Figure 4) from glycan array data of Siglec-F, which belongs to the Siglec family that are well known to bind to sialic acids [15]. As shown in this figure, different glycan structures bound to Siglec-F with various binding affinities, which are

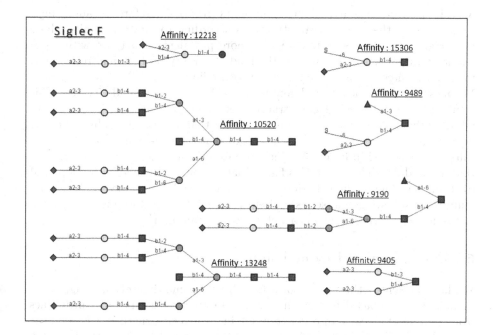

Fig. 4. Input data of Siglec-F with corresponding binding affinity values in relative fluorescence units (RFU) of each glycan structure. Each glycan structure was repeated in the data set according to these values (see text for details).

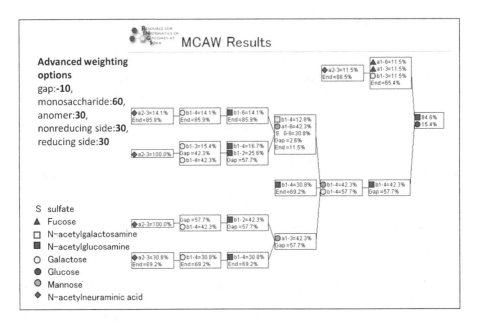

Fig. 5. Resulting glycan profile for glycans with high binding affinity to Siglec-F

provided as average relative fluorescence units (RFU). Therefore, we added multiple copies of the same glycan structures according to binding affinity; those with higher affinities were made to be more prevalent than those with lower affinities. In particular, we divided the binding affinity by 1000 and rounded to the nearest integer. Thus, the structure with affinity 10520 RFU was repeated 11 times, and the structure with affinity 15306 RFU was repeated 15 times. This resulted in a total of 87 glycan structures in the input data set. Moreover, we adjusted the scoring option configurations such that sialic acid residues (NeuAc, purple diamonds in CFG notation) are aligned at the non-reducing end. Our results are illustrated in Figure 5, which also lists the advanced weighting options that we used. It is clear from this figure that not only NeuAc a2-3, but also Gal b1-4 was very highly aligned. Moreover, we find that N-acetyglucosamine comes quite often following this series of NeuAc a2-3 Gal b1-4, forming a sialylated lactosamine structure, which is a well-known glycan motif.

5   Discussion and Conclusions

We have presented the first algorithm to perform multiple glycan alignments as well as a web-based tool so that users can quickly visualize glycan profiles from a group of glycans. In order to weigh input glycan structures based on the strength of binding affinity, weights can be incorporated into the calculation by repeating higher-affinity glycan structures in the input.

We have shown that biologically significant glycan patterns could be extracted from our tool by illustrating that sialic acids as well as other related monosaccharides could be extracted from our Siglec-F experimental results. By performing further experiments with other Siglecs and various other lectins, more patterns in glycan structure recognition can be obtained. Further work will focus on finding relationships between these patterns and protein sequence/structure. We also plan on studying the most appropriate parameters for the advanced weighting options such that users can select predefined sets of parameters that are most appropriate for their input data.

With the development of this algorithm and tool, we can also compute glycan scoring matrices [3] to analyze similarities in terms of physico-chemical properties of monosaccharides and glycosidic linkages, and further work is now possible for state model determination of Profile PSTMM and Profile OTMM. Therefore, this work is a fundamental step towards glycan recognition analysis to progress glycoinformatics research.

References

1. Akune, Y., Hosoda, M., Kaiya, S., Shinmachi, D., Aoki-Kinoshita, K.F.: The RINGS resource for glycome informatics analysis and data mining on the Web. OMICS 14(4), 475–486 (2010)
2. Alvarez, R.A., Blixt, O.: Identification of ligand specificities for glycan-binding proteins using glycan arrays. Methods Enzymol. 415, 292–310 (2006)
3. Aoki, K.F., Mamitsuka, H., Akutsu, T., Kanehisa, M.: A score matrix to reveal the hidden links in glycans. Bioinformatics 21(8), 1457–1463 (2005)
4. Aoki-Kinoshita, K.F., Ueda, N., Mamitsuka, H., Kanehisa, M.: ProfilePSTMM: capturing tree-structure motifs in carbohydrate sugar chains. Bioinformatics 22, e25–e34 (2006)
5. Aoki-Kinoshita, K.F.: Glycome Informatics: Methods and Applications. CRC Press (2009)
6. Aoki, K.F., Yamaguchi, A., Ueda, N., Akutsu, T., Mamitsuka, H., Goto, S., Kanehisa, M.: KCaM (KEGG carbohydrate matcher): a software tool for analyzing the structures of carbohydrate glycans. Nucleic Acids Research 32, 267–272 (2004)
7. Eddy, S.R.: Profile hidden Markov models. Bioinformatics 14, 755–763 (1998)
8. Fitch, W.M., Margoliash, E.: Construction of phylogenetic trees. Science 155, 279–284 (1967)
9. Fukui, S., Feizi, T., Galustian, C., Lawson, A.M., Chai, W.: Oligosaccharide microarrays for high-throughput detection and specificity assignments of carbohydrate-protein interactions. Nat. Biotechnology 20(10), 1011–1017 (2002)
10. Hashimoto, K., Aoki-Kinoshita, K.F., et al.: A new efficient probabilistic model for mining labeled ordered tree. In: Proc. KDD, pp. 177–186 (2006)
11. Ohtsubo, K., Marth, J.: Glycosylation in cellular mechanisms of health and disease. Cell 126(5), 85–867 (2006)
12. Ramakrishnan, S., Lang, W., Raguram, S., Raman, R., Venkataraman, M., Sasisekharan, R.: Advancing glycomics: Implementation strategies at the Consortium for Functional Glycomics. Glycobiology 16, 82–90 (2006)
13. Thompson, J.D., Higgins, D.G., Gibson, T.J.: Clustal W: improving the sensitivity of progressive multiple sequence alignment through sequence weighting, position-specific gap penalties and weight matrix choice. Nucleic Acids Research 22(22), 4673–4680 (1994)

14. Ueda, N., Aoki-Kinoshita, K.F., Yamaguchi, A., Akutsu, T., Mamitsuka, H.: A probabilistic model for mining labeled ordered trees: capturing patterns in carbohydrate sugar chains. IEEE Transactions on Knowledge and Data Engineering 17(8), 1051–1064 (2005)
15. Varki, A., et al. (eds.): Essentials of Glycobiology second edition. Cold Spring Harbor Laboratory Press (2009)
16. Bille, P.: A survey on tree edit distance and related problems. Theoretical Computer Science 337(1-3), 217–239 (2005)
17. Shatsky, M., Nussinov, R., Wolfson, H.J.: A method for simultaneous alignment of multiple protein structures. Proteins: Structure, Function, and Bioinformatics 56(1), 143-156, 1097-0134(2004)
18. RINGS, http://www.rings.t.soka.ac.jp
19. Consortium for Functional Glycomics, http://www.functionalglycomics.org

A Unified Adaptive Co-identification Framework for High-D Expression Data[*]

Shuzhong Zhang[1], Kun Wang[3], Cody Ashby[3],
Bilian Chen[2], and Xiuzhen Huang[3],[**]

[1] University of Minnesota, Minneapolis, MN 55455, USA
zhangs@umn.edu
[2] Xiamen University, Xiamen 361000, China
chenbilian_158@hotmail.com
[3] Arkansas State University, Jonesboro, AR 72467, USA
{kun.wang,cody.ashby}@smail.astate.edu, xhuang@astate.edu

Abstract. High-throughput techniques are producing large-scale high-dimensional (e.g., 4D with genes vs timepoints vs conditions vs tissues) genome-wide gene expression data. This induces increasing demands for effective methods for partitioning the data into biologically relevant groups. Current clustering and co-clustering approaches have limitations, which may be very time consuming and work for only low-dimensional expression datasets. In this work, we introduce a new notion of "co-identification", which allows systematical identification of genes participating different functional groups under different conditions or different development stages. The key contribution of our work is to build a unified computational framework of co-identification that enables clustering to be high-dimensional and adaptive. Our framework is based upon a generic optimization model and a general optimization method termed Maximum Block Improvement. Testing results on yeast and *Arabidopsis* expression data are presented to demonstrate high efficiency of our approach and its effectiveness.

1 Introduction

While genome data is relatively static, gene expression, which reflects gene activity, is highly dynamic. Gene expression of the cell could be used to infer the cell type, state, stage, and cell environment and may indicate a homeostasis response or a pathological condition and thus relate to development of new medicines, drug metabolism, and diagnosis of diseases [33,27,8]. High-throughput gene expression techniques, such as microarray and next-generation sequencing, are generating huge amounts of high-dimensional genome-wide expression data (e.g., data in 2D matrices: genes vs conditions, or in 3D, 4D, or 5D: genes vs time points vs conditions vs tissues vs development stages vs stimulations). While the availability of these data presents unprecedented opportunities, it also

---

[*] This research is supported by grants from NIH NCRR (5P20RR016460-11) and NIGMS (8P20GM103429-11).
[**] Corresponding author.

presents major challenges for extractions of biologically meaningful information from the large data sets. In particular, it calls for effective computational models, equipped with efficient solution methods, to categorize gene expression data into biologically relevant groups in order to facilitate further functional assessment of important biological and biomedical processes. Classical clustering and co-clustering analysis is a worthy approach in this endeavor.

Clustering is usually applied to partition expression data into groups. A lot of research has been conducted in clustering. Cf. [12] for classical clustering, where the author discussed two classes of clustering: hierarchical clustering and partitioning, and three popular clustering methods: hierarchical clustering [15], k-means clustering [35] and the self organizing map (SOM) method [34]. The classical clustering methods cluster genes into groups based on their similar expression on all the considered conditions. The concept of *co-clustering* was introduced to 2D expression data analysis by Cheng and Church [7]. The co-clustering method can cluster genes and conditions simultaneously. Subsequently, many co-clustering algorithms were developed, such as the plaid model approach [23], xMotif[28], BiMax [29], OPSM [3], Bicluster [9], BCC[2], and ROCC [11]. Different techniques improving co-clustering approaches were also developed [1,38,39]. Readers may refer to [26,17,9,11] for the ideas of different co-clustering algorithms and techniques and [29] for a comprehensive comparison of the popular co-clustering approaches. Recently there are approaches developed for 3D expression data clustering analysis [31,25,41,21]. However, for current clustering and co-clustering approaches, there are important issues to address:

How to develop a systematic method to be able to associate one item to multiple co-groups under different conditions or different development stages of high-dimensional gene expression data? Most classical clustering and co-clustering methods assign one element to one specific cluster or co-clusters. For gene expression analysis, it is important to associate genes/conditions with multiple clusters or co-clusters, inducing the concept of "soft" clustering which allows elements to be members of multiple groups. In [30], soft clustering is represented by a probabilistic distribution. There are methods considering co-cluster overlapping such as the ROCC approach in [11], which, however, only tries to merge some related co-clusters as a post-processing step, and the approach in [7], which allows overlaps but has introduced the masking problem (where the elements in a previously-discovered co-cluster are replaced by random numbers). Refer to [26] for different additive and multiple overlapping models for co-clustering.

In this work, our co-identification approach is different from the previous overlapping bicluster approaches: Our approach does not aim to overlap the biclusters as a post-processing step. Our approach will systematically identify the genes involved in different functional groups at different time points or conditions while the biclusters are being built all at the same time. Note that Lazzeroni and Owen [23] attempted to discover one co-cluster at a time in an iterative process where a plaid model is obtained [26].

How to naturally determine the number of clusters and co-clusters? Classical clustering and co-clustering methods usually rely on the predetermined numbers as the numbers of clusters and co-clusters. There are methods for estimating the number of clusters or co-clusters in a data set, such as the SVD method [9], the gap statistic or similarity matrix [4,37,14,24], which are, however, not related to the clustering process. In this work we develop an adaptive method to determine the number of co-clusters while the co-clusters are being formed.

2   Methods

We build the computational approach for co-identification based on block optimization, and develop new algorithms from a general scheme which we termed as Maximum Block Improvement (MBI), for naturally growing the size of co-groups and encouraging the degree of co-identification.

The Co-clustering Problem. To illustrate the ideas, consider the conventional co-clustering formulation [40]. Suppose that $A \in \Re^{n_1 \times n_2 \times \cdots \times n_d}$ is an d-dimensional tensor. Let $I_j = \{1, 2, \cdots, n_j\}$ be the set of indices on the j-th dimension, $j = 1, 2, ..., d$. We wish to find a p_j-partition of the index set I_j, say $I_j = I_1^j \cup I_2^j \cup \cdots \cup I_{p_j}^j$, where $j = 1, 2, ..., d$, in such a way that each of the *sub-tensor* $A_{I_{i_1}^1 \times I_{i_2}^2 \times \cdots \times I_{i_d}^d}$ is as tightly packed up as possible, where $1 \leq i_j \leq n_j$ and $j = 1, 2, ..., d$. The notion that plays an important role in our model is the so-called *mode product* between a tensor X and a matrix P. Suppose that $X \in \Re^{p_1 \times p_2 \times \cdots \times p_d}$ and $P \in \Re^{p_i \times m}$. Then, $X \times_i P$ is a tensor in $\Re^{p_1 \times p_2 \times \cdots \times p_{i-1} \times m \times p_{i+1} \times \cdots \times p_d}$, whose $(j_1, j_2, \cdots, j_{i-1}, j_i, j_{i+1}, \cdots, j_d)$-th component is defined by

$$(X \times_i P)_{j_1, j_2, \cdots, j_{i-1}, j_i, j_{i+1}, \cdots, j_d} = \sum_{\ell=1}^{p_i} X_{j_1, j_2, \cdots, j_{i-1}, \ell, j_{i+1}, \cdots, j_d} P_{\ell, j_i}.$$

Let $X_{j_1, \cdots, j_{i-1}, j_i, j_{i+1}, \cdots, j_d}$ be the value of the co-cluster $(j_1, \cdots, j_{i-1}, j_i, j_{i+1}, \cdots, j_d)$ with $1 \leq j_i \leq p_i$, $i = 1, 2, ..., d$. Let an assignment matrix $Y^j \in \Re^{n_j \times p_j}$ for the indices for j-th array of tensor A be:

$$Y_{ik}^j = \begin{cases} 1, & \text{if } i \text{ is assigned to the } k\text{-th partition } I_k^j; \\ 0, & \text{otherwise.} \end{cases}$$

Then, we introduce a *proximity* measure $f(s) : \Re \to \Re_+$, with the property that $f(s) \geq 0$ for all $s \in \Re$ and $f(s) = 0$ if and only if $s = 0$. The co-clustering problem can be formulated as

$$(CC) \min \sum_{j_1=1}^{n_1} \sum_{j_2=1}^{n_2} \cdots \sum_{j_d=1}^{n_d}$$
$$f\left(A_{j_1, \cdots, j_d} - (X \times_1 Y^1 \times_2 \cdots \times_d Y^d)_{j_1, \cdots, j_d}\right)$$
$$\text{s.t. } X \in \Re^{p_1 \times p_2 \times \cdots \times p_d}, Y^j \in \Re^{n_j \times p_j}$$
$$\text{is a row assignment matrix, } j = 1, 2, ..., d$$

We may consider a variety of proximity measures. For instance, if $f(s) = |s|^2$ then (CC) can be written as

$$(CC_1) \min \left\| A - X \times_1 Y^1 \times_2 Y^2 \times_3 \cdots \times_d Y^d \right\|_F$$
$$\text{s.t. } X \in \Re^{p_1 \times p_2 \times \cdots \times p_d}, Y^j \in \Re^{n_j \times p_j}$$
$$\text{is a row assignment matrix, } j = 1, 2, ..., d,$$

A well-known approach to the above problem is the *block descent method* [5], which, though simple to implement, fails to converge to a stationary point (local optimum). Recently this issue of convergence was resolved in [6] and the authors proposed an enhanced search algorithm termed the *maximum block improvement* (MBI) method. This method is highly effective and easy to implement, according to our experience in the co-clustering analysis for gene expression data, alongside its excellent theoretical convergence properties [40,6].

An Adaptive Co-identification Model. The power of the MBI method is now extended to solve a much more complex model - the co-identification model, where even the size of a block becomes a variable. This degree of flexibility is exactly needed in the analysis of the gene expression data, since any presumed knowledge, such as the total number of co-clusters and the number of times a gene is allowed to assign to co-clusters, would risk the blockage of key information from being revealed. By introducing the needed flexibility one has to deal with the newly introduced complications: optimization will naturally select only *one* assignment in a group, and will like to have *as many as possible* groups, notwithstanding the flexibility. To circumvent the difficulty, an enhanced model can be as follows:

$$
(CI) \quad \min \sum_{j_1=1}^{n_1} \sum_{j_2=1}^{n_2} \cdots \sum_{j_d=1}^{n_d}
$$
$$
f\left(A_{j_1,\cdots,j_d} - (X \times_1 Y^1 \times_2 \cdots \times_d Y^d)_{j_1,\cdots,j_d}\right)
$$
$$
+ \lambda(p_1, p_2, ..., p_d) - \mu\left(\sum_{i,k} Y_{ik}^1, ..., \sum_{i,k} Y_{ik}^d\right)
$$
$$
\text{s.t.} \quad X \in \Re^{p_1 \times p_2 \times \cdots \times p_d},
$$
$$
Y^j \in \{0,1\}^{n_j \times p_j} \text{ with } \sum_{i,k} Y_{ik}^j \geq 1, \ j = 1, ..., d,
$$

where $\lambda(p_1, p_2, ..., p_d)$ is a penalty function, intended to punish the possible abuse of more groups for identification, and $\mu(\sum_{i,k} Y_{ik}^1, ..., \sum_{i,k} Y_{ik}^d))$ is an *incentive* function, intended to encourage the identification of similar data in a group, without restricting to only one data per row. Some immediate choices of a penalty function include $\lambda(p_1, p_2, ..., p_d) = c_1 p_1 \cdots p_d$ or $\lambda(p_1, p_2, ..., p_d) = c_1 \sum_{i=1}^{d} p_i$, where c_1 is a positive constant. Similarly, choices of incentive function include: $\mu(y_1, ..., y_d) = c_2 \sum_{i=1}^{d} y_i$, where $c_2 > 0$ is another parameter. The purpose of introducing such penalty and incentive functions is to: (a) encourage the adaptiveness in the choices of the groups; and (b) avoid the introduction of too many unnecessary groups. Notice that in the new model, even the dimensions $(p_1, ..., p_d)$ become a part of the decision. Such optimization models have rarely been studied in the optimization literature; however, they perfectly fit in the realm where the power of the MBI method would extend, and they are very relevant for the gene expression data analysis. The features of the new model are summarized in the following.

– By replacing co-clusters, we work with the new notion of *co-groups*, which will lead to the new *co-identification* model, allowing assigning one element to multiple co-groups under different conditions.

   The new model will characterize and model the information of different groups of genes being regulated by different transcription factors at different conditions, or the same group of genes at different conditions being regulated by a different group of transcription factors, or the same group of genes involving in different networks and pathways.

– We develop an adaptive scheme to naturally grow the size of *co-groups*.

Our model will naturally systematically search for the number of co-clusters for every specific gene expression datasets. One idea is to apply the MBI approach [6,40] to conduct local search for the values of the parameters $p_1, ..., p_d$ to control the number of co-groups. Methods like higher-order principal components analysis [36] and higher-order singular value decomposition (HOSVD) [22] can be applied to set the initial values of the parameters $p_1, ..., p_d$ as the start point of the local search.

– We develop a general optimization scheme for the co-identification model.

Please refer to Figure 1 for our generic algorithm for the co-identification model based on the MBI method. The general optimization scheme of MBI is suitable for not only 2D but also multiple- and high- dimensional (3D, 4D, 5D) gene expression data.

Our co-identification model could accommodate different evaluation and objective functions. Therefore, different co-clustering approaches previously developed in the literature could be considered as special cases of our approach. Besides L_1, L_2, L_∞ [40], our model could use the Bergman divergence functions [2], where the authors chose the appropriate Bregman divergence based on the underlying data generation process or noise model. For classical clustering, Euclidean distance and Pearson correlation are both reasonable distance measures, with Euclidean distance being more appropriate for log ratio data, and Pearson correlation working better for absolute-valued data [16,10,12].

3   Results and Discussion

To simplify the testing, in the following experiments, we separate the determination of the number of co-groups from the co-identification analysis. Our approach is implemented using C++. The figures are generated using MATLAB. The testing is mainly performed on a regular PC (3GB Mem, 64bit Windows7). The running-time testing is conducted on a server (PowerEdge 2950III, 32GB Mem). We use both synthetic and real datasets to validate and evaluate the co-identification model and the generic MBI algorithm. We give a brief description of the real datasets we use to test our algorithm in this section. The 2D dataset is the yeast gene expression dataset with 2884 genes and 17 conditions. The detailed information about this dataset can be found in [7,35]. The 3D dataset is the Arabidopsis thaliana abiotic stress gene expression from [18,32]. We extract a file which has 2395 genes, 5 conditions (cold, salt, drought, wound, and heat), with each condition containing 6 time points. Due to space limit, some detailed testing results on synthetic datasets and some identified co-groups from other real datasets are not shown here.

Testing Results for Determining the Number of Co-groups. We test the MBI approach for determining the number of co-groups: We first randomly

Generic co-identification algorithm

Input: $A \in \Re^{n_1 \times n_2 \times \cdots \times n_d}$ is an d-dimensional tensor, which holds the d-dimensional gene expression data set. Parameters p_1, p_2, ..., p_d, are all positive integers, $0 < p_i \leq n_i$, $1 \leq i \leq d$.
Output: $p_1 \times p_2 \times \cdots \times p_d$ co-groups of A.

Main Variables: A non-negative integer k as the loop counter;
A $p_1 \times p_2 \times \cdots \times p_d$-tensor X with each entry a real number as the artificial central point of each of the co-groups;
A $n_i \times p_i$-matrix Y_i as the assignment matrix with $\{0,1\}$ as the value of each entry, $1 \leq i \leq d$.

Begin
[A.] Start with some initial values for p_1, p_2, ..., p_d in order to control the number of co-groups, conduct local search on each p_i, $1 \leq i \leq d$.

[A.0] *(Initialization)*. $Y^0 = X$; Choose a feasible solution $(Y_0^0, Y_0^1, Y_0^2, \cdots, Y_0^d)$ and compute the initial objective value $v_0 := f(Y_0^0, Y_0^1, Y_0^2, \cdots, Y_0^d) + c(p_1, p_2, ..., p_d; Y_0^1, Y_0^2, \cdots, Y_0^d)$. Set the loop counter $k := 0$.
[A.1] *(Block Improvement)*. For each $i = 0, 1, 2, \ldots, d$, solve

$$(G_i) \max f(Y_k^0, Y_k^1, \cdots, Y_k^{i-1}, Y^i, Y_k^{i+1}, \cdots, Y_k^d) + c(p_1, ..., p_d; Y_k^1, \cdots, Y_k^{i-1}, Y^i, Y_k^{i+1}, ..., Y_k^d)$$
$$\text{s.t.} \quad Y^i \in \Re^{n_j \times p_j} \text{ is an assignment matrix,}$$

and let

$$y_{k+1}^i := \arg\max f(Y_k^0, Y_k^1, ..., Y_k^{i-1}, Y^i, Y_k^{i+1}, ..., Y_k^d) + c(p_1, ..., p_d; Y_k^1, ..., Y_k^{i-1}, Y^i, Y_k^{i+1}, ..., Y_k^d)$$
$$w_{k+1}^i := f(Y_k^0, Y_k^1, ..., Y_k^{i-1}, y_{k+1}^i, Y_k^{i+1}, ..., Y_k^d) + c(p_1, ..., p_d; Y_k^1, ..., Y_k^{i-1}, Y^i, Y_k^{i+1}, ..., Y_k^d).$$

[A.2] *(Maximum Improvement)*. Let $w_{k+1} := \max_{1 \leq i \leq d} w_{k+1}^i$ and $i^* = \arg\max_{1 \leq i \leq d} w_{k+1}^i$. Let

$$Y_{k+1}^i := Y_k^i, \ \forall \ i \in \{0, 1, 2, \cdots, d\} \backslash \{i^*\}$$
$$Y_{k+1}^{i^*} := y_{k+1}^{i^*}$$
$$v_{k+1} := w_{k+1}.$$

[A.3] *(Stopping Criterion)*. If $|v_{k+1} - v_k| > \epsilon$, set $k := k + 1$, and go to Step 1; Otherwise, set $V_{p_i} = v_{k+1}$.

[B.] According to the assignment matrices $Y_{k+1}^1, Y_{k+1}^2, \cdots, Y_{k+1}^d$ corresponding to the maximum V_{p_i}, $1 \leq i \leq d$, print the $p_1 \times p_2 \times \cdots \times p_d$ co-groups of A.
End

Fig. 1. Co-identification Algorithm Based on Maximum Block Improvement

generate some starting points, say the values of $(p_1, ..., p_d)$, and then we conduct a local improvement strategy, meaning that we try to increase or decrease each p_i value until no more improvement is possible locally. We refer the reader to Table 1 for our testing on the effectiveness of the proposed local search strategy.

Table 1. Testing of the Maximum Block Improvement strategy on the 2D yeast dataset from [35]. The initial objective function value is -25900; the first column: the initial p values; the second column: the new p values after the local search; the third column: the objective function values with the initial p values and with the new p values respectively; and the last column: the running time for the local search.

Initial $p_{1,0}$, $p_{2,0}$	New p_1, p_2	Obj-value (Initial value -25900)	Run Time (seconds)
20,10	25,16	-7192.42, -6810.89	646.91
5,8	13,9	-9291.28, -7498.96	341.67
97,10	96,11	-6706.33, -6220.54	364.64
68,8	69,11	-6763.02, -6337.49	441.18
32,9	35,15	-6967.74, -6591.19	624.38
20,11	25,16	-7202.46, -6810.89	609.70
19,4	28,6	-7480.57, -7041.12	487.72
51,3	50,5	-7157.43, -6808.80	224.95
65,9	64,11	-6736.08, -6413.91	349.50
43,1	42,5	-7888.83, -6832.58	271.23
6,3	11,5	-8918.90, -7677.44	202.40
2,4	11,5	-15973.50, -7677.44	280.21

Coherent Groups from 3D Arabidopsis Gene Expression Data. To demonstrate the effectiveness of our algorithm in search for coherent patterns from gene expression datasets, Figure 2 provides several exemplary 3D co-groups identified from the 3D Arabidopsis dataset. We present here the co-groups with a small number of genes, which shows clear coherent expression pattern over a series of time points and under different conditions. These co-groups clearly facilitate further functional analysis of the genes. The analysis of 3D *Arabidopsis* dataset in [32] has generated three biologically relevant co-cluster module types: 1) modules with genes that are co-regulated under several conditions are the most prevalent ones, 2) Coherent modules with similar responses under all conditions occurred frequently, too, 3) A third module type, which covers a response specific to a single condition was also detected, but rarely. Especially for the third module type, refer to the top-left pattern of Figure 2, which shows the two Arabidopsis genes are co-regulated at all 3 conditions: cold, salt, drought, but are differently expressed at the condition heat. The two genes are: 250296_at and 245955_at. The following information of the two genes is from http://www.arabidopsis.org/: gene 250296_at: 17.6 kDa class II heat shock protein (HSP17.6-CII), identical to 17.6 kDa class II heat shock protein SP:P29830 from (Arabidopsis thaliana); gene 245955_at: glycosyl hydrolase family 1 protein, contains Pfam PF00232 : Glycosyl hydrolase family 1 domain, TIGR-FAM TIGR01233: 6-phospho-beta-galactosidase, similar to beta-glucosidase 1

(GI:12043529) (Arabidopsis thaliana). *Gene 250296_at (green-colored on the figure) with significantly high expression at the heat condition, which is identified by our approach, is confirmed coding for a heat-shock protein.*

Results from Yeast Gene Expression Data. We apply our co-identification approach to the analysis of the Saccharomyces cerevisiae gene expression data collected in [35]. This data contains the expression of 2884 genes under 17 conditions. From our co-identification analysis, we identify genes that are co-listed in two or more co-groups, such as YER068W, YDR103W, YGL130W, YJR129C, YLR425W, YOR383C and YLL004W. This information could be used to predict the functions of unknown genes from the known functions of the genes in the same co-groups. This information could also lead to identification of previously undetected novel functions of genes. Specifically we have checked the function information (http://www.yeastgenome.org) of the following two genes which are involved in more than one co-group.

Gene ORC3/YLL004W: Subunit of the origin recognition complex, which directs DNA replication by binding to replication origins and is also involved in transcriptional silencing. We find out three pathways from KEGG Pathway Database (http://www.genome.jp/kegg/) in which this gene are involved.

Gene STE5/YDR103W: Pheromone-response scaffold protein that controls the mating decision; binds Ste11p, Ste7p, and Fus3p kinases, forming a MAPK cascade complex that interacts with the plasma membrane and Ste4p-Ste18p; allosteric activator of Fus3p. Our approach identifies two co-groups in which Gene STE5/YDR103W is involved. The two co-groups are biologically significant with low p-values: Co-group#41 (6 genes: STE5 ADA2 AFG3 MOT2 PHO23 DSS4), zinc ion binding, with p-value 0.001735, Co-group#62 (4 genes: STE5 MOT2 ORC3 RAD52), pheromone response, mating-type determination, sex-specific proteins, with p-value 0.007773 (The p-value information is obtained from the website of Funcspec: http://funspec.med.utoronto.ca/).

Running Time Analysis. Our approach is highly efficient and could be applied to 2D, 3D and higher-dimensional gene expression data. When testing on 3D datasets (genes vs time points vs conditions), the running time of our approach increases linearly with the number of genes, the number of time points, or the number of conditions. We conduct our running-time testing on the Arabidopsis thaliana abiotic stress 3D gene expression datasets from [32]. We use a file which has 2395 genes, 5 conditions (cold, salt, drought, wound, and heat), with each condition containing 6 time points. Especially when we keep the number of genes (2395) and increase the second dimension for the number of time points, or the third dimension for the number of the conditions, we increase the size of the dataset significantly, however, the running time of our algorithm still has only a linear increase (Figure 3). The performance of our algorithm is very robust. The testing results demonstrate the high efficiency of our algorithm. In contrast, other existing methods for 3D co-clustering such as *TriCluster* [41], the running time is exponential with the number of time points, or the number of conditions. Other existing methods usually do not work for gene expression data of four- or higher- dimensions.

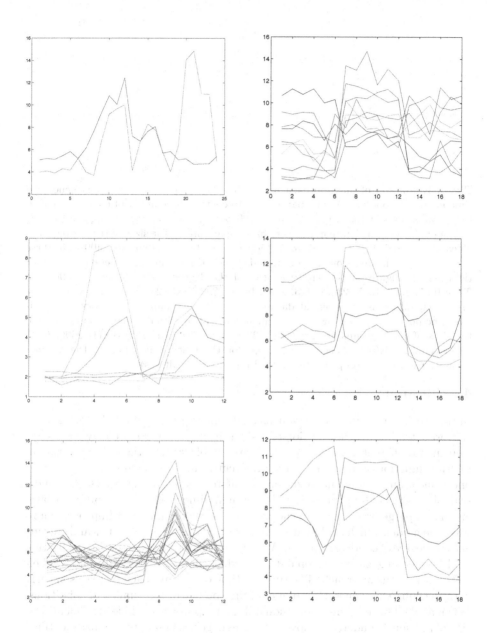

Fig. 2. Exemplary 3D co-groups (with no. of genes x 6 time points x no. of conditions) generated from the 3D Arabidopsis dataset. Genes have different expression patterns at different conditions. The x-axis represents the different number of time points (with every 6 time-points in one condition), while the y-axis represents the values of the gene expression level. Each curve corresponds to the expression of one gene. For example, the co-group at the top-left shows the clear expression patterns of 2 genes at 4 conditions (cold, salt, drought and heat).

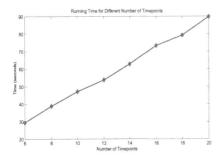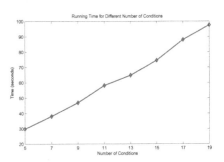

Fig. 3. Evaluation of our approach on the 3D Arabidopsis dataset (2396 genes x 6 timepoints x 5 conditions). The parameters to control the number of co-groups for all these evaluations are set the same: $p_1 = 100, p_2 = 2, p_3 = 3$. For testing on different number of genes (this figure not shown due to space limit), the sizes of the 8 groups are 300x6x5, 600x6x5, 900x6x5, 1200x6x5, 1500x6x5, 1800x6x5, 2100x6x5, 2400x6x5 (these datasets are truncated from the original dataset). For testing on different number of time points (Figure on the left), the sizes of the 8 groups are 2396x6x5, 2396x8x5, 2396x10x5, 2396x12x5, 2396x14x5, 2396x16x5, 2396x18x5, 2396x20x5 (except for the first group which is the original dataset, the other 7 groups contain added repetitive time points). For testing on different number of conditions (Figure on the right), the sizes of the 8 groups are 2396x6x5, 2396x6x7, 2396x6x9, 2396x6x11, 2396x6x13, 2396x6x15, 2396x6x17, 2396x6x19 (except for the first group which is the original dataset, the other 7 groups contain added repetitive conditions).

4 Summary

In this work, for complex high-dimensional gene expression data clustering analysis, we introduce the new notion of co-identification, so that we may assign one element to different *groups* or *co-groups* to enable systematical identification of multiple functions of one gene or the involvement in multiple functional groups of one element under different conditions or different development stages. We build a scalable model not only for 2D but also for high-dimensional gene expression data, and develop a general adaptive scheme based on Maximum Block Improvement to solve the model, which could naturally grow the size of the co-groups and encourage the degree of co-identification. We apply the unified adaptive co-identification analysis to real gene expression datasets, which shows the high efficiency and effectiveness of the approach. The running time of our approach increases linearly with the number of genes, the number of time points, or the number of conditions. When applied to real gene expression data, our approach could lead to identification of previously undetected novel functions of genes. Our approach has identified a differentially expressed heat-shock gene that are co-regulated with other genes under three other conditions (cold, salt, and drought) of 18 time points from the *Arabidopsis* dataset (this type of patterns are considered important and rare by the EDSIA method [32]), and also identified genes that participate different biologically significant functional groups from the yeast dataset. The co-identification analysis could possibly enrich further functional study of important biological processes, which may lead to new insights into genome-wide gene expression of the cell.

Our approach is a unified systematic approach, which could be used for high-dimensional gene expression data analysis (as well as for high-dimensional data analysis for applications of other fields). There are few current approaches which efficiently work for datasets with dimensions greater than 3. Our approach is general enough to embrace many other clustering and biclustering methods proposed in the literature as special cases. Especially our framework could apply as evaluation or objective functions the 6 different schemes listed in [2]. Our co-identification model provides the framework for incorporating additional ideas from approximation [20,19], parameterization [13], randomization and probabilistic analysis, or approaches combined with statistic and greedy strategies.

References

1. Aguilar-Ruiz, J.S.: Shifting and scaling patterns from gene expression data. Bioinformatics 21, 3840–3845 (2005)
2. Banerjee, A., et al.: A generalized maximum entropy approach to bregman coclustering and matrix approximation. JMLR 8, 1919–1986 (2007)
3. Ben-Dor, A., et al.: Discovering local structure in gene expression data: the order-preserving submatrix problem. In: RECOMB 2002, pp. 49–57 (2002)
4. Ben-Hur, A., et al.: A stability based method for discovering structure in clustered data. In: Proc. of PSB (2002)
5. Bertsekas, D.P.: Nonlinear Programming. Athena Scientific, Belmont (1999)
6. Chen, B., et al.: Maximum block improvement and polynomial optimization. SIAM Journal on Optimization 22, 87–107 (2012)
7. Cheng, Y., Church, G.M.: Biclustering of expression data. In: Proc. Int. Conf. Intell. Syst. Mol. Biol., vol. 8, pp. 93–103 (2000)
8. Cheung, A.N.: Molecular targets in gynaecological cancers. Pathology 39, 26–45 (2007)
9. Cho, H., et al.: Minimum sum-squared residue co-clustering of gene expression data. In: Proc. SIAM on Data Mining, pp. 114–125 (2004)
10. Costa, I.G., et al.: Comparative analysis of clustering methods for gene expression time course data. Genet. Mol. Biol. 27, 623–631 (2004)
11. Deodhar, M., et al.: Hunting for Coherent Co-clusters in High Dimensional and Noisy Datasets. In: IEEE Intl. Conf. on Data Mining Workshops (2008)
12. D'haeseleer, P.: How does gene expression clustering work? Nature Biotechnology 23, 1499–1501 (2005)
13. Downey, R.G., Fellows, M.R.: Parameterized Complexity. Springer (1999)
14. Dudoit, S., Fridlyand, J.: A prediction based resampling method for estimating the number of clusters in a data set. Genome Biology 3, 1–21 (2002)
15. Eisen, M.B., et al.: Cluster analysis and display of genome-wide expression patterns. Proc. Natl. Acad. Sci. 95, 14863–14868 (1998)
16. Gibbons, F.D., Roth, F.P.: Judging the quality of gene expression-based clustering methods using gene annotation. Genome Res. 12, 1574–1581 (2002)
17. Hochreiter, S., et al.: FABIA: factor analysis for bicluster acquisition. Bioinformatics 26, 1520–1527 (2010)
18. Kilian, J., et al.: The AtGenExpress global stress expression data set: protocols, evaluation and model data analysis of UV-B light, drought and cold stress responses. The Plant Journal 2, 347–363 (2007)

19. Kolda, T.G., Bader, B.W.: Tensor decompositions and applications. SIAM Review 51, 455–500 (2009)
20. Jegelka, S., Sra, S., Banerjee, A.: Approximation Algorithms for Tensor Clustering. In: Gavaldà, R., Lugosi, G., Zeugmann, T., Zilles, S. (eds.) ALT 2009. LNCS, vol. 5809, pp. 368–383. Springer, Heidelberg (2009)
21. Jiang, D., et al.: Mining coherent gene clusters from gene-sample-time microarray data. In: Proc. ACM SIGKDD, pp. 430–439 (2004)
22. Lathauwer, D., et al.: A multilinear singular value decomposition. SIAM J. Matrix Anal. Appl. 21, 1253–1278 (2000)
23. Lazzeroni, L., Owen, A.B.: Plaid models for gene expression data. Statistica Sinica 12, 61–86 (2002)
24. Lee, M., et al.: Biclustering via Sparse Singular Value Decomposition. Biometrics 66, 1087–1095 (2010)
25. Li, A., Tuck, D.: An Effective Tri-Clustering Algorithm Combining Expression Data with Gene Regulation. Gene Regulation and Systems Biology 3, 49–64 (2009)
26. Madeira, S.C., Oliveira, A.L.: Biclustering algorithms for biological data analysis: a survey. IEEE/ACM Trans. Comput. Biology Bioinform. 1, 24–45 (2004)
27. Magic, Z., et al.: cDNA microarrays: identification of gene signatures and their application in clinical practice. J. BUON 12(suppl.1), S39–S44 (2007)
28. Murali, T., Kasif, S.: Extracting conserved gene expression motifs from gene expression data. In: Pacific Symposium on Biocomputing, vol. 8, pp. 77–88 (2003)
29. Prelic, A., et al.: A systematic comparison and evaluation of biclustering methods for gene expression data. Bioinformatics 22, 1122–1129 (2006)
30. Snider, N., Diab, M.: Unsupervised Induction of Modern Standard Arabic Verb Classes. In: HLT-NAACL, New York (2006)
31. Strauch, M., et al.: A Two-Step Clustering for 3-D Gene Expression Data Reveals the Main Features of the Arabidopsis Stress Response. J. Integrative Bioinformatics 4, 54–66 (2007)
32. Supper, J., et al.: EDISA: extracting biclusters from multiple time-series of gene expression profiles. BMC Bioinformatics 8, 334–347 (2007)
33. Suter, L., et al.: Toxicogenomics in predictive toxicology in drug development. Chem. Biol. 11, 161–171 (2004)
34. Tamayo, P., et al.: Interpreting patterns of gene expression with self-organizing maps: methods and application to hematopoietic differentiation. Proc. Natl. Acad. Sci. USA 96, 2907–2912 (1999)
35. Tavazoie, S., et al.: Systematic determination of genetic network architecture. Nat. Genet. 22, 281–285 (1999)
36. Tucker, L.R.: Some mathematical notes on three-mode factor analysis. Psychometrika 31, 279–311 (1966)
37. Tibshirani, R., et al.: Estimating the Number of Clusters in a Dataset via the Gap Statistic. J. Royal Stat. Soc. B 63, 411–423 (2001)
38. Wang, H., et al.: Clustering by pattern similarity in large data sets. In: Proc. KDD 2002, pp. 394–405 (2002)
39. Xu, X., et al.: Mining shifting-and-scaling co-regulation patterns on gene expression profiles. In: Proc. ICDE 2006, pp. 89–98 (2006)
40. Zhang, S., Wang, K., Chen, B., Huang, X.: A New Framework for Co-clustering of Gene Expression Data. In: Loog, M., Wessels, L., Reinders, M.J.T., de Ridder, D. (eds.) PRIB 2011. LNCS, vol. 7036, pp. 1–12. Springer, Heidelberg (2011)
41. Zhao, L., Zaki, M.J.: Tricluster: an effective algorithm for mining coherent clusters in 3D microarray data. In: Proc. ACM SIGMOD, pp. 694–705 (2005)

Protein Clustering on a Grassmann Manifold

Chendra Hadi Suryanto[1], Hiroto Saigo[2], and Kazuhiro Fukui[1]

[1] Graduate School of Systems and Information Engineering,
Department of Computer Science, University of Tsukuba, Japan
http://www.cvlab.cs.tsukuba.ac.jp/
[2] Department of Bioscience and Bioinformatics,
Kyushu Institute of Technology, Japan
http://www.bio.kyutech.ac.jp/~saigo/

Abstract. We propose a new method for clustering 3D protein structures. In our method, the 3D structure of a protein is represented by a linear subspace, which is generated using PCA from the set of synthesized multi-view images of the protein. The similarity of two protein structures is then defined by the canonical angles between the corresponding subspaces. The merit of this approach is that we can avoid the difficulties of protein structure alignments because this similarity measure does not rely on the precise alignment and geometry of each alpha carbon atom. In this approach, we tackle the protein structure clustering problem by considering the set of subspaces corresponding to the various proteins. The clustering of subspaces with the same dimension is equivalent to the clustering of a corresponding set of points on a Grassmann manifold. Therefore, we call our approach the *Grassmannian Protein Clustering Method (GPCM)*. We evaluate the effectiveness of our method through experiments on the clustering of randomly selected proteins from the Protein Data Bank into four classes: alpha, beta, alpha/beta, alpha+beta (with multi-domain protein). The results show that GPCM outperforms the k-means clustering with Gauss Integrals Tuned, which is a state-of-the-art descriptor of protein structure.

Keywords: protein structure clustering, k-means, Mutual Subspace Method, Grassmann manifold, Gauss Integrals.

1 Introduction

Since there are numerous proteins whose functions are yet to be understood, accurately predicting protein structure and function is a main issue in structural bioinformatics. One important task in such computations is the clustering of 3D protein structures. In the clustering process, a distance metric is required to calculate the similarity between two proteins. The metric mostly used to measure the similarity between two protein structures is based on the root mean square deviation (RMSD) calculated from the coordinates of protein backbones. However RMSD raises problems in finding the best alignment and requires the superposition of two target proteins, which can be especially difficult when the shapes of the proteins are substantially different.

T. Shibuya et al. (Eds.): PRIB 2012, LNBI 7632, pp. 71–81, 2012.

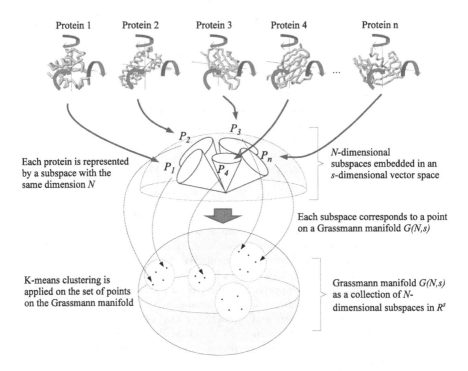

Fig. 1. The framework of the proposed method (GPCM)

There have been many attempts to establish an optimal alignment of protein structures based on RMSD [1][2][3], but there are few effective protein structure descriptors that overcome the limitations of RMSD. For example, when the 3D structure of a protein is represented by an oriented open curve in 3D space, a compact descriptor in the form of a 30-dimensional vector has been defined using two geometric measures, writhe and average crossing number [4]. This idea has been extended to a more robust descriptor, called Gauss Integrals Tuned (GIT) [5].

We propose a new method for clustering 3D protein structures in which the 3D structure of a protein is represented by a linear subspace (see Figure 1). Each subspace is generated using PCA from the set of synthesized multiple view images of the protein, as shown in Figure 2. The similarity of two protein structures is defined by the canonical angles between the corresponding subspaces. The advantage of this similarity measure [6] is that it does not rely on the precise alignment and geometry of the protein structures, so we can avoid the difficulties of the protein structure alignment. In this approach, we tackle the clustering problem of protein structure by considering the set of subspaces corresponding to the proteins. The clustering of subspaces with the same dimension is equivalent to the clustering of a corresponding set of points on a Grassmann manifold. Therefore, we call our approach the *Grassmannian Protein Clustering Method (GPCM)*.

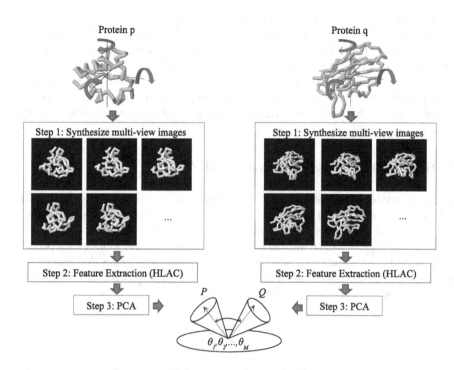

Fig. 2. Calculating the similarity of two protein structures based on canonical angles

The validity of the proposed method is demonstrated through experiments on the clustering of randomly selected proteins from the Protein Data Bank (PDB) into four protein fold classes: alpha (alpha-helices), beta (beta-sheets), alpha/beta (beta-alpha-beta motifs, mainly parallel beta-sheets), and alpha+beta (segregated alpha and beta regions, mainly anti parallel beta-sheets with some multi-domain proteins). The results show that our clustering method outperforms the conventional k-means clustering method with GIT [11], which is a state-of-the-art protein descriptor.

The organization of this paper is as follows. In Section 2 we provide a short description of the Gauss integrals descriptor. Then, in Section 3, we explain our approach which uses canonical angles to define protein structure similarity. Next, in Section 4 we outline the method of clustering subspaces using k-means on a Grassmann manifold. The experimental results are described and discussed in Section 5. Finally, in Section 6 we give some conclusions.

2   Protein Descriptor Based on Gauss Integrals

In the methodology of Gauss integrals as a protein descriptor, a protein backbone which is a trace of C_α carbon atoms is considered as an oriented open curve in space [4]. A series of 29 first, second, and third-order invariants, based on the generalized Gauss integrals for the writhe and average crossing number, are computed over the curve. This set is considered as a vector $\in \mathcal{R}^{29}$ for interpreting the topology of 3-dimensional protein structure [5]. Including the number of residues, the final descriptor is a compact 30-dimensional feature vector of Gauss integrals. In order to make the Gauss integral based descriptor more robust to perturbations of protein structure, it has been extended to the Gauss Integrals Tuned (GIT) descriptor which uses a 31-dimensional vector [5]. In this paper, we focus on the k-means clustering method using Euclidean distances in the 31-dimensional GIT vector space.

3   Similarity Based on Canonical Angles

Canonical angles are also used as the similarity measure for image sets in the Mutual Subspace Method (MSM) [9][10]. The general procedure for using MSM to determine 3D protein structure similarity is as follows.

Let $\mathbf{x}_{i(i=1,\ldots,n)}$ be an f-dimensional feature vector that belongs to protein p, where n is the number of samples. The basis vectors of an N-dimensional subspace $\mathcal{P}$ corresponding to protein p can be computed as the eigenvectors $[\phi_1, \ldots, \phi]_N$ of the correlation matrix $\mathbf{A}$ [12]:

$$\mathbf{A} = \frac{1}{n} \sum_{i=1}^{n} \mathbf{x}_i \mathbf{x}_i^T \quad . \tag{1}$$

M canonical angles $(0 \leq \theta_1 \leq \ldots \leq \theta_M \leq \frac{\pi}{2})$ between an M-dimensional subspace $\mathcal{Q}$ and an N-dimensional subspace $\mathcal{P}$ $(M \leq N)$ are defined as follows [7].

$$\cos \theta_i = \max_{\boldsymbol{u}_i \in \mathcal{Q}} \max_{\boldsymbol{v}_i \in \mathcal{P}} \boldsymbol{u}_i^{\mathrm{T}} \boldsymbol{v}_i \quad , \tag{2}$$

s.t. $\boldsymbol{u}_i^{\mathrm{T}} \boldsymbol{u}_i = \boldsymbol{v}_i^{\mathrm{T}} \boldsymbol{v}_i = 1, \boldsymbol{u}_i^{\mathrm{T}} \boldsymbol{u}_j = \boldsymbol{v}_i^{\mathrm{T}} \boldsymbol{v}_j = 0, i \neq j.$

In practice, we can obtain $cos^2\theta_i$ from the singular value of $\mathbf{P}^T\mathbf{Q}$, where $\mathbf{P} = [\phi_1, \ldots, \phi_N]$, $\mathbf{Q} = [\psi_1, \ldots, \psi_M]$. Here ϕ_i and ψ_i are the orthogonal basis vectors of the subspace $\mathcal{P}$ and $\mathcal{Q}$ respectively. The final similarity between two subspaces is given by

$$Sim = \frac{1}{M} \sum_{i=1}^{M} cos^2\theta_i \quad . \tag{3}$$

4   Algorithm of Clustering on a Grassmann Manifold

The standard k-means algorithm [14] attempts to partition a set of observation data $(\mathbf{d}_1, \mathbf{d}_2, \ldots, \mathbf{d}_n)$ into k clusters $C_{i(i=1,\ldots,k)}$ such that the sum of the distances among the data within each cluster is minimum:

$$\underset{C}{\arg min} \sum_{i=1}^{k} \sum_{\mathbf{d}_j \in C_i} ||\mathbf{d}_j - \mu_i||^2 \quad , \tag{4}$$

where μ_i is the mean of the data within cluster C_i.

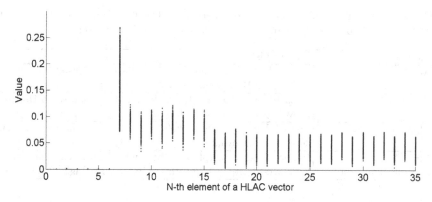

Fig. 3. Plot of 35-dimensional normalized HLAC feature vectors from various kinds of protein multiple view images. We removed the first 6 HLAC elements (the zeroth- and first-order correlations) which are close to zero.

In our problem, since each protein is represented by a linear subspace, we need to calculate the mean of multiple subspaces under the condition that the similarity of canonical angles should be regarded as a geodesic distance. Assume that an N-dimensional subspace $\mathcal{P}$ which lies in an s-dimensional vector space $\mathcal{R}^s$ corresponds to a point on the Grassmann manifold $G(N, s)$, and the subspace $\mathcal{P}$ is spanned by the columns of the $s \times N$ matrix $\mathbf{U}$. The mean of the points corresponding to subspaces on the Grassmann manifold can be obtained by using Algorithm 1[13], where $\mathrm{Glog}\,(\mathbf{X}, \mathbf{Y})$ can be calculated using Algorithm 2.

Algorithm 1. Computation of Karcher Mean on a Grassmann manifold [13]

1: Let $\mathbf{U}_{i(i=1,\dots,n)} \in G(N, s)$ be the points on Grassmann manifold, and choose an error precision ϵ which is small enough (close to zero).
2: Initialize $\mu = \mathbf{U}_1$.
3: **repeat**
4: $\delta = \frac{1}{N} \sum_{i1}^{N} \mathrm{Glog}\,(\mu, \mathbf{U}_i)$
5: Update $\mu = \mu\mathbf{V}\cos{(\mathbf{S})} + \mathbf{U}\sin{(\mathbf{S})}$, where $\mathbf{USV}^T = \delta$
6: **until** $||\delta|| < \epsilon$

In summary, the flow of our proposed method is as follows.

Algorithm 2. $\mathrm{Glog}\,(\mathbf{X}, \mathbf{Y})$ [13]

1: $\mathbf{USV}^T = (\mathbf{I} - \mathbf{XX}^T)\mathbf{Y}(\mathbf{X}^T\mathbf{Y})^{-1}$
2: $\mathbf{\Theta} = \tan^{-1}(\mathbf{S})$
3: $\mathrm{Glog}\,(\mathbf{X}, \mathbf{Y}) = \mathbf{U\Theta V}^T$

Table 1. List of proteins used in the experiment. The first four characters are the PDB code and the last character indicates the chain ID of the protein.

Class	Protein List
Alpha (α)	1bbha,1hbga,1i3ea,1me5a,1qc7a,1s56a,1sr2a,2ccya,256ba,1tlha, 1jr5a,1c75a,1b7va,1k3ha,1k3ga,1enha,1hdpa,1ocpa,1b72a,1pufa
Beta (β)	1bioa,1d1ia,1exha,1ifca,1k1ja,1lcla,1mdca,1nsba,1rsub,1bwwa, 1b0wa,1b4ra,1ncia,1ncga,1op4a,1eeqa,1qaca,1ap2a,1cd0a,1pw3a
Alpha/Beta (α/β)	1aaza,1abaa,1g4ta,1hfra,1kofa,1mxia,1p2va,1rnha,1tcaa,1zona, 2foxa,3adka,3chya,2tpsa,1spqa,1v7za,1j2ta,1btaa,1h4xa,1h4za
Alpha+Beta ($\alpha+\beta$ and multi-domain proteins)	1apme,1atpe,8cata,1pfma,1r28a,1pu3a,178la,1hlea,1jtia,1as4a, 1qmna,1szqa,1qlpa,1opha,1hp7a,1bsca,1lxya,1ag2a,2baaa,3lzta

Step 1: For each protein synthesize multi-view backbone images of size 128×128 pixels by rotating the 3D model of the protein randomly around its viewing axes, using 3D molecular graphics software, Jmol [8].

Step 2: Extract a position-invariant feature vector, HLAC [15], from each multiple view image of the protein. Although the original HLAC is a 35-dimensional feature vector, which consists of several orders of local correlations, in this process we use a 29-dimensional HLAC starting from the second-order correlation. This is because the zeroth- and first-order elements of HLAC are almost zero, as shown in Figure 3. To deal with the diversity in the appearance and size of proteins, we empirically change the range for calculating local correlations from 1, 2, 3, 4, 5, 7, and 8 pixels, so that seven 29-dimensional HLACs are produced. Finally, we concatenate all the HLACs into a 203-dimensional HLAC feature vector.

Step 3: For each protein apply PCA to the set of 203-dimensional HLAC feature vectors to generate a subspace.

Step 4: Apply k-means clustering to the set of points on the Grassmann manifold corresponding to the set of subspaces.

5   Experiment

In the experiment, we randomly collected 80 proteins from the PDB site [16]. The test data are listed in Table 1. We applied our proposed clustering method, GPCM, and the conventional k-means clustering with GIT descriptor to this dataset. First we explain the details of the experimental conditions in Section 5.1. Then, the results of the clustering are discussed in Section 5.2. In Section 5.3, we discuss the results of an additional experiment.

5.1   Experimental Conditions

Since one protein may contain more than one chain, we first removed the unnecessary chain from each protein. Then, by following the flow of our framework from Step 1 to Step 3, as described in Section 4, we collected 3000 synthesized

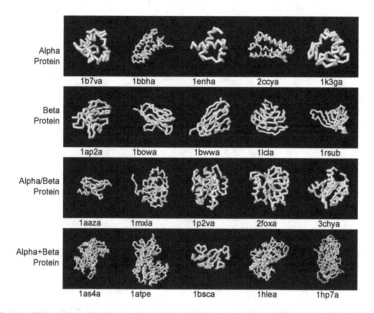

Fig. 4. Examples of the synthesized protein images used in the experiment

Table 2. Clustering results for the 80 proteins

Method	Measurement	Average (%)	Worst (%)	Best (%)
k-means with GIT	Accuracy	80.29	65.63	85
	Sensitivity	60.58	31.25	70
	Specificity	86.86	77.08	90
GPCM	Accuracy	**81.52**	**71.88**	**86.88**
	Sensitivity	**63.04**	**43.75**	**73.75**
	Specificity	**87.68**	**81.25**	**91.25**

protein images of size 128×128 pixels from each protein backbone visualization by using Jmol [8] which is included in the Matlab Bioinformatics Toolbox. Figure 4 shows some of the synthesized protein images. Next, we extracted HLAC vectors from these images to obtain 203-dimensional feature vectors. Finally, we constructed a subspace by applying PCA to each HLAC feature set. The dimension of the subspace was set to 4. Considering the randomness of the k-means clustering result, we repeated the clustering experiment 5000 times for both the proposed method and the k-means with GIT descriptor. The number of clusters was set to 4 ($k = 4$).

5.2   Clustering Results

Figure 5 shows box plots which summarize the accuracy of the clustering results from the experiment. Here, the clustering accuracy was defined as $(TP + TN)/(TP+TN+FP+FN)$, where TP is true positive, TN is true negative, FP

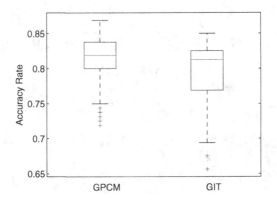

Fig. 5. Box plots of the 5000 experimental results using the proposed method and k-means with GIT descriptor

Table 3. The average true positive rate of each cluster from the 5000 repeated experiments. The columns indicate the clustering result. The rows indicate the ground truth label ($\alpha + \beta^*$ includes multi-domain proteins).

(a) GPCM

Class	α (%)	β (%)	α/β (%)	$\alpha+\beta^*$ (%)
α	71.15	0.13	22.03	6.69
β	0.3	71.98	7.77	19.95
α/β	22.75	8.19	51.84	17.22
$\alpha+\beta^*$	16.35	8.48	17.96	57.21

(b) k-means with GIT

Class	α (%)	β (%)	α/β (%)	$\alpha+\beta^*$ (%)
α	93.67	0	5.11	1.21
β	1.95	73.31	1.17	23.58
α/β	39.84	0.12	56.12	3.93
$\alpha+\beta^*$	28.6	51.12	1.05	19.22

is false positive, and FN is false negative. First, all possible combinations of the class labels for the clustering result were listed. Next, we computed the accuracy rate for each combination of class labels. Finally, the class label which produced the best accuracy rate was considered to be the correct label. The average, worst, and best clustering results of the GIT and the proposed method are shown in Table 2. The sensitivity (true positive rate) is defined as $TP/(TP + FN)$. The specificity (true negative rate) is defined as $TN/(FP + TN)$. We see that our proposed method is able to cluster the proteins more accurately than the conventional method. The proposed method could achieve up to 86.88% accuracy, 73.75% sensitivity, and 91.25% specificity. On the other hand, the conventional method achieved up to 85% accuracy, 70% sensitivity, and 90% specificity.

For further analysis, we examined the clustering results for each protein in both methods. Table 3 shows the average sensitivity (true positive rate) for each protein. These results show that the GIT descriptor is good at separating the alpha-helices and beta-sheets; however, it has serious difficulty clustering the overlapped structures of the fourth class which contains the alpha+beta proteins

Table 4. Clustering results for 400 proteins

Method	Measurement	Average (%)	Worst (%)	Best (%)
k-means with GIT	Accuracy	75	68	77
	Sensitivity	50.06	36	54
	Specificity	83.35	78.67	84.67
GPCM1	Accuracy	**75.08**	**71.13**	76.38
	Sensitivity	**50.16**	**42.25**	52.75
	Specificity	**83.39**	**80.75**	84.25
GPCM2	Accuracy	74.11	69	**77.88**
	Sensitivity	48.22	38	**55.75**
	Specificity	82.74	79.33	**85.25**

and the complicated multi-domain proteins. On the other hand, our proposed method has a more consistent performance across the categories. These results imply that there is room to improve the performance of our method by considering more effective features and tuning parameters, while the method with GIT may have some fundamental problems when dealing with overlapped and complicated protein structures. Moreover, the incapability of GIT to describe a protein which contains more than three consecutive missing carbon atoms is also a drawback of that method.

In terms of the computational speed, k-means with GIT is much faster than the proposed method. When using an Intel Xeon E5506 2.13Ghz and the Matlab statistical toolbox, the average execution time for the built-in k-means function with GIT descriptor was 0.0066s. On the other hand, our proposed method had an average execution time of 1.7s. However, it is worth noting that we have not optimized our Matlab implementation code to benefit from parallel processing, as we wrote our own implementation of k-means on the Grassmann manifold.

5.3   Additional Experiment

We conducted an additional experiment using 400 proteins. As in the previous experiment, we repeated the clustering 5000 times. However, in this experiment the fourth class of the protein does not contain multi-domain proteins (only alpha+beta proteins were used) to reduce the difficulty of classification. The experimental results for the clustering of the 400 proteins are shown in Table 4. GPCM1 used the same experimental parameters that were used in the experiment with 80 proteins. In GPCM2, the HLAC parameters were set to 2, 4, 6, and 8, and the subspace dimension was set to 5. Although the performance of the proposed method is quite similar to that of the conventional method in both cases, this experiment demonstrates that the performance of the proposed method can be improved by tuning the parameters of the subspace and having better feature extraction for the protein visualization images.

6   Conclusion and Future Work

In this paper, we have proposed a novel framework, called *Grassmannian Protein Clustering Method (GPCM)*, for solving the protein clustering problem. In GPCM, a 3D protein structure is represented by a linear subspace generated by applying PCA to the multiple-view of synthesized protein images. The similarity of two protein structures is defined by the canonical angles between the corresponding subspaces. The advantage of this approach is that it does not require precise alignment of the proteins. Since the protein is represented by a subspace, we regarded the protein clustering problem as a subspace clustering problem, and we applied the k-means algorithm for subspace clustering on a Grassmann manifold.

The experimental results demonstrated that the proposed method is superior to the conventional k-means with GIT approach, especially in identifying the overlapped structure of alpha+beta proteins which results a higher clustering accuracy. The GIT descriptor has a compact protein representation so the k-means computation is very fast. However, as shown by our experimental results, it may have a problem separating overlapped and complicated structures in which both alpha-helix and beta-sheet motifs exist.

Since this research is still in its early stages, we will conduct further experiments using more of protein data available from the PDB site. We will also consider different methods for extracting features from the protein visualization images and different classifiers, such as the nonlinear constrained subspace[17], with the aim of improving the performance of our method.

References

1. Holm, L., Sander, C.: DALI: a network tool for protein structure comparison. Trends Biochem. Sci. 20, 478–480 (1995)
2. Shindyalov, I., Bourne, P.: Protein structure alignment by incremental combinatorial extension (CE) of the optimal path. Protein Engineering 11, 739–747 (1998)
3. Orengo, C.A., Taylor, W.R.: SSAP: Sequential structure alignment program for protein structure comparison. Methods in Enzymology 266, 617–635 (1996)
4. Røgen, P., Bohr, H.G.: A new family of global protein shape descriptors. Mathematical Biosciences 182(2), 167–181 (2003)
5. Røgen, P.: Evaluating protein structure descriptors and tuning Gauss Integrals based descriptors. Journal of Physics Condensed Matter 17, 1523–1538 (2005)
6. Suryanto, C.H., Jiang, S., Fukui, K.: Protein structures similarity based on multi-view images generated from 3D molecular visualization. In: International Conf. on Pattern Recognition, ICPR 2012 (to appear 2012)
7. Chatelin, F.: Eigenvalues of matrices. John Wiley & Sons, Chichester (1993)
8. Jmol: an open-source Java viewer for chemical structures in 3D, http://www.jmol.org/
9. Yamaguchi, O., Fukui, K., Maeda, K.: Face recognition using temporal image sequence. In: International Conf. on Face and Gesture Recognition, pp. 318–323 (1998)

10. Fukui, K., Yamaguchi, O.: Face recognition using multi-viewpoint patterns for robot vision. In: 11th International Symposium of Robotics Research, pp. 192–201 (2003)
11. Harder, T., Borg, M., Boomsma, W., Røgen, P., Hamelryck, T.: Fast large-scale clustering of protein structures using Gauss Integrals. Journal of Bioinformatics, 510–515 (2012)
12. Oja, E.: Subspace Methods of Pattern Recognition. Research Studies Press, England (1983)
13. Begelfor, E., Werman, M.: Affine invariance revisited. In: Proceedings of International Conf. on Computer Vision and Pattern Recognition, pp. 2087–2094 (2006)
14. Lloyd, S.P.: Least squares quantization in PCM. IEEE Trans. Information Theory 28, 129–137 (1982)
15. Otsu, N., Kurita, T.: A new scheme for practical flexible and intelligent vision systems. In: Proc. of IAPR Workshop on CV, pp. 431–435 (1988)
16. Berman, H.M., Westbrook, J., Feng, Z., Gilliland, G., et al.: The Protein Data Bank. Nucleic Acids Research 28, 235–242 (2000)
17. Fukui, K., Stenger, B., Yamaguchi, O.: A Framework for 3D Object Recognition Using the Kernel Constrained Mutual Subspace Method. In: Narayanan, P.J., Nayar, S.K., Shum, H.-Y. (eds.) ACCV 2006. LNCS, vol. 3852, pp. 315–324. Springer, Heidelberg (2006)

Improving the Portability
and Performance of jViz.RNA –
A Dynamic RNA Visualization Software

Boris Shabash, Kay Wiese, and Edward Glen

Simon Fraser University, School of Computing Science
8888 University Drive, Burnaby, BC, Canada
`http://www.sfu.ca`

Abstract. In this paper, four methods were explored for improving the performance of jViz.RNA's structure drawing algorithm when dealing with large sequences; First, the approximation based Barnes-Hut algorithm was explored. Second, the effects of using multithreading were measured. additionally, dynamic C libraries, which integrate C code into the Java^{TM} environment, were investigated. Finally, a technique termed structure recall was examined.

The results demonstrated that the use of the Barnes-Hut algorithm produced the most drastic improvements in run-time, but distorts the structure if too crude of an approximation is used. Multithreading and integration of C code proved to be favorable approaches since these improved the speed at which calculations are done, without distorting the structures.

jViz.RNA is available to download from http://jviz.cs.sfu.ca/.

1 Introduction

jViz.RNA is an RNA visualization software developed at Simon Fraser University [5,12,13]. jViz.RNA draws its strength from three major components. First, since it is written in Java^{TM}, it is platform independent and can run on any machine that has the Java Virtual Machine (JVM) installed. Second, unlike other RNA visualization tools [4,6,7], jViz.RNA offers a dynamic model that users can interact with. This gives users the ability to modify the layout generated by jViz.RNA if they feel they need to. Third, jViz.RNA offers the ability to inspect several different aspects of the RNA molecule explored, as well as the ability to compare two RNA structures. This aspect is particularly useful when working alongside algorithms designed to predict RNA structure.

In the research proposed in this paper, our aim was to improve the run-time performance of jViz.RNA's structure drawing algorithm, in order to allow faster rendering of large RNA sequences. During the course of this research, jViz.RNA's portability was also extended by incorporating the RNAML and FASTA file formats. However, the details of these file formats' integration are omitted for brevity purposes.

T. Shibuya et al. (Eds.): PRIB 2012, LNBI 7632, pp. 82–93, 2012.

Unlike other visualization tools that allow for manipulation of the resulting layout [2,3,9,8], the RNA drawing approach jViz.RNA employes relies on simulating the natural forces that are at work on RNA inside the cell, thus simulating its natural folding process[1]. Furthermore, jViz.RNA renders the folding while it calculates it, thus giving the user an interactive model. However, a major drawback of such an approach is that it is more time consuming than preparing an immutable layout discretely. The calculations involved propose a running time of, at worst, $O(n^2)$ for each iteration. This work aimed at reducing the run-time required for jViz.RNA's drawing algorithm by using both algorithmic methods aimed to try and approximate the interactions involved at each iteration in exchange for a faster theoretical run time, as well as software engineering approaches aimed at decreasing the run time of individual instructions.

The remainder of this article is organized as follows; Section 2 presents insight into the different approaches employed to improve jViz.RNA's drawing algorithm, as well as the results, Section 3 discusses the significance of the results and possible extensions, and finally Section 4 provides a concluding summary.

2   jViz.RNA's Drawing Algorithm and Potential Improvements

2.1   Approach

jViz.RNA aims at presenting the user with a dynamic, interactive, model. The model would behave according to the forces acting on each nucleotide, which were estimated by incorporating Newtonian mechanics and electrostatics into the simulation; Each nucleotide would experience attraction forces ($F_{attraction}$) from nucleotides that it is bonded to (two nucleotides from the backbone and another one if it is involved in base pairing), as well as repulsion forces ($F_{repulsion}$) from all other nucleotides in the simulation, simulating electrostatic repulsion. The structure would be in a stable conformation when the two forces for each nucleotide cancel out, i.e. when $F_{attraction} = F_{repulsion}$.

To simulate the forces acting on each nucleotide, a spring based model was developed. The attraction force between two nucleotides is given by

$$F_{attraction} = k\Delta d \tag{1}$$

where k is a spring coefficient and Δd is the distance between two nucleotides. The repulsion force between two nucleotides is given by

$$F_{repulsion} = \frac{q}{\Delta d^2} \tag{2}$$

where q is the repulsion coefficient and Δd is again the distance between the two nucleotides. To draw the RNA structure's layout, the sequence is first laid in a

---

[1] This is not to imply jViz.RNA predicts folding, it simply simulates it after other software have established the correct base pairing of the RNA molecule.

circle, and the forces acting on the nucleotides are calculated iteratively until $F_{attraction} = F_{repulsion}$ for all nucleotides. The calculation of attracting forces for all nucleotides at a given iteration is in order of $O(n)$, since each nucleotide is bonded to at most three other nucleotides. However, calculation of repulsive forces can be in order of $O(n^2)$ if all interactions are considered.

The most recent iteration of jViz.RNA employed the idea of a "neighborhood" of nucleotides. The neighborhood of each nucleotide increased throughout the iterations, until eventually, for every nucleotide, all nucleotides were considered for calculating repulsion forces. However, for large sequences of size 1,000+ nucleotides (nt), this approach would still lag and large structures would take up to half an hour to collapse into a stable conformation.

The Barnes-Hut Algorithm. The Barnes-Hut algorithm [1] was originally employed in astrophysical simulations to simulate the movements of planets within galaxies. However, it showed potential to be useful in the context of bioinformatics as well.

The main premise of the Barnes-Hut algorithm is that if a group of bodies (nucleotides, in the case of jViz.RNA), e.g. B_1, B_2 and B_3, are far enough away from a body B_0 and close enough together, then the forces they exert on B_0 can be approximated by creating a virtual body, B_v, which combines all of their properties and lies in the center of the body group. In essence, if the group of bodies mentioned exerts repulsive forces over B_0 by Equation 2 described above, then instead of calculating $F_{repulsion_{1 \to 0}} + F_{repulsion_{2 \to 0}} + F_{repulsion_{3 \to 0}}$, one can calculate $F_{repulsion_{v \to 0}}$. The virtual body B_v combines the properties of B_1, B_2 and B_3 into one body, and so fewer calculation need to be made (Figure 1).

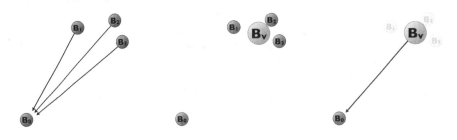

(a) B_0 experiences repulsion forces by B_1, B_2, and B_3

(b) B_1, B_2 and B_3 can be approximated by a virtual body, B_v which encompasses all their properties and lies in the centre of the body group

(c) The virtual body B_v can be used to approximate the repulsion forces exerted on B_0, thus reducing calculations

Fig. 1. The basic premise for the Barnes-Hut algorithm is that if a group of bodies is close enough together, and far enough from another body, B_0, then their effects on it can be approximated by a virtual body

The algorithm accomplishes this simplification by first constructing a quad-tree (or oct-tree for 3D) where all the leaves are either bodies in the simulation or empty spaces and all internal nodes are groups of bodies represented by a virtual body.

Then each body in the simulation traces through the quad-tree and calculates which internal nodes simulate a group of bodies far enough away from it and close enough together such that their forces can be approximated. In theory, this results in a cruder approximation of the forces acting in the simulation but performs calculations in the order of $O(n \log n)$ rather than $O(n^2)$.

Multithreading. Multi-threading can be employed if there are independent calculations that can be distributed. In the case of force calculation in jViz.RNA, the set of force calculations can be divided by assigning each thread a set of nucleotides and having it calculate the forces acting on each nucleotide in the set.

For our purposes we chose to implement multithreading such that each of the m threads receives a set of $\frac{n}{m}$ nucleotides under its care, where n is the number of nucleotides in the structure in total. Then, each of these threads would calculate the forces acting on each of its nucleotides by all other $n - 1$ nucleotides. In theory, the running time of the algorithm when multithreading is employed should be in the order of $O(\frac{n^2}{m})$.

There is the possibility that for smaller structures the time involved in managing the threads may add an increase in run-time greater than any time gained by splitting the calculations. In addition, larger structures may benefit from a different number of threads than smaller structures, and as a result, we have tested several different structures of varying lengths and saw how they behaved under an increasing number of threads.

The goal of this research was to deliver a proof of principle regarding the advantages multithreading can provide to jViz.RNA. As such, we tested multithreading on the brute-force approach (without the use of a neighborhood), and compared it with jViz.RNA's neighborhood algorithm. If multithreading shows promising results, it can then be implemented in jViz.RNA's neighborhood based algorithm.

Native C Code. Java^{TM} offers the advantage of platform independence due to the presence of the JVM. However, sometimes the use of the JVM can cause slowdowns since code needs to be interpreted through the JVM, adding an extra step for execution. To compensate, Java^{TM} offers the potential to integrate C code through dynamic libraries. However, unlike the Java^{TM} code, C code is not platform independent and so a library must be compiled for each separate architecture the code is to be executed on. In addition, since calling a C library from the Java^{TM} environment introduces an overhead, the integration of C libraries into Java^{TM} is most fruitful when large sets of computations are executed by the C code.

In this research as well, our purpose was to provide a proof of principle that the C code would work faster than the Java^{TM} code when dealing with medium-size and large-size sequences. If the improvement found would be great enough, it will be safe to assume that the use of native C code can be scaled to jViz.RNA's neighborhood based algorithm.

Structure Recall. The final approach for improving the performance of jViz.RNA is a concept we denote as 'structure recall'. In this setup, jViz.RNA would recall structures it has previously seen and lay them out exactly as they were last viewed. Since structures are often viewed more than once, it would save researchers and users the time spent waiting for the structure to collapse into a stable conformation.

In order to achieve this form of 'memory', jViz.RNA has to keep the information of each RNA structure previously viewed in files. The main decision that motivated the design of these files was whether the data should be stored in a file containing binary data (flat files) or files that contain the (x, y) coordinates of each nucleotide (structured files). Flat files offer the advantage of quick access time, since once the data would be read, simply a reference to the right type of data would have to be assigned to it. Alternatively, structured files would take longer to access the data, but could be employed by other applications.

Since users and programmers would not be able to visualize how an RNA molecule is laid out given the coordinates, and other applications may display an RNA molecule in a different way, it was decided to use the flat file format.

2.2   Results – Implementing and Testing Run-Time Optimization Methods

For the experiments in this section, the following 16 sequences were used to test the performance of different methods (Table 1). The sequences were selected to represent a wide spectrum of sizes that could be of interest to researchers in the natural sciences.

The experiments involving these 16 sequences were all performed on a Mac-Book Pro with a 2.4 GHz Intel Core 2 Duo and 2GB of RAM running Mac OS X Version 10.5.8.

The Barnes-Hut Algorithm. The first step in implementing the Barnes-Hut algorithm was defining a quad-tree and how it should be built. The tree was built anew at every iteration since updating it could be more costly in computation time than building it. At every iteration, the maximum and minimum x and y coordinates were found in order to establish the enclosing area of the quad-tree. Each internal node in the tree would represent a space, and would store its center, as well as its center of mass (which need not be the same as the center).

Table 1. The 16 sequences used for the experiments and their corresponding accession numbers

Structure Name	Accession Number	Size (nt)
Bacillus stearothermophilus 5S ribosomal RNA	AJ251080	117
Saccharomyces cerevisiae 5S ribosomal RNA	X67579	118
Agrobacterium tumefaciens 5S ribosomal RNA	X02627	120
Arthrobacter globiformis 5S ribosomal RNA	M16173	122
Deinococcus radiodurans 5S ribosomal RNA	AE000513:254392-254515	124
Metarhizium anisopliae var. anisopliae strain 33 28S ribosomal RNA group IB intron	AF197122	436
Tetrahymena thermophila 26S ribosomal RNA fragment (with intron)	V01416	517
Acomys cahirinus mitochondrial 12S ribosomal RNA	X84387	940
Xenopus laevis mitochondrial 12S ribosomal RNA	M27605	945
Homo sapiens mitochondrial 16S ribosomal RNA	J01415:648- 1601	954
Ailurus fulgens mitochondrial 16S ribosomal RNA	Y08511	964
Sulfolobus acidocaldarius 16S ribosomal RNA	D14876	2080
Aureoumbra lagunensis 18S ribosomal RNA	U40258	2236
Hildenbrandia rubra 18S ribosomal RNA (with intron)	L19345	2283
Porphyra leucosticta 18S ribosmal RNA (with intron)	AF342746	2404
Chlorella saccharophila 18S ribosomal RNA	AB058310	2510

Each node would also store a **mass** attributes denoting the mass of the body it represents. Leafs would have a **mass** of either 0 (for empty spaces), or 1 (nucleotides). Internal nodes would have their **mass** set equal to the sum of their four children's masses, and the center of mass for each node would be the weighted average of the centers of mass of its children. The nodes also each had a **size** attribute which indicated their circumference.

With the nodes defined, the tree itself was relatively straightforward to define; the only crucial parameter to define for the tree was the ratio parameter θ. When a nucleotide traverses the tree it inspects every internal node for its size, s, and the distance between the nucleotide and the node's center, Δd. If Equation 3 holds true,

$$\frac{s}{\Delta d} < \theta \tag{3}$$

then that internal node can be used as a virtual body to approximate the entire subtree beneath it. Otherwise, the nucleotide must examine each of the node's children and so on.

For the purposes of our experiments, we have experimented with θ values of 0.5, 1.0 and 5.0, as well as a set of experiments where the value of θ varied according to the sequence length. The run-time results can be seen in Figure 2.

It may seem as though the Barnes-Hut algorithm performs better as the value of θ increases. When using a θ value of 5.0, the running time of jViz.RNA dropped over 90%. However, it was important to also explore the visual effects that the Barnes-Hut algorithm delivered. Figure 3 demonstrates the results for

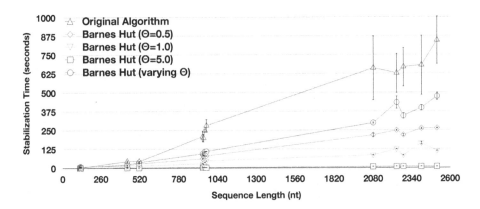

Fig. 2. The stabilization times (in seconds) for the different RNA sequences (as the number of nucleotides increases) using the Barnes-Hut algorithm under different θ values. 99% confidence intervals are used as error bars

a particular sequence when using the Barnes-Hut algorithm, and similar results were observed for the other sequences. When the value of θ is 1.0, the resulting visualization becomes very crude. Especially for long stems of RNA base pairs, which seem to thin out as they extend (Figure 3(c)). These effects extend to even short helices as the value of θ increases to 5.0. However, when using a θ value of 0.5 (Figure 3(b)), the RNA sequence maintains a conformation similar to its original one (Figure 5(a)), while still demonstrating a run time drop of between 30% and 70%. The case where θ can be related to the sequence length was also considered, decreasing as the sequence length increases. Figure 3(d) demonstrates the results of employing such an approach, while the run time results can be seen in Figure 2. The value of θ was calculated using Equation 4[2];

$$\theta = 0.8936 \times n^{-0.1649} \tag{4}$$

where n is the length of the given sequence.

In addition, future iterations of jViz.RNA could combine the Barnes-Hut algorithm with the original neighborhood based algorithm. Under this setup, once the sequence achieves force equilibrium under the Barnes-Hut algorithm, the neighborhood based algorithm would take over, and calculate the finer conformation of the sequence. This may result in diminished performance improvements compared to using only the Barnes-Hut algorithm, but overall giving aesthetically pleasing layouts for the RNA sequences in less time than the neighborhood based algorithm alone.

---

[2] The equation was developed using a regression that aimed to ensured that the θ value would be ≈ 0.41 for 117nt, and ≈ 0.25 for 2510nt.

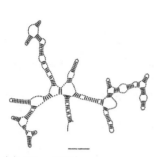

(a) The RNA sequence as drawn by the original algorithm

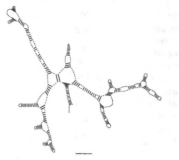

(b) The RNA sequence as drawn by the Barnes-Hut algorithm where $\theta = 0.5$

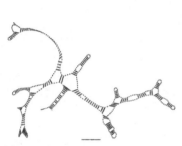

(c) The RNA sequence as drawn by the Barnes-Hut algorithm where $\theta = 1.0$

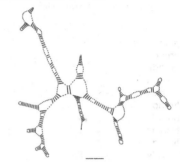

(d) The RNA sequence as drawn by the Barnes-Hut algorithm where θ varies ($\theta = 0.31$ for this sequence)

Fig. 3. The RNA sequence for *Tetrahymena thermophila* 26S ribosomal RNA fragment (with intron) (V01416) as drawn by jViz.RNA's original algorithm and the Barnes-Hut algorithm

Multithreading. Multithreading in jViz.RNA was implemented as a `Searcher` class using the `Runnable` interface Java™offers [11]. The `Runnable` interface allows a class to run on its own thread. The `Searcher` class would search through a subset of the nucleotides and check their interactions with all other nucleotides.

The results in Figure 4 show that as the number of threads increases, calculations for large sequences become faster and faster. However, even for large sequences there is a diminishing return as the number of threads increases. Furthermore, for medium and small size sequences the ideal number of threads is not the same as for large ones. For smaller sequences, an increase in the number of threads past four or five can actually cause a slowdown in the run-time. The reason for these observations is that although an increasing number of threads can allow CPU cores to process more data in parallel, they also require additional management. In addition, the number of available CPU cores would greatly effect these results since more cores would be able to handle more threads in parallel.

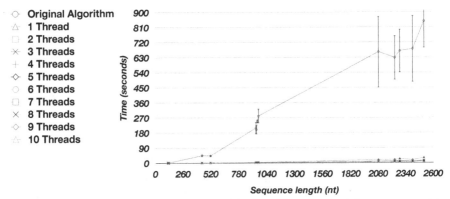

(a) The running times of the original, neighborhood based, algorithm versus multi-threading

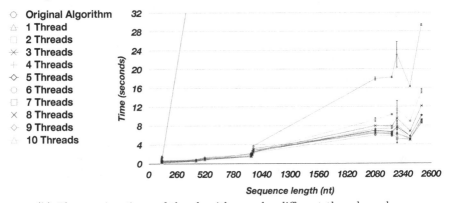

(b) The running times of the algorithm under different thread numbers

Fig. 4. The stabilization times (in seconds) for the different RNA sequences (as the number of nucleotides increases) using multithreading and the brute-force algorithm with different thread numbers versus the original algorithm emloyed by jViz.RNA. 99% confidence intervals are used as error bars

What is even more interesting is that even the brute force implementation, when employing multithreading, far outperforms jViz.RNA's current algorithm. There could be many factors accounting for this difference. However, it appears that dedicating more than one CPU core for the processing of the structure's layout drastically improves jViz.RNA's performance. With these results in mind, it is still worth noticing that the size of the sequence greatly effects the number of threads that allows for optimal performance insofar as run-time goes.

Native C Code. Keeping the overhead involved in calling C code from a Java^{TM}environment in mind, it was very interesting to find that even for the short sequences, the running times were faster when C based code was employed

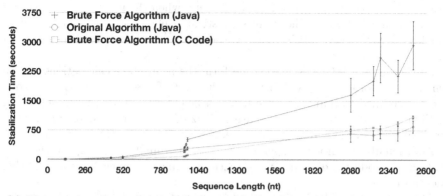

(a) The running times of the original, neighborhood based, algorithm versus the brute-force algorithm using both Java and C code

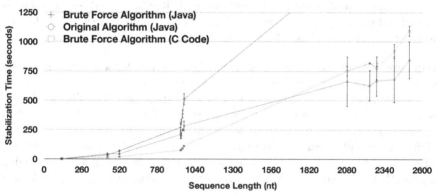

(b) The running times of the original, neighborhood based, algorithm versus the brute-force algorithm using C code

Fig. 5. The stabilization times (in seconds) for the different RNA sequences (as the number of nucleotides increases) using the original algorithm, and the brute force algorithm implemented in both C and Java[TM] code. 99% confidence intervals are used as error bars

than Java[TM] code, as Figure 5 demonstrates. It isn't surprising to see that the brute-force C algorithm is outperformed by jViz.RNA's original algorithm as the sequences tested get larger. However, the fact that the C algorithm performs better over the smaller sequences shows that the overhead involved in calling C code does not contribute to a slowdown in performance. Had the overhead for calling C code accounted for a major slowdown, one would expect to see the C code perform slower than the Java[TM] code over the smaller sequences, since the amount of time required for calculations is smaller, but the overhead involved in sending the data is fairly consistent. The results seen here, however, are just the opposite. For smaller sequences, the C code outperforms even the use of the neighborhood based approach in Java^{TM}. Only when the number of nucleotides increases, and the use of the more sophisticated neighborhood approach gives the Java^{TM}code a competitive edge, is the Java^{TM}code performing better than the C based code.

Furthermore, the C code used to perform the brute force calculations consistently outperforms the Java^{TM}code for the brute force approach. The C code can be modified to implement the use of a neighborhood, and the results seen in Figure 5 demonstrate that the use of jViz.RNA's current algorithm in a C environment can improve performance for many, if not all, RNA sequences.

Additionally, an important point to make is that multithreading and the integration of C code do not distort the structures in any way since no approximation is entailed.

The main drawback of this method is the fact that for each different architecture, the dynamic library needs to be recompiled. However, this does not present a major problem since the calls to load the library and execute the native method can be put in blocks of defensive programming (`try`/`catch` blocks) and the original Java^{TM}code be used as default.

Structure Recall. Structure recall is a simple method for improving run-time, but has proven very effective. It is reminiscent of RNAViz's use of 'skeleton files' [4], but differs in its interaction with users since structure recall files are loaded automatically, if they are available. Structure recall employs flat files to store coordinate information regarding a sequence's nucleotides' positions.

Often groups of researchers are interested in a particular group of structures and would view a small set of structures many times. It is therefore only logical for jViz.RNA to avoid redundant calculations regarding structure conformation.

3   Discussion

This paper described several extensions and improvements for jViz.RNA, All of which provided promising results. The Barnes-Hut algorithm proved to improve the performance greatly, but some of the improvement needs to be traded in favor of aesthetic and clear layouts of the RNA sequences. On the other hand, software engineering approaches such as multithreading and C based execution of repetitive code, yielded very promising results both run-time wise and display wise, without the use of any approximation.

Furthermore, these approaches lend themselves to integration such that future iterations of jViz.RNA would use the Barnes-Hut algorithm, traversed in parallel by several threads, in a C environment for all force calculations, as well as integrate structure recall for sequences that have been previously viewed. Moreover, since multithreading benefits from having more cores, jViz.RNA could explore the possibility of General Purpose GPU (GPGPU) programming. GPUs are designed to perform calculations in parallel for large arrays of data, and are becoming increasingly popular with NVIDIA's introduction of the C based CUDA programming API [10], which would be ideal for jViz.RNA's required calculations

4 Conclusion

This paper describes four extensions to jViz.RNA which show potential for further experimentation and future inclusion in jViz.RNA. With these extensions, jViz.RNA overcomes the performance limitation for larger sequences. Additionally, as was mentioned in Section 1, jViz.RNA was also extended by the addition of functionality to process FASTA and RNAML files, allowing it to interact better with the input and output files of other programs and contributing to the standardization of RNAML.

References

1. Barnes, J., Hut, P.: A hierarchical O(N log N) force-calculation algorithm. Nature 324(4), 446–449 (1986)
2. Broccoleri, R.E., Heinrich, G.: An Improved Algorithm for Nucleic Acid Secondary Structure Display. Bioinformatics 4(1), 167–173 (1988)
3. Darty, K., Denise, A., Ponty, Y.: Varna: Interactive drawing and editing of the rna secondary structure. Bioinformatics 25(15) (2009)
4. De Risjk, P., De Wachter, R.: Rnaviz, a program for the visualisation of rna secondary structure. Nucleic Acids Research 25(22), 4679–4684 (1997)
5. Glen, E.: JVIZ.RNA - A Tool for Visual Comparison and Analysis of RNA Secondary Structures. Master's thesis, Simon Fraser University (2007)
6. Han, K., Byun, Y.: PseudoViewer3: generating planar drawings of large-scale RNA structures with pseudoknots. Bioinformatics 25(11), 1435–1437 (2009)
7. Hofacker, I.L.: Vienna RNA secondary structure server. Ivo L. Hofacker 31(13), 3429–3431 (2003)
8. Jossinet, F., Ludwig, T.E., Westhof, E.: Assemble: an interactive graphical tool to analyze and build RNA architectures at the 2D and 3D levels. Bioinformatics 26(16), 2057–2059 (2010)
9. Jossinet, F., Westhof, E.: The RnamlView Project. Institut de biologie moleculaire et cellulaire du CNRS
10. NVIDIA®. CUDA™ Parallel Programming Made Easy (2011),
    http://www.nvidia.com/object/cuda_home_new.html
11. Oracle. Interface Runnable, http://download.oracle.com/javase/1.4.2/docs/api/java/lang/Runnable.html
12. Wiese, K.C., Glen, E.: jViz.Rna -a java tool for RNA secondary structure visualization. IEEE Transactions on NanoBioscience 4(3), 212–218 (2005)
13. Wiese, K.C., Glen, E.: jViz.Rna - An Interactive Graphical Tool for Visualizing RNA Secondary Structure Including Pseudoknots. In: CBMS, pp. 659–664. IEEE Computer Society (2006)

A Novel Machine Learning Approach for Detecting the Brain Abnormalities from MRI Structural Images

Lavneet Singh, Girija Chetty, and Dharmendra Sharma

Faculty of Information Sciences and Engineering
University of Canberra, Australia
{Lavneet.singh,Girija.chetty,Dharmendra.sharma}@canberra.edu.au

Abstract. In this study, we present the investigations being pursued in our research laboratory on magnetic resonance images (MRI) of various states of brain by extracting the most significant features, and to classify them into normal and abnormal brain images. We propose a novel method based ondeep and extreme machine learning on wavelet transform to initially decompose the images, and then use various features selection and search algorithms to extract the most significant features of brain from the MRI images. By using a comparative study with different classifiers to detect the abnormality of brain images from publicly available neuro-imaging dataset, we found that a principled approach involving wavelet based feature extraction, followed by selection of most significant features using PCA technique, and the classification using deep and extreme machine learning based classifiers results in a significant improvement in accuracy and faster training and testing time as compared to previously reported studies.

Keywords: Deep Machine Learning, Extreme Machine Learning, MRI, PCA.

1   Introduction

Magnetic Resonance Images (MRI) is an advance technique used for medical imaging and clinical medicine and an effective tool to study the various states of human brain. MRI images provide the rich information of various states of brain which can be used to study, diagnose and carry out unparalleled clinical analysis of brain to find out if the brain is normal or abnormal. However, the data extracted from the images is very large and it is hard to make a conclusive diagnosis based on such raw data. In such cases, we need to use various image analysis tools to analyze the MRI images and to extract conclusive information to classify into normal or abnormalities of brain. The level of detail in MRI images is increasing rapidly with availability of 2-D and 3-D images of various organs inside the body.

Magnetic resonance imaging (MRI) is often the medical imaging method of choice when soft tissue delineation is necessary. This is especially true for any attempt to classify brain tissues [1]. The most important advantage of MR imaging is that it is non-invasive technique [2]. The use of computer technology in medical decision support is now widespread and pervasive across a wide range of medical area, such as

T. Shibuya et al. (Eds.): PRIB 2012, LNBI 7632, pp. 94–105, 2012.

cancer research, gastroenterology, heart diseases, brain tumors etc. [3, 4].Fully automatic normal and diseased human brain classification from magnetic resonance images (MRI) is of great importance for research and clinical studies. Recent work [2, 5] has shown that classification of human brain in magnetic resonance (MR) images is possible via machine learning and classification techniques such as artificial neural networks and support vector machine (SVM) [2] and unsupervised techniques such as self-organization maps (SOM) [2] and fuzzy c-means combined with appropriate feature extraction techniques [5]. Other supervised classification techniques, such as k-nearest neighbors (k-NN), which group pixels based on their similarities in each feature image [1, 6, 7, 8] can be used to classify the normal/pathological T2-wieghted MRI images.

Out of several debilitating ageing related health conditions, white matter lesions (WMLs) are commonly detected in elders and in patients with multiple brain abnormalities like Alzheimer's disease, Huntington's disease and other neurological disorders. According to previous studies, it is believed that total volume of the lesions (lesion load) and their progression relate to the aging process as well as disease process. Therefore, segmentation and quantification of white matter lesions via texture analysis is very important in understanding the impact of aging and diagnosis of various brain abnormalities. Manual segmentation of WM lesions, which is still used in clinical practices, shows the limitation to differentiate brain abnormalities using human visual abilities. Such methods can produce a high risk of misinterpretation and can also contribute to variation in correct classification. Automated texture analysis algorithms have been developed to detect brain abnormalities using image segmentation techniques and machine learning algorithms. The signal of homogeneity and heterogeneity of abnormal areas in Region of Interest (ROI) in white matter lesions of brain in T2-MRI images can be quantified by texture analysis algorithms [reference]. The ability to measure small differences in MRI images is essential and important to reduce the diagnosis errors of brain abnormalities. The supervised feature classification from T2 MRI images, however, suffers from two problems. First, because of the large variability in image appearance between different datasets, the classifiers need to be retrained from each data source to achieve good performances. Second, these types of algorithms rely on manually labeled training datasets to compute the multi-spectral intensity distribution of the white matter lesions making the classification unreliable. Inspired by new segmentation algorithms in computer vision and machine learning, we propose an efficient semi-automatic and deep learning algorithm for white matter (WM) lesion segmentation around ROI based on extreme and deep machine learning. Further, we compare this novel approach with some of the other supervised machine learning techniques reported previously.

Rest of the paper is organized as follows. Next Section gives a brief background of materials and methods used in Section 2. The details of the feature extraction, and feature selection, and other classifiers techniques used is described in same Section 2, 3 and Section 4 presents some of the experimental work carried. The paper concludes with in section 5 with some outcomes of the experimental work using proposed approach, and outlines plans for future work.

2 Materials and Methods

2.1 Coarse Image Segmentation

Color image segmentation is useful in many applications. From the segmentation results, it is possible to identify regions of interest and objects in the scene, which is very beneficial to the subsequent image analysis or annotation. However, due to the difficult nature of the problem, there are few automatic algorithms that can work well on a large variety of data. The problem of segmentation is difficult because of image texture. If an image contains only homogeneous color regions, clustering methods in color space are sufficient to handle the problem. In reality, natural scenes are rich in color and texture. It is difficult to identify image regions containing color-texture patterns. The approach taken in this work assumes the following:

- Each region in the image contains a uniformly distributed color-texture pattern.
- The color information in each image region can be represented by a few quantized colors, which is true for most color images of natural scenes.
- The colors between two neighboring regions are distinguishable - a basic assumption of any color image segmentation algorithm.

2.2 K-Means Clustering Based Coarse Image Segmentation

K-Means clustering algorithm is a well-known unsupervised clustering technique to classify any given input dataset. This algorithm classifies a given dataset into discrete k-clusters using which k-centroids are defined, one for each cluster. The next step is to take each point in the given input data set and associate it to the possible nearest centroid. This process is repeated for all the input data points, based on which next level of clustering and the respective centroids are obtained. This procedure is iterated until it converges. This algorithm minimizes the following objective function.

$$J = \sum_{j=1}^{k} \sum_{i=1}^{k} \left\| x_i{}^j - c^j \right\|^2 \tag{1}$$

Where $\left\| x_i{}^j - c^j \right\|^2$ is a chosen distance measure between a data point $(x_i)^{\,j}$ and the cluster centre, c_j is an indicator of the distance of the k data points from their respective cluster centers. The proposed unsupervised segmentation algorithm uses the principle of K-means clustering.

The proposed technique segments the region of interest (ROI) of an input image (input_img) by an interactive user defined shape of square or rectangle to obtain select_img. Then, the number of bins for coarse data computation (bin size), the size of overlapping kernel to partition (w-size) and the maximum number of clusters for segmentation (max_class) are fed as input data for the computation of coarse data. The coarse data identified by each kernel is aggregated to form the final_coarse_data which is further clustered using the principle of K-means clustering in order to produce the segment_img. The algorithmic description of the proposed technique is given herein under:

2.3   Algorithm

1. Read a grayscale image as input_img
/* Define the area to be segmented as a runtime interactive input. The shape of the selection can either be a square or a rectangle */
2. Let select_img is the selected subimage of input_img
3. Assign:
a. binsize=5
/* number of bins for coarse data computation */
b. wsize= 7
/* wsize is the size of overlapping kernel to partition the select_img */
c. max_class= 3
/* maximum number of clusters for segmentation */
4. Repeat step 5 and 6 until the select_img is read
5. Read select_img in the order of (wsize*wsize) as window_img
6. Compute coarse_img for window_img as coarse_win_data
7. Aggregate coarse_win_data for select_img as final_coarse_data
8. Cluster final_coarse_data using K-means clustering technique using max_class in order to obtain segment_img
9. Stop

This algorithm can segment an object either fully or partially based on user's choice. If the image has a background and objects then it partitions the object from the background and displays its coarse image. If the image has no background, then the segmented image reveals the inner details of the object. This technique finds application in image processing as well as image analysis. Figure 1 shows the segmented coarse image using above mentioned algorithm.

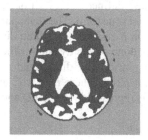 

Fig. 1. (a) Coarse Segmented MRI Image based on above algorithm (b) ROI segmented image of White Lesions

2.4   Datasets

The input dataset consists of axial, T2-weighted, 256 X 256 pixel MR brain images (Fig. 2). These images were downloaded from the (Harvard Medical School website (http:// med.harvard.edu/AANLIB/) [9]. Only those sections of the brain in which lateral ventricles are clearly seen are considered in our study. The number of MR brain images in the input dataset is 60 of which 6 are of normal brain and 54 are of abnormal brain. The abnormal brain image set consists of images of brain affected by Alzheimer's and other diseases. The remarkable feature of a normal human brain is the symmetry that it exhibits in the axial and coronal images. Asymmetry in an axial MR brain image strongly indicates abnormality. A normal and an abnormal T2-weighted MRI brain image are shown in Fig. 2(a), 2(b) and 2(c), respectively. Indeed, for multilayer learning models like deep and extreme machine learning algorithms needed big datasets for training, however due to lack of availability of proper datasets in MRI imaging, we used this dataset for examining the performance of proposed approaches for this paper, but acquiring other suitable datasets for future studies.

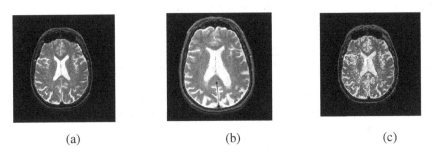

(a)   (b)   (c)

Fig. 2. (a) T2, weighted an axial MRI Brain Image; (b) T2, weighted an axial MR brain image as abnormal brain; (c) T2, weighted an axial MR brain image as normal brain after Wavelets Decomposition and denoising

2.5   Discrete Wavelets Transform (DWT)

The DWT is an implementation of the wavelet transform using a discrete set of the wavelet scales and translation obeying some defined rules. For practical computations, it is necessary to discretize the wavelet transform. The scale parameters are discretized on a logarithmic grid. The translation parameter (τ) is then discretized with respect to the scale parameter, i.e. sampling is done on the dyadic (as the base of the logarithm is usually chosen as two) sampling grid. The discretized scale and translation parameters are given by, s = 2-m and t = n2-m, where m, n $\in$ Z, the set of all integers. Thus, the family of wavelet functions is represented in Eq. (2) and (3),

$$\psi_{m,n}(t) = 2^{\frac{m}{2}}\psi(2^m t - n) \qquad (2)$$

$$W\psi(a,b) = \int_{-\infty}^{\infty} f(x) * \psi_{a,b}(t)dx \qquad (3)$$

In case of images, the DWT is applied to each dimension separately. This result in an image Y is decomposed into a first level approximation component Y_a^1 and detailed components Y_h^1 Y_v^1 and Y_d^1 corresponding to horizontal, vertical and diagonal details. Fig.1 depicts the process of an image being decomposed into approximate and detailed components.

The approximation component (Y_a) contains low frequency components of the image while the detailed components (Y_h, Y_v and Y_d) contain high frequency components. Thus,

$$Y = Y_a{}^1 + \{ \ Y_h{}^1 + Y_v{}^1 + Y_d{}^1 \} \tag{4}$$

At each decomposition level, the length of the decomposed signals is half the length of the signal in the previous stage. Hence the size of the approximation component obtained from the first level decomposition of an NXN image is N/2 X N/2, second level is N/4 X N/4 and so on. As the level of decomposition is increased, compact but coarser approximation of the image is obtained. Thus, wavelets provide a simple hierarchical framework for interpreting the image information.

2.6   Deep Belief Nets

DBNs[10] are multilayer, stochastic generative models that are created by learning a stack of Restricted Boltzmann Machines (RBMs), each of which is trained by using the hidden activities of the previous RBM as its training data. Each time a new RBM is added to the stack, the new DBN has a better variation lower bound on the log probability of the data than the previous DBN, provided the new RBM is learned in the appropriate way [11].

A Restricted Boltzmann Machine (RBMs) is a complete bipartite undirected probabilistic graphical model. The nodes in the two partitions are referred as hidden and visible units. An RBM is defined as

$$p(v,h) = \frac{e^{-E(v,h)}}{\sum_u \sum_g e^{-E(u,g)}} \tag{6}$$

Where v ϵ V are the visible nodes and h ϵ H are the latent random variables. The energy function E (v,h,W) is described as

$$E = -\sum_{i=1}^{D} \sum_{j=1}^{K} v_i W_{ij} h_j \tag{7}$$

Where W ϵ R^{DXK} are the weights on the connections, and where we assume that the visible and hidden units both contain a node with value of 1 that acts to introduce bias. The conditional distribution for the binary visible and hidden units are defined as

$$p(v_i = 1/h, W) = \sigma(\sum_{j=1}^{K} W_{ij} h_j) \tag{8}$$

$$p(h_j = 1/v, W) = \sigma(\sum_{i=1}^{D} W_{ij} v_i) \tag{9}$$

Where σ is the sigmoid function. Using above equations, it easy to go back and forth between the layers of RBM. While training, it consists of some input to the RBM on the visible layer, and updating the weights and the biases such that p(v) is high. In generalized way, in as set of C training cases $\{v^c |c \in \{1,....,C\}\}$, the objective is to maximize the average log probability defined as

$$\sum_{c=1}^{C} \log p(v^c) = \sum_{c=1}^{C} \log \frac{\sum_g e^{-E(v^c,g)}}{\sum_u \sum_g e^{-E(u,g)}} \tag{10}$$

The whole training process involves updating the weights with several numbers of epochs and the data is split in 20 batches which we take it randomly and the weights are update at the end of every batch. We use the binary representation of hidden units activation pattern for classification and visualization. The autoencoder with N_h hidden nodes is trained and fine-tuned using back-propagation to minimize squared reconstruction error, with a term encouraging low average activation of the units.

2.7 Extreme Machine Learning

The Extreme Learning Machine [12, 13, 14] [15] [18] is a Single hidden Layer Feed forward Neural Network (SLFN) architecture. Unlike traditional approaches such as Back Propagation (BP) algorithms which may face difficulties in manual tuning control parameters and local minima, the results obtained after ELM computation are extremely fast, have good accuracy and has a solution of a system of linear equations. For a given network architecture, ELM does not have any control parameters like stopping criteria, learning rate, learning epochs etc., and thus, the implementation of this network is very simple. Given a series of training samples $(x_i, y_i)_{i=1, 2 ...N}$ and $\hat{N}$ the number of hidden neurons where $x_i = (x_{i1},....x_{in}) \in R^n$ and $y_i = (y_{i1},....y_{in}) \in R^m$, the actual outputs of the single-hidden-layer feed forward neural network (SLFN) with activation function g(x) for these N training data is mathematically modeled as

$$\sum_{k=1}^{\hat{N}} \beta_k g\big((w_k, x_i) + b_k\big) = 0_i , \forall = i = 1, ...,N \tag{11}$$

Where $w_k = (w_{k1},.....,w_{kn})$ is a weight vector connecting the kth hidden neuron, $\beta_k = (\beta_{k1},...... \beta_{km})$ is the output weight vector connecting the kth hidden node and output nodes. The weight vectors w_k are randomly chosen. The term (w_k, x_i) denotes the inner product of the vectors w_k and x_i and g is the activation function. The above N equations can be written as $H\beta = O$ and in practical applications $\hat{N}$ is usually much less than the number N of training samples and $H\beta \neq Y$, where

$$H = \begin{bmatrix} g\big((w_1,x_1) + b_1\big) & \cdots & g\big((w_{\hat{N}}, x_1) + b_{\hat{N}}\big) \\ \vdots & \ddots & \vdots \\ g\big((w_1, x_{1N}) + b_1\big) & \cdots & g\big((w_{\hat{N}}, x_N) + b_{\hat{N}}\big) \end{bmatrix}_{N \times \hat{N}} \tag{12}$$

The matrix H is called the hidden layer output matrix. For fixed input weights $w_k = (w_{k1},.....,w_{kn})$ and hidden layer biases b_k, we get the least-squares solution $\hat{\beta}$ of the linear system of equation $H\beta = Y$ with minimum norm of output weights β, which gives a good generalization performance. The resulting $\hat{\beta}$ is given by $\hat{\beta} = H +$ Y where matrix H^+ is the Moore-Penrose generalized inverse of matrix H [14].

2.8 Trained Classifiers and Feature Selection Evaluators

In this study, apart from deep learning based on Restricted Boltzmann machines and extreme machine learning based on Single hidden Layer Feed forward Neural Network (SLFN) architecture as classifiers, several other classifiers are also examined in terms of accuracy and performance, including K-nearest neighbor, SVM , Naive Bayes, MultiboostAB, RotationForest, VFI, J48 and Random Forest.

To reduce the dimensionality of the large set of features of dataset, in our study, we propose the use of three optimal attribute selection algorithms: correlation based feature selection (CFS) method, which evaluates the worth of a subset of attributes by considering the individual predictive ability of each feature along with the degree of redundancy between them, secondly an approach based on wrappers which evaluates attribute sets by using a learning scheme. Also in this study, three search methods are also examined: the Best First, Greedy Stepwise and Scatter Search algorithms. These search algorithms are used with attribute selector's evaluators to process the greedy forward, backward and evolutionary search among attributes of significant and diverse subsets. In total, these feature selection algorithms were tested to select nearly 10 optimal and significant features out of 1024 features. The whole proposed method is implemented using Weka 3.6 platform.

3 Experiments and Results

3.1 Level of Wavelet Decomposition

We obtained wavelet coefficients of 60 brain MR images, each of whose size is 256 X 256. Level-1 HAR wavelet decomposition of a brain MR image produces 16384 wavelet approximation coefficients; while level-2 and level-3 produce 4096 and 1024 coefficients, respectively. The preliminary experimental analysis of the wavelet coefficients through simulation in Matlab 7.10., we showed that level-2 features are the best suitable for different classifiers, whereas level-1 and level-3 features results in lower classification accuracy. We also use the DAUB-4 (Daubachies) as mother wavelets to get decomposition coefficients of MRI images at Level 2 for comparative evaluation of two wavelets decomposition methods in terms of classification accuracy.

3.2 Attribute Selection and Classification

In our study, three attribute or feature selection algorithms are used: correlation based feature selection (CFS) method which evaluates the worth of a subset of attributes by considering the individual predictive ability of each feature along with the degree of redundancy between them, secondly an approach based on a wrapper which evaluates attribute sets by using a learning scheme. Also in this study, three search methods are also examined: the Best First, Greedy Stepwise and Scatter Search algorithms. These search algorithms are used with attribute selector's evaluators to process the greedy forward, backward and evolutionary search among attributes of significant and diverse subsets. Table 1 shows the accuracy of classification (percentage of correctly classified samples), True Positive Rate (TP), False Positive Rate (FP) and Average Classification Accuracy (ACC) over all pair-wise combination with different feature evaluators and search algorithms with respect to multi-class classification.

Table 1 shows the performance of several learning classifiers, including K-nearest neighbor, SVM, Naive Bayes, MultiboostAB, Rotation Forest, VFI, J48 and Random Forest. Among the pair-wise classification, the lowest accuracy is observed for the classification VFI classifiers of 74.16% and the highest accuracy for the classification by Rotational forest of 97.06%. Moreover, the combination of CFS feature evaluator with the of Best First search algorithm gives the highest classification accuracy.

While Table 1 shows the performance of indivual classifiers, Table 2 defines the comparative results of various combined search techniques and feature evaluators using above prescribed classifiers. Table 3 compares the proposed method against a popular dimensionality reduction method, known as Principal Component Analysis (PCA). PCA applies an orthogonal linear transformation that transforms data to a new coordinate system of uncorrelated variables called principal components. We have applied PCA to reduce the number of attributes or feature to 18 attributes and plotted the ROC curves using several above mentioned learning classifiers in terms of True Positive and False Positive Rate, as seen in figure 3. As can be seen in figure 3, ROC curves for all the trained learning classifiers examined in this study, the curves lie above the diagonal line describing the better classification rather than any other random classifiers. The optimal points of various trained classifiers are indicated by bold solid circles as False Positive rate (FP) and True Positive rate (TP). These optimal points in ROC curves show the maximum optimal value (FP, TP) of all trained classifiers.

Table 1. Various Classifiers comparision with respect Average Classification Accuracy(%) and other parameters

Classifiers	TP Rate	FP Rate	Precision	Recall	F-Measure	(ACC %)
KNN	0.935	0.917	0.826	0.853	0.839	91.04
SVM	0.912	0.912	0.831	0.912	0.87	91.17
Naive Bayes	0.868	0.916	0.828	0.868	0.847	86.76
MultiboostAB	0.91	0.91	0.829	0.91	0.868	91.04
Rotation Forest	0.971	0.285	0.971	0.971	0.968	97.06
VFI	0.742	0.049	0.93	0.742	0.796	74.16
J48	0.96	0.314	0.958	0.96	0.957	95.98
Random Forest	0.97	0.271	0.97	0.97	0.968	97.01

Table 2. Comparison of pair wise combination of various Attribute Selectors and classifiers with respect to ACC (%)

Evaluator	Search Algorithm	Classifier	N	ACC (%)
CFS	Best First	K-NN	6	91.04
CFS	Greedy Stepwise	K-NN	2	89.70
CFS	Scatter Search	K-NN	4	88.23
Wrapper	Best First	K-NN	5	89.32
Wrapper	Greedy Stepwise	K-NN	4	87.56
Wrapper	Scatter Search	K-NN	4	88.20
CFS	Best First	SVM	6	91.17
CFS	Greedy Stepwise	SVM	6	89.23
CFS	Scatter Search	SVM	4	91.04
Wrapper	Best First	SVM	2	90.65
Wrapper	Greedy Stepwise	SVM	2	90.65
Wrapper	Scatter Search	SVM	5	89.56
CFS	Best First	Naive Bayes	8	86.76
CFS	Greedy Stepwise	Naive Bayes	8	82.78
CFS	Scatter Search	Naive Bayes	7	82.12
Wrapper	Best First	Naive Bayes	4	85.44
Wrapper	Greedy Stepwise	Naive Bayes	2	85.44
Wrapper	Scatter Search	Naive Bayes	2	80.12
CFS	Best First	MultiboostAB	5	91.04
CFS	Greedy Stepwise	MultiboostAB	5	91.04
CFS	Scatter Search	MultiboostAB	4	86.54
Wrapper	Best First	MultiboostAB	5	89.39
Wrapper	Greedy Stepwise	MultiboostAB	5	90.45
Wrapper	Scatter Search	MultiboostAB	4	88.76

Table 2. (*Continued*)

CFS	Best First	Rotation Forest	9	97.06
CFS	Greedy Stepwise	Rotation Forest	9	96.21
CFS	Scatter Search	Rotation Forest	8	91.66
Wrapper	Best First	Rotation Forest	5	93.78
Wrapper	Greedy Stepwise	Rotation Forest	6	93.78
Wrapper	Scatter Search	Rotation Forest	6	89.54
CFS	Best First	VFI	3	74.16
CFS	Greedy Stepwise	VFI	2	71.01
CFS	Scatter Search	VFI	4	71.01
Wrapper	Best First	VFI	3	72.22
Wrapper	Greedy Stepwise	VFI	2	72.85
Wrapper	Scatter Search	VFI	4	72.85
CFS	Best First	J48	7	95.98
CFS	Greedy Stepwise	J48	7	95.98
CFS	Scatter Search	J48	6	91.41
Wrapper	Best First	J48	7	95.98
Wrapper	Greedy Stepwise	J48	7	95.98
Wrapper	Scatter Search	J48	6	91.41
CFS	Best First	Random Forest	8	97.01
CFS	Greedy Stepwise	Random Forest	8	95.47
CFS	Scatter Search	Random Forest	8	95.47
Wrapper	Best First	Random Forest	5	96.25
Wrapper	Greedy Stepwise	Random Forest	6	96.25
Wrapper	Scatter Search	Random Forest	5	90.01

Table 3. Comparison using PCA and other feature attribute evaluators in terms of ACC (%)

Classifier	PCA (%)	CFS-Best First (%)	Wrapper-Best First (%)
KNN	91.38	91.04	89.32
SVM	96.24	91.17	90.65
Naive Bayes	85.63	86.76	85.44
MultiboostAB	94.52	91.04	89.39
Rotation Forest	97.06	97.06	93.78
VFI	77.12	74.16	72.22
J48	95.34	95.98	95.98
Random Forest	97.34	97.01	96.25

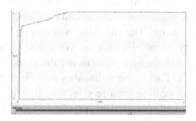

Fig. 3. Shows the ROC curve of the above mentioned trained classifiers

Table 4 describes the classification results using Extreme Machine Learning and Deep Machine Learning. In table 4, we compared the training time, testing time and classification error using extreme and deep machine Learning. As we can see in the table both learning algorithms are processed to many hidden layers and their evaluations is done in terms of various factors. As depicted in Table 4, it clearly shows that deep machine learning plays a major role in reducing the classification error. As Deep and extreme machine learning are designed to work on large datasets for it is difficult to compare the performance. However, they result in acceptable accuracy levels, and we are currently examining several other publicly available large MRI datasets for enhancing the performance of these two novel approaches (Deep learning and Extreme machine learning approaches).

Table 4. Classification results using Extreme Machine Learning and Deep Machine Learning

Hidden Layers	Training Time(s)			Testing Time(s)			Classification Error		
	10	15	20	10	15	20	10	15	20
Deep Learning	0.56	0.47	0.72	0.51	0.34	0.64	0.083	0.065	0.071
Extreme Learning	0.31	0.31	0.61	0.41	0.31	0.56	0.042	0.042	0.061

However, the deep learning networks do not need any particular feature reduction algorithms because of the inherent capability for feature reduction in terms of deep learning (learning through multiple layers). In case of extreme machine learning, the learning proceeds through random assignment of weights and hidden nodes (unlike gradient descendent based techniques). Due to this, there is a significant improvement in training and testing time as depicted in Table 4.

4   Conclusions

In this study, we have presented a principled approach for investigating brain abnormalities based on wavelet based feature extraction, PCA based feature selection and deep and extreme machine learning based classification comparative to various others classifiers. Experiments on a publicly available brain image dataset show that the proposed principled approach performs significantly better than other competing methods reported in the literature and in the experiments conducted in the study. The classification accuracy of more than 93% in case of deep machine learning and 94% in case of extreme machine learning demonstrates the utility of the proposed method. In this paper, we have applied this method only to axial T2-weighted images at a particular depth inside the brain. The same method can be employed for T1-weighted, proton density and other types of MR images. With the help of above approaches, one can develop software for a diagnostic system for the detection of brain disorders like Alzheimer's, Huntington's, Parkinson's diseases etc. Further, the proposed approach uses reduced data by incorporating feature selection algorithms in the processing loop and still provides an improved recognition and accuracy. The training and testing time for the whole study used by deep and extreme machine learning is much less as compared to SVM and other traditional classifiers reported in the literature. Further work will be pursued to classify different type of abnormalities, and to extract new features from the MRI brain images on various parameters as age, emotional states and their feedback.

References

1. Fletcher-Heath, L.M., Hall, L.O., Goldgof, D.B., Murtagh, F.R.: Automatic segmentation of non-enhancing brain tumors in magnetic resonance images. Artificial Intelligencein Medicine 21, 43–63 (2001)
2. Chaplot, S., Patnaik, L.M., Jagannathan, N.R.: Classification of magnetic resonance brain images using wavelets as input to support vector machine and neuralnetwork. Biomedical Signal Processing and Control 1, 86–92 (2006)
3. Gorunescu, F.: Data Mining Techniques in Computer-Aided Diagnosis: Non-InvasiveCancer Detection. PWASET 25, 427–430 (2007) ISSN 1307-6884
4. Kara, S., Dirgenali, F.: A system to diagnose atherosclerosis via wavelet transforms,principal component analysis and artificial neural networks. Expert Systems with Applications 32, 632–640 (2007)
5. Maitra, M., Chatterjee, A.: Hybrid multi-resolutionSlantlet transform and fuzzy c-means clustering approach for normal-pathological brain MR image segregation. Med. Eng. Phys. (2007), doi:10.1016/j.medengphy.2007.06.009
6. Abdolmaleki, P., Mihara, F., Masuda, K., DansoBuadu, L.: Neural networks analysis of astrocyticgliomas from MRI appearances. Cancer Letters 118, 69–78 (1997)
7. Rosenbaum, T., Engelbrecht, V., Krolls, W., van Dorstenc, F.A., Hoehn-Berlagec, M., Lenard, H.: MRI abnormalities in neuro-bromatosistype 1 (NF1): a study of men and mice. Brain & Development 21, 268–273 (1999)
8. Cocosco, C., Zijdenbos, A.P., Evans, A.C.: A fully automatic and robust brainMRI tissue classification method. Medical Image Analysis 7, 513–527 (2003)
9. Database taken from, http://med.harvard.edu/AANLIB/
10. Hintonand, G.E., Salakhutdinov, R.R.: Reducing the dimensionality of data with neural networks. Science 313(5786), 504–507 (2006)
11. Hinton, G.E., Osindero, S.: A fast learning algorithm for deep belief nets. Neural Computation 18, 1527–1554 (2006)
12. Lin, M.-B., Huang, G.-B., Saratchandran, P., Sudararajan, N.: Fully complex extreme learning machine. Neurocomputing 68, 306–314 (2005)
13. Huang, G.-B., Zhu, Q.-Y., Siew, C.K.: Extreme Learning Machine: Theory and Applications. Neurocomputing 70, 489–501 (2006)
14. Serre, D.: Matrices: Theory and Applications. Springer Verlag, New York Inc. (2002)
15. Mishra, A., Singh, L., Chetty, G.: A Novel Image Water Marking Scheme Using Extreme Learning Machine. In: Proceedings of IEEE World Congress on Computational Intelligence (WCCI 2012). IEEE Explore, Brisbane (2012)
16. Singh, L., Chetty, G., Sharma, D.: A Hybrid Approach to Increase the Performance of Protein Folding Recognition Using Support Vector Machines. In: Perner, P. (ed.) MLDM 2012. LNCS, vol. 7376, pp. 660–668. Springer, Heidelberg (2012)
17. Singh, L., Chetty, G.: Review of Classification of Brain Abnormalities in Magnetic Resonance Images Using Pattern Recognition and Machine Learning. In: Proceedings of International Conference of Neuro Computing and Evolving Intelligence, NCEI 2012, Auckland, New-Zealand. LNCS Bioinformatics, Springer (2012)
18. Singh, L., Chetty, G.: A Novel Approach for protein Structure prediction Using Pattern Recognition and Extreme Machine Learning. In: Proceedings of International Conference of Neuro Computing and Evolving Intelligence, NCEI 2012, Auckland, New-Zealand. LNCS Bioinformatics. Springer (2012)

An Open Framework for Extensible Multi-stage Bioinformatics Software

Gabriel Keeble-Gagnère[1], Johan Nyström-Persson[2],
Matthew I. Bellgard[1], and Kenji Mizuguchi[2]

[1] Centre for Comparative Genomics, Murdoch University, Australia
[2] National Institute of Biomedical Innovation, Japan

Abstract. In research labs, there is often a need to customise software at every step in a given bioinformatics workflow, but traditionally it has been difficult to obtain both a high degree of customisability and good performance. Performance-sensitive tools are often highly monolithic, which can make research difficult. We present a novel set of software development principles and a bioinformatics framework, Friedrich, which is currently in early development. Friedrich applications support both early stage experimentation and late stage batch processing, since they simultaneously allow for good performance and a high degree of flexibility and customisability. These benefits are obtained in large part by basing Friedrich on the multiparadigm programming language Scala. We present a case study in the form of a basic genome assembler and its extension with new functionality. Our architecture[1] has the potential to greatly increase the overall productivity of software developers and researchers in bioinformatics.

1 Introduction

Bioinformatics poses a particularly difficult challenge for software developers, with constantly changing end-user requirements and the need to interact with an ever-expanding range of tools and data formats. The advent of big data means that the tools and skills required for data manipulation and basic research are now more advanced than before. However, researchers are fundamentally biologists and more interested in the data itself than in addressing technical issues, which traditionally fall into the computer science field. The challenge for software developers is thus to put the maximum amount of power and flexibility in the hands of the users while assuming as little technical knowledge as possible.

When large data volumes are processed, high performance software tools are often used. However, such tools are often highly specialised and optimised for a specific purpose, permitting only limited customisation. This kind of software is often also *monolithic*. Monolithic tools can be efficient for handling big data problems, but such a design often runs counter to a natural research process,

---

[1] Available freely under a dual GPL/MIT open-source license from
https://bitbucket.org/jtnystrom/friedrich/.

T. Shibuya et al. (Eds.): PRIB 2012, LNBI 7632, pp. 106–117, 2012.
© Springer-Verlag Berlin Heidelberg 2012

since researchers often need to make adjustments to various parts of the tools that they work with, particularly in fast-changing fields such as bioinformatics. MacLean and Kamoun [8], reporting on their experience bringing a small bioinformatics laboratory into the age of big data, state that biologists at first tend to regard bioinformatics processes as being monolithic, but once they understand their inner workings generally become more productive, especially if they can take charge of tools and methods themselves to some degree. Clearly, transparent and flexible tools have the potential to play a very important role.

We argue that it is possible to develop software that makes researchers more productive and enables them to ask more questions about their data and their process by adopting a new set of software development principles. In the following, we present the Friedrich architecture (Section 2). We then discuss the Friedrich framework, a toolkit for building bioinformatics applications according to these principles (Section 3). We discuss the implementation of a basic genome assembler based on Friedrich in Section 4. We compare with other tools and frameworks in Section 5, and conclude the paper in Section 6.

2   The Friedrich Software Principles

The Friedrich architecture is a set of interlocking software design principles that, in our view, can support bioinformatics research very effectively.

Expose Internal Structure. Bioinformatics software should expose its internal building blocks and data flow to a high degree, permitting reconfigurability. Bioinformatics computation often consists of sending data through a number of processing stages until the desired output is produced. Frameworks should reflect this by consisting of modules that can easily be rewired - reconnected in different sequences - to represent changing workflows. This is the opposite of a monolithic application, which is effectively a black box.

Conserve Dimensionality Maximally. The processing of a given data set – which can essentially be viewed as a set of points in a mathematical space – to produce a given output, is analogous to a *projection* in geometry. For example, in $\mathbb{R}^3$, the equation

$$x_1^2 + x_2^2 + x_3^2 = 1,$$

defines a sphere of radius 1 centred at the origin. The projection $proj_1$, which sends $(x_1, x_2, x_3) \in \mathbb{R}^3$ to $x_1 \in \mathbb{R}$, when applied to the sphere defined above, yields: $x_1^2 = 1$, which defines the set of two points $\{-1, 1\}$. If $f : \mathbb{R}^3 \to \mathbb{R}^3$ is a mapping, then given a surface in $\mathbb{R}^3$ (such as the sphere defined above), the function $proj_1 \circ f$ returns an answer to the query "At what points does the mapped surface intersect the x_1-axis?" Given an answer to the query, we cannot extract information about the original surface. In an analogous way, raw bioinformatics data contains all possible information from a given experiment. Thus it has *maximum dimensionality*. As various data processing is performed on this data set, its dimensionality is reduced. For example, given a set of reads from a DNA sequencing run, one processing step might be to remove duplicates, to produce a set of non-redundant sequence reads. This would clearly reduce the dimensionality of the resultant data set, since the redundancy information is lost.

Maximal conservation of dimensionality permits users who are applying tools experimentally to go back to previous stages of their computation and attempt different parameters, adding a great deal of flexibility to the experimental process, allowing new questions to be asked, and saving time. It can also be thought of as maximal preservation of the results of intermediate phases in the computation.

Multi-stage Applications. Many tools need to be used in at least two different stages, which may loosely be called *experimentation* and *production*. In the experimental stage, researchers explore newly available data in order to develop methods and a basic understanding of what can be done. It is in this stage that the need for customisation and flexibility is greatest. In the production stage, a repeatable process is extracted and applied systematically a large number of times. In this stage there is less need for flexibility; instead, robustness, reliability, and performance are valued. However, *a given analysis or tool, once developed, often has to move across this boundary from the experimental stage to the production stage.* This transition is often nontrivial given that hitherto, incompatible technologies have often been used in the two stages. In such a situation, one may opt to use experimental stage technologies in both stages, resulting in poor performance. Alternatively, one may use production-stage technologies in both stages, resulting in difficulty of experimentation. Finally, one may re-develop the analysis from scratch once it makes the transition, which would be a large additional effort.

Friedrich software should support a full range of development stages, including experimentation, production, and any intermediate points. Because a single technology framework is used consistently, it becomes easy to move from experimentation to production, and also to move back again. This enables a feedback loop between experimental usage and production usage: when something unexpected occurs in the large scale application of a tool, it can easily be taken back to the workbench for inspection, and any adjustments made can be propagated back again. Table 1 gives a comparison.

Table 1. A comparison of Friedrich's target characteristics with tools designed mainly for either experimentation or production

Context	Necessary flexibility	Typical programming language	Performance	Examples
Experimental stage tools	High	Perl, Python, R, ...	Low/ moderate	BioPerl, BioPython
Production stage tools	Low	C, C++, Java, ...	Very high	Velvet, Abyss, BioJava
Friedrich	High	Scala, Java	High	Section 4

Flexibility with Performance. This is closely related to the previous principle. If programming languages have traditionally been separable into on one hand a category of high-performing but inflexible ones (in that applications written in them are relatively hard to customise) and on the other a category of poorly performing but flexible ones, we believe that the relatively recent language Scala (see Section 3.1) is an outlier that provides for both good performance and high flexibility. This enables flexibility with performance. For many bioinformatics applications, one should not seek extreme performance or extreme flexibility but good levels of both.

Minimal Finality. Monolithic software often makes unsustainable assumptions about data formats, algorithm parameters and data sizes. For example, the so-called next generation of sequencing equipment is expected to render many of the current genome analysis software tools unusuable, largely for the reason that certain quantity and size parameters will change. Friedrich applications should assume a minimum of finality. Software developers should not dictate how the framework or its building blocks should ultimately be used, since they cannot possibly anticipate all the usage scenarios that may eventually appear. MacLean and Kamoun found that reorienting research from a top-down model to a bottom-up model helped increase productivity in the Sainsbury Laboratory [8]. Minimising finality also helps achieve this end.

Ease of Use. Friedrich applications should not be hard for novices to use. They should provide sensible defaults at all times, so that new users can deploy them in common use cases with little effort. Simplicity should not be sacrificed to the other principles.

We have now described the software design principles of the Friedrich architecture. Next, we describe our implementation of the Friedrich framework, as well as an application built on top of it.

3   The Friedrich Framework

The Friedrich framework is implemented in the form of a Scala library that permits users to develop bioinformatics applications easily. In implementing this framework, our aim has been to allow application developers to follow the principles we outlined in the previous section easily. The framework is still under development, and this section describes its current state.

3.1   The Scala Programming Language

An early decision was made to base Friedrich on Scala, a novel programming language for the Java virtual machine, which is being developed by Martin Odersky and others [12] (http://www.scala-lang.org). Programming languages are traditionally classified as *functional* or *imperative*. Functional languages emphasise avoidance of side effects and composition of functions. Imperative languages, such as Perl, C, and Java, have been more widely used in the mainstream, and generally functions in these languages may have side effects. Scala blends these two paradigms. It provides libraries, constructs and idioms for stateless, purely functional programming as well as for stateful, imperative, object-oriented programming. Scala code is often very compact compared with equivalent Java code, and, provided that the programmer is somewhat disciplined, can be highly readable.

Scala brings several important benefits to Friedrich.

- Scala provides for high programmer productivity and is very well suited to big data tasks, performing well [5] even under heavy loads, thanks to the maturity of the underlying Java platform.
- Existing Java libraries for tasks such as graph processing, database access, calculation and so on can be taken advantage of immediately.
- Because of its strong support for functional programming and immutable state, Scala is a foundation that lends itself well to parallel processing, the need for which cannot be ignored in bioinformatics today.

Scala has much of the flexibility and productivity of scripting languages such as Ruby, Python and Perl. For example, Scala has features such as an interactive interpreter with auto-completion, pattern matching and convenient regular expression support. Type inference means that types in many cases do not need to be declared. SBT (Simple Build Tool), which is widely used by the Scala community, permits automatic dependency management and library downloading in a style that resembles Perl's well-known CPAN package repository.

In a survey of software engineering techniques used in 22 different bioinformatics software projects, Rother et al. described 12 practices that were found to be useful [14]. Scala and Friedrich directly support many of these, benefiting both from the mature development tools available for the Java platform and from its own tools. For example, Scala has good support for unit testing and a sophisticated documentation generator, and Friedrich supports practices such as frequent release and feedback cycles, since it enables easy transitions between the experimental and production stages.

3.2   Friedrich Application Components

Friedrich contains the following key components for building applications.

Phases and Pipelines. Friedrich applications are organised as sets of *phases*, according to the model illustrated in Figure 1. Sequences of phases are called *pipelines*. Friedrich provides foundational classes that can be extended to implement new phases, as well as functions for managing and running pipelines.
Data Object Classes. Friedrich phases operate on standardised data objects. For a given application, all experimental data as well as configuration parameters is stored in these objects.
Configuration Management. Pipelines and general application parameters are stored in XML configuration files (Figure 1). Friedrich provides facilities for reading these configurations and automatically creating pipelines from them.

Core Bioinformatics Functionality. Friedrich provides a small library of core bioinformatics algorithms and data representations.

In order to implement a new Friedrich application, one should select a data object type or define a new one, implement the necessary phases, and write a main method that invokes a pipeline using the Friedrich API. As we will see, implementing phases is not difficult.

Phases receive input that they make certain assumptions about (*phase preconditions*), perform some computation on it, and then pass on this data in a new state (*phase postconditions*) as output. For example, our genome assembler makes use of phases such as ScanReads, BuildGraph and FindPaths, among others (shown in Figure 2). Phases can perform almost any functionality. In accordance with our dimensionality principle, phases should add information to the shared data object rather than remove or overwrite. This permits the user to explore and manipulate the data (in interactive mode) in between pipeline phases. Friedrich applications can easily invoke pipelines based on their names only, which means that workflows can be changed without recompiling an application.

```
1  <settings>
2    <pipeline name="default">
3      <phase>miniasm.ScanReadsPhase</phase>
4      <phase>miniasm.BuildGraphPhase</phase>
5      <phase>miniasm.FindTipsPhase</phase>
6      <phase>miniasm.ComputeCoveragePhase</phase>
7      <phase>miniasm.FindPathsPhase</phase>
8    </pipeline>
9  </settings>
```

Fig. 1. An example of a pipeline configuration. The phases will be run in the order shown. 'Miniasm' is the package name of the corresponding classes.

The components we have described support the six principles as much as possible. Phases and pipelines are a natural way to expose structure. When an application is made up of a set of relatively independent phases, it becomes clear what its internal parts are, and the configuration system permits them to be rewired easily. Conservation of dimensionality is not enforced by the framework itself. Phase implementors are recommended to always add data to the shared data object and not overwrite or remove it unless necessary. In the future, we plan to provide automatic data management facilities to assist interactive use. Multi-stage applications and flexibility with performance are benefits that we derive largely from our use of the Scala language, as outlined above. Minimal finality is something we obtain in part from Scala, and in part from the pipeline and phase system, since the overall data flow of an application can be changed at a late stage. Ease of use is a principle to be upheld by application developers.

4   Genome Assembly with Friedrich

Genome assembly refers to the process of turning raw sequence reads – produced from a sequencing run – into contiguous regions of DNA, know as contigs, that represent the original genome being analysed. In particular, *de novo* genome assembly refers to assembling a novel genome for the first time directly from individual reads – that is, without a reference genome to guide it. Assembly methods have evolved from the *overlap-layout-consensus* (OLC) method (employed by early sequencing efforts, including the Human Genome Project, which took advantage of the long reads produced by traditional Sanger technology) to the *de Bruijn* graph methods employed by most assemblers that accept current high-throughput short read data. For the technical details of genome assembly, we refer the reader to [2]. In short, the nodes of the de Bruijn graph are sequences of length k base pairs (known as k-mers); an edge exists between two k-mers if their sequence overlaps by $k-1$ bases. This graph is then processed and contigs read off directly as non-ambiguous paths.

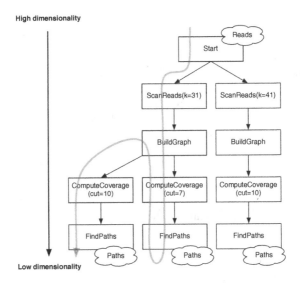

Fig. 2. The internal data flow of a hypothetical genome assembler. The grey path illustrates how a user may wish to try a number of different coverage cutoff values, which involves returning to a previous phase.

One of the early motivations for Friedrich was the desire to investigate in detail the inner workings of this process. Investigating assemblies with commonly used assemblers such as Velvet [17] and ABySS [15], we found that output can vary considerably given the same input data. As well as this, we found that outputs could vary even on very small toy data sets (data not shown). Indeed, anyone who has used these tools will be aware that different assemblers produce different output, but rarely will the user have a clear idea of what exactly has been done differently.

Figure 2 outlines a simplified typical workflow for an assembler. Internally, data is sent through a number of phases in order to produce the final output. As a rule, the output of each phase is less complex than its input, and the final output is much simpler than the initial input. This can be understood as a successive reduction of the dimensionality of the data. Each phase within a tool such as this assembler can be controlled by parameters (for example k, cut), and modifying the parameters of a phase might affect the final output significantly. Thus, researchers might want to traverse what we might call a *phase tree* following the curved arrow in order to compare outputs resulting from various configurations. In a monolithic tool, this is generally not possible, since one cannot return to earlier phases in the pipeline: the tool must be re-run from the starting point even when only parameters of late phases are changed, if they can be changed at all. With Friedrich, it is possible to interrogate the assembly at every step of the way.

The Friedrich-based assembler that we have developed consists of an efficient representation of sequences and reads, 11 processing phases and various utility classes. The source code is about 3000 lines in length.

4.1   Interactive Use

The following is an example of an interactive Friedrich session to process Illumina short read data[2]. We launch the interactive Friedrich console using SBT. If the source code of any phases or libraries being used has changed when Friedrich is launched in this way, they will automatically be recompiled, permitting a smooth development and testing workflow. The interactive Scala environment has features such as tab-completion to show all available alternatives. This environment evaluates Scala expressions as they are typed in, and allows for functions and classes to be defined on the fly.

```
> console
[info] Starting scala interpreter...
scala> import miniasm._
scala> Assembler. <tab>
   T          asInstanceOf      initData          isInstanceOf
   main       runPhases         toShort           toString
   writeContigFile
scala> val asm = Assembler.initData("-input /export/home/staff/
   gkeeble/temp/ERR015569.1in9.fa -k 31")
scala> ScanReadsPhase(asm)
   Fasta format
   miniasm.genome.bpbuffer.BPKmerSet@7a5cf2b8 Cache hits: 160731328
   misses: 15668672 ratio: 0.91, rate: 1221.00/ms
```

---

[2] NCBI SRA experiment ERX005938, run ERR015569. Only 1/9 of the reads were passed to Friedrich.

```
scala> BuildGraphPhase(asm)
  15676904 nodes
  15566712 edges
scala> FindPathsPhase(asm)
  . . . . . . . . . .
scala> contigs.size
  res7: Int = 5686
scala> contigs.toList.sortWith(_.size > _.size).head.size
  res8: Short = 19032
scala> contigs.toList.sortWith(_.size > _.size).head
  res11: Contig = GGAAGCCACAAAGCCTACATAAATATTCATTCCCTCTGGAGGCA...
```

In this interactive session, we first prepare a data object using Assembler.initData. This method takes the same parameters that the Friedrich assembler accepts when run in non-interactive mode. The resulting AssemblyData object asm is then manually passed to different phases by the user. At any time, the user can construct additional data objects and compare them or interrogate them more closely. After FindPathsPhase has finished running, contiguous paths will be available in the asm data object. We can now use the full power of Scala to explore or alter the data that has become available. First we ask for the number of contigs that were found (5686). Then we sort the contigs to have the largest first, defining a sort function on the fly (_.size >_.size, called an *anonymous function*) and asking for its size (19032). Finally we examine the actual base pairs in this long contig.

4.2   Extending the Assembler for Motif Recognition

We now show how to extend the assembler with a phase that detects and displays repeating motifs in the contiguous base pair sequences (contigs) that have been found. In Scala, *traits* are a basic unit of composition. Classes and traits can inherit from multiple traits simultaneously. In this way, one can build up a family of traits, each one representing a functionality, and compose them as needed. Phases must extend a basic Friedrich phase trait called Phase. In our assembler, which is an application that is built on top of Friedrich, we define AsmPhase, which extends Phase and make it the convention that all our assembly phases will extend this new trait. Thus, we now add a phase called FindRepeatsPhase (Figure 3).

The method runImpl implements the concrete functionality of each phase. Lines 7-12 show a generalised *for-comprehension*, a special feature of Scala. The variable c iterates over all contigs that have previously been found in the assembly. These are taken from the data object, which has previously been operated on by other phases. The notation (start, length, pattern) declares a 3-tuple of three variables. These will iterate over the repeated motifs returned by the method c.repeats for each value of c. When repeated motifs are found, they are printed to the console. This short snippet demonstrates that in many cases, Scala code can be considerably more compact than corresponding Java code (not shown). Also note the default values of the parameters: minTotLen: Int = 8, minMotifLen:

```
1   class FindRepeatsPhase [T <: Kmer[T]]( minTotLen : Int = 8,
2       minMotifLen : Int = 3) extends AsmPhase[T] {
3
4       def runImpl(data : AssemblyData [T]): Unit = {
5         println ("Finding repeats ...")
6         for (c <- data . contigSet . contigs ;
7             (start , length , pattern ) <- c . repeats ( minTotLen ,
8               minMotifLen )) {
9           println ("Contig : " + c + " pattern : " +
10            pattern + " start offset : " + start)
11        } }
12  }
```

Fig. 3. The newly added FindRepeatsPhase

Int = 3. These are the two parameters that constructors for this class take, but since they have default values, they can be omitted if needed.

Since the runImpl method can contain any Scala code, it has access to the full range of Java and Scala APIs. Here we obtain the desired repeats by using the method c.repeats, which is defined in Friedrich's contig class (not shown). However, we are in no way limited to using only such built-in methods.

After the new phase has been defined in this fashion, no additional work is needed. It can be included in pipelines, as shown in Figure 1. It can also be used interactively. We have written a small convenience function (not shown) to allow the new phase to be invoked by simply referring to its name. After assembly has been carried out, as shown in our previous interactive example (Section 4.1), we can apply the new phase:

```
scala> FindRepeatsPhase(res1)
(...) TAGACTTATTAGCGACAATAAAGATTATGAGCCTATCAGTCTGGACGGGGAAGATTT
TGAGATGCTTGGTGTAGTTGTAGGCGAGTTTAAAAGAATGGATTAAAATAGACTTAAGAAAAC
TTTAAGT [TGTCTCCTAGTGTCTCCTAG] TGT...
pattern: TGTCTCCTAGTGTCTCCTAG start offset: 299
```

A large number of repeated motifs are found in the contigs that were previously assembled; we show one of them here. At this point, it is possible to retain the data that has been produced so far, make adjustments, and assemble again with different parameters. One can then easily contrast repeated motifs that are produced by different assembly configurations, all without leaving Friedrich.

5   Comparison with Other Tools and Libraries

Friedrich has similarities with many existing tools and frameworks, although we believe that there are no well-known bioinformatics tools precisely filling Friedrich's role at the moment. The pipeline and phase structure is similar to a class of software that might be called *toolkits*. These packages consist of individual specialised programs that operate on a shared file format. The user is free to run the programs in any order and can thus create their own workflow, perhaps through shell scripting. Examples of such toolkits are SAMtools/Picard [7] and GATK [10], for handling nucleotide sequence alignments. While these toolkits come close in spirit to the Friedrich design, one essential difference is that our ability to run Friedrich phases in an interactive Scala environment permits users to very easily inspect and modify data manually in between phases. Unsupported extensions to a toolkit such as SAMtools require first writing a new program from scratch, and interactive experimentation would require even more additional work. Friedrich minimises the cost of free experimentation with data as it is being processed. Note that Picard and GATK provide Java APIs, which could be easily integrated into Friedrich.

There are many general frameworks for bioinformatics, such as BioJava[4], BioPython[1], BioPerl[16], BioScala[13] and BioRuby[11]. These are all utility libraries of varying size and scope, aimed at bioinformatics tasks in the respective programming language. Mangalam provides an informative comparison of the first three [9]. Although this survey is now ten years old, most of the points it makes about programming language differences are still essentially valid. However, its conclusion that BioPerl is sufficient for about 90% of bioinformatics programming needs is now outdated, with the need to process ever larger data sets. In general, the Bio-* toolkits provide useful routines and data models but do not prescribe any specific software development style. Therefore, they are somewhat orthogonal to our effort, which aims to provide both an architectural style and foundational libraries to support it. The Bio-* toolkits can in principle be integrated into Friedrich applications, in particular BioScala and BioJava. Bioinformatics workflow systems, for example those provided in Yabi[6] and Galaxy[3], are user-friendly ways of managing and applying high level computation pipelines. However, in their focus on ready-made, finalised modules they are quite different from what Friedrich seeks to become.

6   Conclusion and Remarks

We have argued for the introduction of a new set of software development principles for bioinformatics software, and we provide a framework that supports application development based on these principles. We have also shown an existing application based on the framework. Principles such as conservation of dimensionality and an exposed internal structure will allow developers to produce software that is more useful to bioinformaticians and better suits the research process. While Friedrich does not aim to provide either the highest performance or the greatest flexibility, with good levels of both it represents a new tradeoff that should be considered an important option for many areas in bioinformatics.

We view the architectural principles presented in Section 2 as essentially complete. However, the Friedrich software framework is still in an early stage of its development and many enhancements and extensions have yet to be implemented. For example, the BioJava[4], and BioScala[13] libraries provide a large amount of functionality for bioinformatics applications, and when doing so is suitable, it would be natural to "wrap" this functionality as Friedrich phases, rather than reimplement the functionality from scratch in Friedrich.

It remains to develop more applications on top of Friedrich, in addition to the genome assembler we have discussed in this work, in order to verify that the design principles hold up across a wider range of tasks in practice.

References

1. Cock, P.J.A., et al.: Biopython: freely available Python tools for computational molecular biology and bioinformatics. Bioinformatics 25(11), 1422–1423 (2009)
2. Compeau, P.E.C., et al.: How to apply de Bruijn graphs to genome assembly. Nature Biotechnology 29(11), 987–991 (2011)
3. Goecks, J., et al.: Galaxy: a comprehensive approach for supporting accessible, reproducible, and transparent computational research in the life sciences. Genome Biology 11(8), R86+ (2010)
4. Holland, R.C.G., et al.: BioJava: an Open-Source Framework for Bioinformatics. Bioinformatics 24(18), 2096–2097 (2008)
5. Hundt, R.: Loop Recognition in C++/Java/Go/Scala. In: Proceedings of Scala Days 2011 (2011)
6. Hunter, A.A., et al.: Yabi: An online research environment for grid, high performance and cloud computing. Source Code for Biology and Medicine 7(1), 1+ (2012)
7. Li, H., et al.: The Sequence Alignment/Map format and SAMtools. Bioinformatics 25(16), 2078–2079 (2009)
8. MacLean, D., Kamoun, S.: Big data in small places. Nature Biotechnology 30(1), 33–34 (2012)
9. Mangalam, H.: The Bio* toolkits–a brief overview. Briefings in Bioinformatics 3(3), 296–302 (2002)
10. McKenna, A., et al.: The Genome Analysis Toolkit: A MapReduce framework for analyzing next-generation DNA sequencing data. Genome Research 20(9), 1297–1303 (2010)
11. Mitsuteru, N.G., et al.: BioRuby: open-source bioinformatics library (2003)
12. Odersky, M.: The Scala Language Specification, Version 2.9 (May 2011), http://www.scala-lang.org/docu/files/ScalaReference.pdf
13. Prins, P.: BioScala (March 2011), https://github.com/bioscala/bioscala
14. Rother, K., et al.: A toolbox for developing bioinformatics software. Briefings in Bioinformatics 13(2), 244–257 (2012)
15. Simpson, J.T., et al.: ABySS: a parallel assembler for short read sequence data. Genome research 19(6), 1117–1123 (2009)
16. Stajich, J.E., et al.: The Bioperl toolkit: Perl modules for the life sciences. Genome research 12(10), 1611–1618 (2002)
17. Zerbino, D.R., Birney, E.: Velvet: Algorithms for de novo short read assembly using de Bruijn graphs. Genome Research 18(5), 821–829 (2008)

An Algorithm to Assemble
Gene-Protein-Reaction Associations for
Genome-Scale Metabolic Model Reconstruction

João Cardoso[1,2], Paulo Vilaça[1,2], Simão Soares[1], and Miguel Rocha[2]

[1] SilicoLife Lda., Spinpark, Avepark, Apart. 4152, 4806-909 Guimarães, Portugal
{jcardoso,pvilaca,ssoares}@silicolife.com
[2] CCTC, School of Engineering, University of Minho
mrocha@di.uminho.pt

Abstract. The considerable growth in the number of sequenced genomes and recent advances in Bioinformatics and Systems Biology fields have provided several genome-scale metabolic models (GSMs) that have been used to provide phenotype simulation methods. Given their importance in biomedical research and biotechnology applications (e.g. in Metabolic Engineering efforts), several workflows and computational platforms have been proposed for GSM reconstruction. One of the challenges of these methods is related to the assignment of gene-protein-reaction (GPR) associations that allow to add transcriptional/ translational information to GSMs, a task typically addressed through manual literature curation. This work proposes a novel algorithm to create a set of GPR rules, based on the integration of the information provided by the genome annotation with information on protein composition and function (protein complexes, sub-units, iso-enzymes, etc.) provided by the UniProt database. The methods are validated by using two state-of-the-art models for *E. coli* and *S. cerevisiae*, with competitive results.

Keywords: Metabolic models, gene-protein-reaction rules, genome annotation.

1 Introduction

Genome-scale metabolic models (GSMs) are being increasingly used tools for the understanding of the metabolic behaviour of micro-organisms, allowing the simulation of their phenotypes in distinct environmental and genetic conditions. They have been used to find genetic modifications able to synthesize desired compounds within the realm of Metabolic Engineering (ME) [8] (e.g. *E. coli* strains have been designed *in silico* to overproduce lactate, ethanol, succinate and aminoacids), but also used to guide biological discovery by comparing predicted and experimental data, to analyse global network properties and to study evolution [2]. So, GSMs have become a core element of biological systems analysis and a common denominator for computational and experimental studies.

T. Shibuya et al. (Eds.): PRIB 2012, LNBI 7632, pp. 118–128, 2012.

GSMs gather information regarding different cellular entities. All models have basic information on the portfolio of metabolic reactions and the metabolites involved (including stoichiometry and reversibility information), and in many cases the compartment where reactions occur. Most GSMs also include information on the transcriptional/ translational level, including the enzymes that catalyse the reactions, information on the peptides making the protein complexes and, finally, the genes encoding those peptides [4]. The relationship between genes, proteins and reactions is usually represented using logical rules, commonly called Gene-Protein-Reaction (GPR) rules. These rules represent these relationships using the logical operators AND and OR at two levels: the former states how proteins are encoded by their genes and the latter how the reactions depend on the enzymes.

The inclusion of GPRs within GSMs is essential to allow the phenotype prediction of the cell under different genetic conditions, e.g. gene knockouts or over/underexpression. The capability of performing these predictions is fundamental, for instance in determining gene essentiality and in strain optimization efforts, where the best set of genetic modifications to impose over the the wild type strain is sought, for a given industrial application related to the overproduction of a given compound [13]. In this last case, it has been shown in previous work that the ability to perform simulations of gene knockouts, instead of reaction deletions used in earlier approaches, is essential to obtain more robust and biologically meaningful solutions [10].

The reconstruction of GSMs is being increasingly automated by structured pipelines [5,4] and making use of several Bioinformatics tools, related to genome annotation and re-annotation, homology searches, database integration, protein localization, among others [12,1]. However, in spite of the growing availability of such tools, some of the steps in GSM reconstruction are still done by semi-automated processes with need for manual curation by experts. The determination of the GPRs associated to each metabolic reaction is one of these steps, where the lack of computational tools for the automation of their generation is particularly felt being this task typically conducted by a laborious and time consuming literature search [12].

Therefore, the main aim of this work consists in developing an algorithm that allows to fully automate the process of adding GPR rules to GSMs in the context of their reconstruction process. Thus, the objective is to discover the best GPR rule for each reaction in a GSM, taking as input the information connecting genes and metabolic activities resulting from the genome annotation. The result of this work will be a computational tool to address this task that makes use of existing information in Bioinformatics databases, mainly UniProt [6].

This task is not absent from important hurdles, being the first the inherent complexity of these GPRs. Indeed, two main factors contribute to this complexity: on one hand, different enzymes can have the same metabolic activity (iso-enzymes) and, on the other hand, an enzyme can be a protein complex formed by different sub-units encoded by different genes. Figure 1 illustrates the distinct cases and the corresponding representation in terms of Boolean rules, using examples from the iJR904 model for *Escherichia coli* [9].

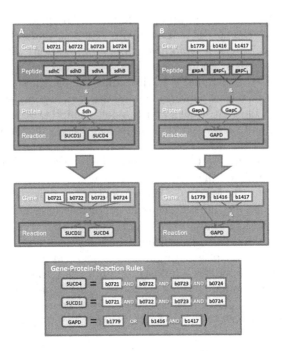

Fig. 1. Illustration of the different cases of GPRs: a) the Sdh enzyme is built from 4 sub-units and catalyses two reactions SUCD4 and SUCD1i; b) GAPD reaction is catalysed by two iso-enzymes (GapA and GapC); GapC is composed of two sub-units encoded by distinct genes.

The most recently published models include, as expected, GPR rules. This is the case with the iAF1260 model for *Escherichia coli* [3] and iMM904 for *Saccharomyces cerevisiae* [7] that will be used in this work to validate our approach.

The remaining of the paper is organized as follows: first, a detailed description of the proposed algorithm is given; next, the results obtained in the two case studies are provided and analysed; finally, conclusions and directions for further work are outlined.

2   Algorithm

An outline of the approach followed in this work is provided in Figure 2. The basic steps of this approach will be explained next with a high-level view. Specific details of each step will follow, organized in sub-sections.

The input for this process is an annotated genome of an organism, assumed in this work as a table containing a gene identifier, one or more Enzyme Commission (EC) numbers with (a list of) assigned metabolic functions and a textual definition of the function of the gene. EC numbers are a recommendation created in order to ensure a systematic organization to define the known metabolic conversions [14].

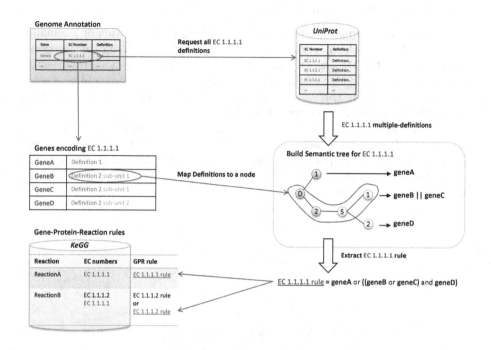

Fig. 2. Overall scheme of the approach followed in this work

For each EC number collected from the genome annotation, a search is conducted in SwissProt, the manually curated database from UniProtKB collection [6]. In each case, a list of matching entries in SwissProt is collected and a semantic tree is created from the definitions included in this list. The aim of this tree is to semantically represent the structure of proteins and their sub-units associated to the respective metabolic function.

In the next step of this process, the aim will be to associate the list of genes associated to that specific EC number to nodes in the semantic tree previously built. This will be done by matching definitions of specific genes to the definitions associated to the tree nodes. Once this association is complete, it is possible to infer a GPR rule by traversing the tree and gathering the linked genes, outputting a rule in the form of a Boolean function. Based on the biological meaning of each tree node, the algorithm can infer the biological association of the genes using the AND or OR logical operators and, thus, create a GPR rule for each EC number.

The final step of the algorithm is to create GPR rules for each reaction. From the GSM reconstruction process, a table is provided containing the list of reactions and their associated EC numbers (e.g. this information can be obtained from databases such as KEGG http://www.genome.jp/kegg). The GPR rule for a given reaction is obtained by the rules from the assigned EC numbers. If more than one EC number is assigned to a reaction, the respective rule will be created by joining the rules from the EC numbers using the operator OR.

2.1   Building the Semantic Tree

One of the most important steps of this algorithm is the creation of a semantic tree for each metabolic function (EC number). This tree is a n-ary tree structure, similar to a suffix tree, where the values are textual expressions representing biological definitions for functional roles. The input for this step will be a set of textual definitions, in this case from the list of entries as a result of a search for a specific EC number in the SwissProt database.

The first step is to create a matrix from the list of definitions, where each row is a definition and each column is obtained by splitting the expressions using white spaces and parenthesis as separators. The matrix is composed of all possible definitions available in the database. Figure 3 describes how the matrix is built from a definition set.

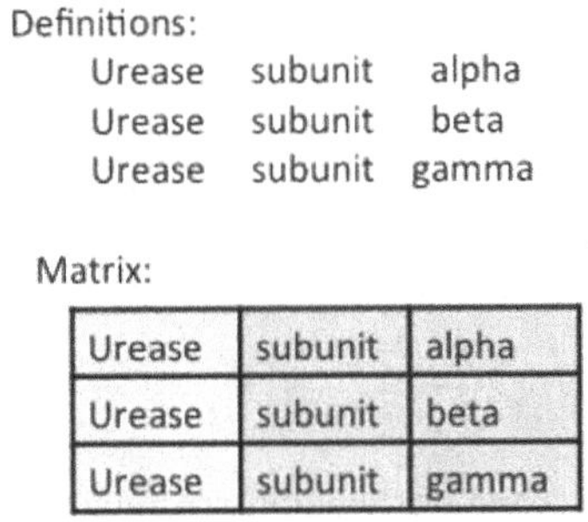

Definitions:
Urease	subunit	alpha
Urease	subunit	beta
Urease	subunit	gamma

Matrix:

Urease	subunit	alpha
Urease	subunit	beta
Urease	subunit	gamma

Fig. 3. Description of the matrix assembly process. The expressions are split into the terms and placed in the matrix resulting in 3x3 matrix structure.

To avoid problems with mismatches caused by synonyms of protein names or functional definitions, a dictionary is created for each EC number. This dictionary is filled with information from UniProt regarding synonyms or alternative names. The strategy is to keep in the matrix only one recommended name in each case and this strategy is applied to all definition rows.

Also, to prevent mismatches caused by typos or other small differences in terms, a global dictionary is used with common terms. A Levenshtein Automaton [11] is applied to every word, finding the closest word in the dictionary. If the distance is equal to 1, the word is replaced by the dictionary word. This allows to correct misspelled words, such as "putativ" or "cmponent", instead of "putative" or "component" respectively.

In order to reduce the information noise, some expressions were defined as useless to the definition match process and these terms are removed from the expressions. In this list the following are included: cellular localization terms, as the definition is the same; organs or organism structure, such as "leaf", "liver", etc; synonyms of homology or same function, such as "like" or "isoenzyme" are also not required. Those words are removed from the expressions before building the matrix.

The semantic tree is created by traversing the matrix row by row. The algorithm used to build the tree follows the ones used to build suffix trees, i.e. when a new row is considered the algorithm will match its words with the nodes in the tree, starting by the root and following the respective branches. When, at a certain level, the branch for that term does not exist, a new branch is created. The tree is composed by two types of nodes: terms, that represent each unique word available in the definition; and the genes that are associated to the last word in the expression.

To create the Boolean rules, it is required to identify how the components are assembled together. Thus, it is necessary to identify for each branch if it will associated to an AND or to an OR relationship. This process will take into account the semantics of the terms found in the annotations. The gene nodes associated connected to the same root are associated by an OR expression. Terms such as "subunit", "chain", "component", "peptide" or any synonym to these words identify the existence of a complex structure and therefore will be associated to an AND relationship. The remaining terms under the same node are also related with an OR relationship.

There are also distinct identifiers for different types of substructure: Greek alphabet characters, Roman numerals, digits and Latin alphabet characters. In some cases, the complex is made by a pair of a "small" and a "large" or "heavy" and "light" chain or units described by their molecular weight. Figure 4 exemplifies the generated semantic tree for an example.

2.2 Matching the Tree with the Genome Annotation

Given a table with the annotated genome, containing for each metabolic gene a set of EC numbers and the textual definition of its functional role, the next step will be to map the genes onto the trees created in the previous step.

For each EC number, a tree is created as explained in the previous section. Also, a list of genes related to that EC number is extracted from the genome annotation table. Each of these genes will then be mapped to the tree by matching its definition text with the one on the tree nodes. The matching algorithm is similar to the one using in the construction of the tree explained above. The gene will be linked to the deepest node in the tree where the matching process is possible.

When all genes for a given EC number are matched onto the tree, it is possible to create a rule for this EC number. The tree is traversed generating a string; each branch has a Boolean function, i.e. the nodes in that branch are connected by either "AND" or "OR". Sub-trees without genes are disregarded and nodes with genes will add the gene identifier to the string.

Figure 5 shows an example, based on the tree shown in the previous subsection. In this case, the generated GPR will be the following: *BCE_3662 AND BCE_3663 AND BCE_3664*.

The last step is to generate rules for the reactions in the target model. Assuming there is information available on the set of EC numbers for each rule, this step is achieved by joining together the rules for the set of EC numbers through an OR operator.

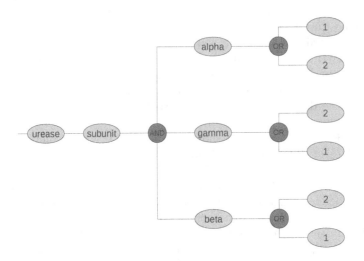

Fig. 4. Illustration of an example semantic tree

2.3 Implementation

The previous algorithm was implemented using the Java programming language, being the software available on demand to the authors. To collect all information from the SwissProt database, the UniProtJAPI provided by European Bioinformatics Institute (EBI) (`http://www.ebi.ac.uk/uniprot/remotingAPI`) has been used. The mappings of reactions to EC numbers can be taken from the KEGG database. In the experiments, this information is available from the models.

3 Results

In order to validate the proposed algorithm, two existing GSMs were used: the iAF1260 model for *Escherichia coli* [3] and iMM904 for *Saccharomyces cerevisiae* [7]. Since these methods have a set of GPR rules associated to most reactions, in both cases as a result of thorough literature curation process Table 1 shows basic statistics of both models, including the number of reactions with an assigned EC number, the ones with GPR rules available and the intersection of both sets. These last sets will be the ones of interest in the analysis of the results, to provide a fair comparison with the proposed method.

By running the methods described in the previous section in the provided case studies, the following number of GPR rules were created (showing also the percentage over the total number of reactions with GPR and EC number):

- *Escherichia coli*: 674 (71%)
- *Saccharomyces cerevisiae*: 535 (70%)

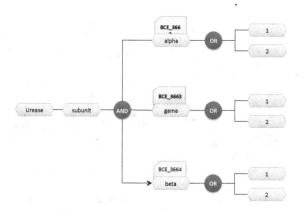

Fig. 5. Illustration of the process of mapping genes onto the semantic tree

Table 1. Model statistics

	E.coli	S.cerevisiae
Reactions with EC number	955	760
Reactions with GPR	1944	1043
Reactions with GPR and EC number	932	693

To provide an analysis of the results by comparing the rules obtained with the ones in the models, the Jaccard coefficient (J) will be defined to compare GPR rules for the same reaction. For each rule, the sets of genes used in the target rule (T) and the proposed rule (P) are taken and J is calculated as follows:

$$J(T,P) = \frac{|P \cap T|}{|P \cup T|} \tag{1}$$

Figure 6 shows the distribution of J values over all reactions for both case studies. The values are divided into four categories: $J = 1$ (perfect match), $J \geq 0.5$ (considered a good match), $J < 0.5$ (partial match) and $J = 0$ (no match). In both cases, the large majority of the rules obtain a good match with the rules in the model, with over 50% with a perfect match and more than half of the remaining with a match over 50%. It is important to notice that about half of the cases where there is no match are situations where the proposed method provides a rule and the model does not have one.

These results show the high correspondence between both data. However, since GPR rules are Boolean functions it is important not only to check the correspondence of the variables used, but also to compare the results of the function. This analysis was conducted for the cases where there was a full match of the sets of variables used. A truth table with all possible values for the genes was created in each case, where each row stands for a possible combination of the values of the genes involved. The output of the GPR rule was compared in

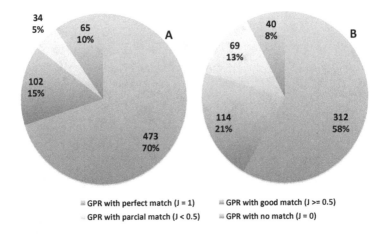

Fig. 6. Results of the proposed methods applied to the E. coli (a) and S. cerevisiae (b) models. Pie charts show the distribution of the Jaccard indexes (J) calculated over the GPR rules created using the proposed methods and obtained from the models

each case between the proposed rule and the existing one. It was verified that the results are 99.8% identical in *E. coli* and 100% in *S. cerevisiae*.

It is also important to notice that models are also composed by transport reactions that have a specific annotation - the Transport Commission numbers - and their semantic composition is more complex. For that reason, this algorithm is not suitable for assemble GPR for those conversions.

Other discrepancies between the existing rules have been found. For instance, the EC 1.2.1.3 (*aldehyde dehydrogenase*) is associated to *b1300* in the model, however the UniProt database describes it with EC number 1.2.1.5 . This mismatch can be either explained by two reasons: the annotation was reviewed and associated with a new function or the manual curation and literature mining process during the reconstruction determined that the gene is also related to the function. Another issue identified is the lack of EC function associated with the gene (e.g. *b3610* in the *E. coli* model). Although the model has an association with the metabolic function EC 1.8.4.2, there is no evidence at the UniProt database.

4   Conclusions and Further Work

In this work, a novel algorithm and computational tool has been proposed to address the task of gene-protein-reaction rule inference from the genome annotation. This is an important task within the larger effort of genome-scale metabolic model reconstruction that has been traditionally performed using laborious manual literature curation. Although the results are still preliminary and the methods can be improved, this contribution already shown interesting results when applied to well known and validated models from *E. coli* and *S. cerevisiae*.

Some issues are still preventing more accurate results. One one hand, the models used as case studies were built over the last decade in a process of iterative refinement involving huge resources and extensive manual curation. Also, in many cases, divergences on the EC number annotations between the models and the UniProt database are the reason for many mismatches. This should be further examined in posterior work, namely by considering the use of additional databases complementing UniProt.

Also, the approach proposed here is not able to encompass an important class of reactions that handle the transport of metabolites from the exterior of the cell and between cell compartments. Since these are mostly not covered by EC number nomenclature, a distinct approach needs to be developed, for instance based on TC numbers from the TCDB database (http://www.tcdb.org/). This will be a major task in future work, together with other possible improvements in the proposed methodology.

Acknowledgements. The work is partially funded by ERDF - European Regional Development Fund through the COMPETE Programme (operational programme for competitiveness) and by National Funds through the FCT (Portuguese Foundation for Science and Tech-nology) within projects ref. COMPETE FCOMP-01-0124-FEDER-015079 and PEst-OE/EEI/UI0752/2011.

References

1. Dias, O., Rocha, M., Ferreira, E., Rocha, I.: Merlin: Metabolic models reconstruction using genome-scale information. Computer Applications in Biotechnology 11, 120–125 (2010)
2. Feist, A.M., Palsson, B.: The growing scope of applications of genome-scale metabolic reconstructions using escherichia coli. Nature Biotechnology 26(6), 659–667 (2008)
3. Feist, A.M., Henry, C.S., Reed, J.L., Krummenacker, M., Joyce, A.R., Karp, P.D., Broadbelt, L.J., Hatzimanikatis, V., Palsson, B.Ø.: A genome-scale metabolic reconstruction for escherichia coli k-12 mg1655 that accounts for 1260 orfs and thermodynamic information. Molecular Systems Biology 3 (2007)
4. Feist, A.M., Herrgård, M.J., Thiele, I., Reed, J.L., Ø Palsson, B.: Reconstruction of biochemical networks in microorganisms. Nature Reviews. Microbiology 7(2), 129–143 (2009)
5. Henry, C.S., DeJongh, M., Best, A.A., Frybarger, P.M., Linsay, B., Stevens, R.L.: High-throughput generation, optimization and analysis of genome-scale metabolic models. Nature Biotechnology 28(9), 977–982 (2010)
6. Magrane, M., Uniprot Consortium: Uniprot knowledgebase: a hub of integrated protein data. Database: the Journal of Biological Databases and Curation 2011:bar009 (January 2011)
7. Mo, M., Palsson, B.Ø., Herrgård, M.J.: Connecting extracellular metabolomic measurements to intracellular flux states in yeast. BMC Systems Biology 3, 37 (2009)
8. Nielsen, J.: Metabolic engineering. Appl Microbiol Biotechnol. 55, 263–283 (2001)
9. Reed, J.L., Vo, T.D., Schilling, C.H., Palsson, B.Ø.: An expanded genome-scale model of escherichia coli k-12 (ijr904 gsm/gpr). Genome Biology 4, R54 (2006)

10. Rocha, M., Maia, P., Mendes, R., Pinto, J.P., Ferreira, E.C., Nielsen, J., Patil, K.R., Rocha, I.: Natural computation meta-heuristics for the in silico optimization of microbial strains. BMC Bioinformatics 9 (2008)
11. Schulz, K.U., Mihov, S.: Fast string correction with levenshtein automata. International Journal on Document Analysis and Recognition 5, 67–85 (2002)
12. Thiele, I., Palsson, B.Ø.: A protocol for generating a high-quality genome-scale metabolic reconstruction. Nature Protocols 5, 93–121 (2010)
13. Vilaça, P., Maia, P., Rocha, I., Rocha, M.: Metaheuristics for Strain Optimization Using Transcriptional Information Enriched Metabolic Models. In: Pizzuti, C., Ritchie, M.D., Giacobini, M. (eds.) EvoBIO 2010. LNCS, vol. 6023, pp. 205–216. Springer, Heidelberg (2010)
14. Webb, E.C.: International Union of Biochemistry, and Molecular Biology. In: Enzyme nomenclature 1992. Recommendations of the Nomenclature Committee of the International Union of Biochemistry and Molecular Biology on the Nomenclature and Classification of Enzymes, 6th edn., Academic Press (1992)

A Machine Learning and Chemometrics Assisted Interpretation of Spectroscopic Data – A NMR-Based Metabolomics Platform for the Assessment of Brazilian Propolis

Marcelo Maraschin[1,3], Amélia Somensi-Zeggio[1], Simone K. Oliveira[1],
Shirley Kuhnen[1], Maíra M. Tomazzoli[1], Ana C.M. Zeri[2],
Rafael Carreira[3], and Miguel Rocha[3,*]

[1] Plant Morphogenesis and Biochemistry Laboratory, Federal University of Santa Catarina,
Florianópolis, SC, Brazil
[2] National Laboratory of Bioscience, Campinas, SP, Brazil
[3] CCTC, School of Engineering, University of Minho, Campus Gualtar, Braga, Portugal
mrocha@di.uminho.pt

Abstract. In this work, a metabolomics dataset from ^{1}H nuclear magnetic resonance spectroscopy of Brazilian propolis was analyzed using machine learning algorithms, including feature selection and classification methods. Partial least square-discriminant analysis (PLS-DA), random forest (RF), and wrapper methods combining decision trees and rules with evolutionary algorithms (EA) showed to be complementary approaches, allowing to obtain relevant information as to the importance of a given set of features, mostly related to the structural fingerprint of aliphatic and aromatic compounds typically found in propolis, e.g., fatty acids and phenolic compounds. The feature selection and decision tree-based algorithms used appear to be suitable tools for building classification models for the Brazilian propolis metabolomics regarding its geographic origin, with consistency, high accuracy, and avoiding redundant information as to the metabolic signature of relevant compounds.

Keywords: Supervised classification techniques, evolutionary algorithms, Random Forest, PLS-DA, wrapper methods, NMR-based metabolomics.

1   Introduction

One and two dimensional NMR spectroscopy (1D-, 2D-NMR) has increasingly been used for complex matrix analysis such as plant extracts and biofluids in metabolomics studies. From a ^{1}H-NMR spectrum, a set of peaks, or features, indicative of the metabolite signatures and chemical composition of the sample is obtained and may be used as a basis to build descriptive and predictive models (e.g. for classification tasks). In this context, feature selection may be employed to improve classification accuracy or

---

* Corresponding author.

T. Shibuya et al. (Eds.): PRIB 2012, LNBI 7632, pp. 129–140, 2012.
© Springer-Verlag Berlin Heidelberg 2012

aid model explanation by establishing a subset of class discriminating features. Factors such as experimental noise and threshold selection may adversely affect the set of selected features. Furthermore, the high dimensionality and multi-collinearity inherent to ^{1}H-NMR signals may increase discrepancies between the set of features retrieved and those required to provide a complete explanation of metabolite signatures. Thus, previously to classification of metabolomics data, it is interesting to perform descriptive studies, e.g. using principal component analysis (PCA) [1].

Discriminant analyses such as soft independent modelling by class analogy (SIMCA), support vector machine (SVM), partial least squares discriminant analysis (PLS-DA), and more recently random forests (RF) have also been used within the metabolomics domain.

Feature selection may be employed to improve a classification model in terms of generalization, performance, and accuracy by eliminating non-informative features, as well as to gain deeper insights into the rationale underlying class divisions within a particular domain. In the context of metabolomics, retrieving the set of class discriminating features may aid in the identification of the class determining metabolites. However, features selected on the basis of classification accuracy, i.e. features that are sufficient to separate classes, may not to always be the best approach due to the redundancy of information. This is typically found in high dimensional NMR-based metabolomics studies, where a metabolite may be represented by one or more spectral features as only a part of the metabolite signature identification may be enough to provide a perfect classification model.

In this work, to overcome such constraints we have adopted an approach where accuracy based approaches are complemented with feature selection methods less prone to the bias effects of multi-collinear features, including those based on variable influence on the projection (VIP) values, derived from PLS-DA and variable importance produced by a RF classifier. Indeed, contrarily to PLS-DA, RF is a non-parametric technique unaffected by feature scale so that the techniques seem to be somewhat complementary.

PLS extracts the set of latent variables which model the data, but which are also correlated to the class membership vector. Once a PLS model has been built the influence of individual features is captured by measuring the VIP scores derived from the PLS coefficients for the optimal set of features. After that, features are ranked by these scores and selected considering the choice of an appropriate threshold (usually $\alpha \geq 1$), a step that may greatly affect the set of retrieved features. Finally, PLS-DA is also a scale dependent technique as the choice of scaling factor affects the features selected [2].

In its turn, RF is a classification technique based on growing many classification trees, in which feature values are used to build a model that enables the classification of unlabeled samples. RF allows assigning importance values to features resulting from their influence on the classification accuracy of the forest, aiding feature selection, and allowing gaining further insights into the data. The importance of a particular feature is determined by randomly permuting the feature over samples in each tree's 'out-of-bag' test set, followed by the reclassification of the samples using the RF. Such a calculation approach is advantageous for feature selection because it covers both the impact of each feature individually and its multivariate interactions with other features. Besides, as RF is a decision tree-based technique it also deals well with differently scaled features [3], a relevant trait for NMR-based metabolomics where the peaks vary greatly in intensity.

An alternative approach for feature selection is the use of wrapper methods. In wrapper approaches, the feature selection processes are performed by optimization algorithms that search the space of possible subsets of attributes, to find the best alternative. These approaches train the classifier with a subset of the available attributes and estimate its generalization error. These methods are dependent on the classifier that is used. Indeed, there is no guarantee that an optimal subset of attributes chosen for one classifier will be the optimal one when used with another algorithm.

The wrapper approach followed in this work is based on two components: the use of classifiers implemented by the open-source data mining software Weka [4] for the inner layer (decision trees and rule set induction methods will be used), and the use of Evolutionary Algorithms (EAs) as the optimization engine. Together, these techniques may allow extracting relevant features from a given dataset, minimizing the redundant information as to metabolite signature identification. This work aimed at proving the later assertion as our scientific hypothesis, using a high dimensional, multi-collinear metabolomics dataset (80 samples x 81675 variables) of Brazilian propolis NMR spectra as a study model.

Propolis has been chosen because it has long been recognized as a useful source of valuable compounds for human health, but due to its huge chemical heterogeneity, the production of standardized and homogeneous extracts is a difficult task. This is due to the fact that chemical characterization and standardization of propolis extract is technically tedious, time expensive, and non-cost effective as one adopt traditional analytical selective techniques such as high performance liquid chromatography. Besides, the effect of flora composition on the propolis' chemical profile is well known and considering the huge biodiversity of plant species found in some producer regions [5], e.g., Atlantic Rainforest in Santa Catarina State, southern Brazil, one could expect a high chemical heterogeneity among samples from distinct geographic regions where propolis has been collected; an important underlying assumption addressed in this study.

On the other hand, over the past years nuclear magnetic resonance (NMR) spectroscopy has been recognized as a powerful tool as one aims at characterizing chemically complex matrices. Indeed, NMR spectroscopy is a fast, robust, and non-selective analytical technique able to detect virtually any molecule in a solution, given a minimum value of concentration (detection limit, ug/ml). However, the amount of information afforded by NMR analysis is huge as a typical high dimensional ^{1}H-NMR spectrum easily contains 32.000 or 64.000 data points. The analysis of such an amount of information is unthinkable without the aid of powerful computational tools, but one should bear in mind this scenario for metabolomics studies.

In order to deal with large NMR datasets, data mining techniques have been adopted to build descriptive and predictive models. Here, machine learning and chemometrics techniques are thought to be a suitable approach to gain insights as to important spectroscopic features associated to the chemical composition and geographic origin of propolis produced in Santa Catarina state, southern Brazil. For that, emphasis will be given to accurate feature selection and classification techniques in order to avoid retrieving redundant information (i.e., multi-collinear features) and overfitting in classification models by using prominent machine learning algorithms.

2   Methods

2.1   Propolis Sample Preparation and NMR Spectroscopy

In autumn, 2010, propolis samples (n=16) were collected from each of the five geographic regions (East, Central, Highlands, North, and West) of Santa Catarina State, southern Brazil. The lyophilized ethanolic extracts (2g/10 ml, EtOH 70%, v/v) were added of 700 μl of CD_3OD, centrifuged (5 krpm/10min), and transferred to 5 mm NMR tubes. The propolis ^{1}H-NMR spectra were acquired on a Varian Inova 600 MHz NMR spectrometer by collecting (time domain) 32,000 data points (32 scans, acquisition time = 4s, delay time = 2s, recycle time = 6s, 25°C) over a spectral window of 8000 Hz, and water signal suppression. The recycle time was considered of sufficient length (e.g. 3 T_1 s) to avoid significant (<10%) peak saturation. Prior to Fourier transformation (FT), the 1D FIDs were zero-filled to 64K data points and a line broadening factor of 0.5 Hz was applied. A routine implemented in the ACD/NMR processor software (v.12.01) consisting of phasing, baseline correction, and calibration ($TSP_{\delta H}$ 0.00ppm) was used for processing all the ^{1}H-NMR spectra. Each relevant peak, i.e., selected feature, in the spectrum was integrated using a quantitation script of the Quanalyst tool of ACD/NMR processor software.

2.2   Metabolomics Data Processing

From the processed full spectra dataset (0.80 – 13.00 ppm) a peak list was extracted using a two-column comma separated values format, where the first column indicates peak position (ppm) and the second one represents peak intensities. A set of 80 samples was used, containing a total of 81675 peaks with an average of 425.4 peaks per sample. Peak alignment grouped proximal peaks together according to their position using a moving window of 0.03ppm and a step of 0.015ppm. Peaks of the same group were aligned to their median positions across all samples and those detected in very few samples (< 50% in both classes) were excluded. Besides, the missing and zero values were replaced with a value of 0.00005, the half of the minimum positive values in the original data, assuming to be the detection limit. Indeed, most missing values are caused by low abundance metabolites with contents lower than the detection limit.

In order to identify and remove variables that are unlikely to be of use when modeling data, a filtering protocol was applied based on interquantile range, affording a 5% reduction in features. No phenotype information was used in the filtering process, allowing the result to be used in any downstream analysis. Such processing step is strongly recommended for datasets with large number of variables (> 250) containing much noise [6] as typically found in NMR-based metabolomics analysis. Taking into account the very distinct orders of magnitude of the variables, quantile normalization within replicates of the dataset was performed [7].

2.3   Statistical and Machine Learning Data Analysis

In order to extract latent information from the ^{1}H-NMR dataset, classification models were built by applying supervised classification and feature selection methods. The MetaboAnalyst 2.0 tool provides a framework for conducting analyses over metabolomics datasets and was used to perform PLS-DA and RF analysis [8]. The wrapper approach was implemented by combining classifiers from the Weka open-source data mining software (v. 3.6.6) [4] and EAs were implemented using the Java open-source library JECoLi (http://darwin.di.uminho.pt/jecoli).

The methods used in this work are described in detail next:

PLS-DA: PLS is a supervised method that uses multivariate regression techniques to extract via linear combination of original variables (X) the information that can predict the class membership (Y). To assess the significance of class discrimination, a permutation test is performed. In each permutation, a PLS-DA model is built between the data (X) and the permuted class labels (Y) using the optimal number of components determined by cross-validation for the model based on the original class assignment. Further variable importance in projection (VIP), a weighted sum of squares of the PLS loadings taking into account the amount of explained Y-variation in each dimension was measured for purpose of calculation of the feature importance.

The PLS regression was performed using the *plsr* function provided by R *pls* package. The classification and cross-validation were performed using the corresponding wrapper function of the *caret* package [9, 10].

Random Forest (RF): Random Forest is a supervised learning algorithm suitable for high dimensional data analysis. It uses an ensemble of classification trees, each of which is grown by random feature selection from a bootstrap sample at each branch. Class prediction is based on the majority vote of the ensemble. RF also provides other useful information such as OOB (out-of-bag) error, variable importance measure, and outlier measures. During tree construction, about one-third of the instances are left out of the bootstrap sample. This OOB data is then used as test sample to obtain an unbiased estimate of the classification error. Variable importance is evaluated by measuring the increase of the OOB error when it is permuted. The outlier measures are based on the proximities during tree construction. RF analysis was performed using the *randomForest* package for R [11].

Wrapper Approach - Weka Classifiers and Evolutionary Algorithms: An EA is used to evolve the best set of attributes for the classification task, using a set-based representation to encode each solution. Regarding the reproduction operators, two types were used: crossover and mutation. The crossover operator used was inspired on uniform crossover and works as follows: the genes that are present in both parent sets are kept in both offspring; the genes that are present in only one of the parents are sent to one of the offspring, selected randomly with equal probabilities. Regarding mutation, the random mutation operator was deployed, replacing a gene in the set by a random value in the allowed range. Both reproduction operators are used with equal probabilities to create new solutions. The operators are implemented taking into consideration the need to comply with the constraints imposed by the minimum and maximum set size and also to avoid repeated elements in the sets. In the experiments reported in this work, the minimum size is always set to 1 and the maximum size to 10.

The selection procedure is a tournament scheme with k=2. In each generation, 50% of the individuals are kept from the previous generation and 50% are bred by the application of the reproduction operators. An elitism value of 1 is used, allowing the best individual of the population to be always kept. The EA's population size is set to 100 and the termination criterion was defined based on a maximum of 100 generations. The EA was executed 30 times for each case.

Each solution in the EA is evaluated by retrieving the attributes encoded in its genome and building classifiers based solely on those attributes. These classifiers are built and evaluated resorting to Weka and therefore it is easy to select different classifiers implementing distinct data mining algorithms. In this work, we used J48, a classification decision tree induction method based on the well known C4.5 algorithm and JRip, a rule set induction method inspired in the RIPPER algorithm. The fitness function of each solution is computed calculating an accuracy estimation of the classifier, obtained by performing 5-fold cross-validation over the available dataset.

3   Results

The dataset for this classification task includes 80 samples with five classes, one per each geographic region. The dataset is balanced since there are 16 samples for each class. The aim is to classify samples regarding their geographic region.

Previously to PLS-DA and RF analyses, a descriptive model was built based on the calculation of the principal components (PCAs) for the ^{1}H-NMR dataset as previously suggested [1]. PC1 and PC2 afforded for 89.9% of the explained variance of the data, but a clear discrimination was not achieved as the samples spread over the PC1 and PC2 axes. These findings prompted us to adopt a classification model in order to gain insights as to the relevant features associated to an eventual discrimination according to the propolis sample chemical composition and its geographic origin. In order to extract relevant but not redundant information we applied PLS-DA and RF to the propolis metabolomics dataset.

A model was built by performing PLS-DA that was able to identify important features to predict the propolis sample classification by measuring the variable importance in projection (VIP) as shown in Table 1. The most important fifteen ^{1}H-NMR resonances, i.e. features, were identified by PLS-DA and most of them (10) resulting from aliphatic compounds, as five features were associated to anomerical ones. Among other, the features detected by PLS-DA were mostly assigned to chemical groups of the alkane moiety (e.g., C-CH$_2$-C, 1.30-1.33 ppm; C-CH-C, 1.21 and 1.47 ppm) or acetyl group (COCH$_3$, 2.07-2.37 ppm) [12, 13, 14] of fatty acids and waxes commonly found in propolis. One-way ANOVA followed by the *post-hoc* Tukey test of the ^{1}H -NMR dataset confirmed the significance of most of the selected features.

Table 1. Important features (^{1}H resonances) ranked according to the VIP score calculated by PLS-DA of propolis samples. The colored boxes on the right indicate the relative concentrations of the corresponding metabolite in each studied group, i.e., geographic region.

Resonances (δH ppm)	VIP score	p-value[1] (-log 10)	^{3}C E H N W
1.30	4.57	47165e-14	
2.29	3.08	2.9594e-10	
1.33	2.88	n.s[2]	
5.26	2.78	5.8257e-19	
1.21	2.74	1.3077e-17	
4.84	2.63	8.9478e-10	
1.47	2.43	5.9462e-14	
4.81	2.21	6.9457e-10	
5.07	2.16	3.7371e-18	
2.12	2.11	2.4206e-09	
4.87	2.09	1.0308e-10	
2.07	2.02	3.1639e-10	
1.80	1.87	1.1753e-17	
1.53	1.82	n.s	
2.37	1.81	n.s	

[1] One-way ANOVA and *post-hoc* Tukey test ($p<0.05$), [2] not significant, [3] geographic regions of Santa Catarina state (southern Brazil): **C** = central, **E** = east, **H** = highlands, **N** = north, and **W** = west.

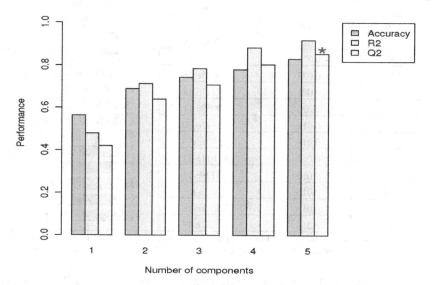

Fig. 1. Performance of the PLS-DA model classification using different numbers of components. The red asterisk indicates the best classifier.

The quantitative measure of the performance for PLS-DA classification model given by the R2, Q2, and accuracy values showed to be higher than 83% for those statistics and reveals a good performance of the method (Fig. 1).

In a second set of experiments, the non parametric RF analysis was applied to the ^{1}H-NMR dataset allowing to selecting extra and non-redundant features for an accurate classification of Brazilian propolis according the geographic origin (Table 2). Eleven out of the top fifteen features identified by RF analysis occur in the spectral window of aliphatic compounds, corroborating the PLS-DA findings, but expanding the metabolite signatures associated to the selected features. Indeed, features associated to saturated (C-CH$_3$, C-CH-C, C-CH$_2$-C) and unsaturated (=C-CH$_3$) alkyl and acetyl (COCH$_3$) groups were predominantly identified by the RF supervised learning algorithm. Preliminary analysis of the features selected by RF, PLS-DA, and also 2D-NMR experiments (data not shown) suggests the presence of long chain fatty acids in propolis samples such as arachidonic, oleic, stearic, and palmitic/palmitoleic acids, associated to the resonances at 1.30, 1.64, 2.04, and 2.76 ppm, for instance [14].

Table 2. Significant features (^{1}H resonances) ranked by the mean decrease in classification accuracy when permuted by RF analysis. The colored boxes on the right indicate the relative effect of the corresponding metabolite in each group of propolis in study, according to their regions of production.

Resonances (δH ppm)	Mean decrease Accuracy	p-value[1] (-log 10)	^{3}C E H N W
2.04	0.035	3.1456e-22	
1.08	0.034	3.2706e-21	
2.46	0.028	9.7457e-25	
1.18	0.026	3.7861e-16	
1.30	0.024	4.7165e-14	
2.56	0.023	3.9608e-16	
5.07	0.021	3.7371e-18	
2.61	0.020	7.9958e-11	
2.49	0.019	6.5947e-20	
1.64	0.018	n.s[2]	
5.20	0.017	5.6611e-21	
6.16	0.016	3.4739e-11	
2.76	0.015	n.s	
2.98	0.014	3.8818e-18	
6.10	0.013	3.6287e-19	

[1] One-way ANOVA and *post-hoc* Tukey test (p<0.05), [2] not significant, [3] geographic regions of Santa Catarina state (southern Brazil): **C** = central, **E** = east, **H** = highlands, **N** = north, and **W** = west.

The confusion matrix revealed a quite interesting performance of the RF supervised learning algorithm, since the classification error found for the predicted class and actual class was zero. Besides, the descriptive model based on the univariate statistics one-way ANOVA followed by the *post-hoc* Tukey test corroborate thirteen out of the top fifteen features selected by RF analysis. Indeed, only two features (1.30 ppm and 5.20 ppm) were simultaneously detected by both PLS-DA and RF, characterizing redundant information.

PLS-DA and RF methods were also able to reveal distinct effects of the selected features regarding the geographic origin of propolis samples (Tables 1 and 2). A quantitative approach was applied to the PLS-DA selected features by calculating the values of their absolute integral (data not shown). Differences in relative concentrations of the corresponding metabolite in each studied group (geographic origin) were detected for all the features, adding extra information to the classification model. Thus, for example, the propolis samples originated from the east region of Santa Catarina state were characterized for their lower content of metabolites comparatively to samples from the other studied regions.

Finally, it is worth mentioning that propolis is a complex matrix well known for its phenolic constituents so that the most interesting spectral windows are 5.50-8.25 ppm, containing mainly the aromatic compound signals, and 8.25-13.00 ppm, where the carbonylic and carboxylic proton signals are found. However, features belonging to those spectral regions did not influence the classification by PLS-DA and RF analysis.

The following task involved the validation of the wrapper approach described in section 2.3. In this study, the coupling of J48, a decision tree-inducing algorithm, and JRip, a rule set induction method, to EA as an optimization engine, in a wrapper approach allowed to identify a certain number of features over the ^{1}H-NMR spectral window as shown in Table 3.

Table 3. J48-EA and JRip-EA wrapper performances and the 15-top ^{1}H-NMR resonances identified taking into account a calculation for 5 and 10 features. The EA was executed 30 times for each case and the prediction accuracy of the classifiers was evaluated using 5-fold cross-validation.

Wrappers	Features	Mean fitness (%)	Standard deviation	Mean cross-validation accuracy (%)	Standard deviation	Resonances (δH ppm)
J48-EA	5	99.84	0.34	93.70	2.61	2.04, **5.61**, **6.64**, 0.89, 4.97, **5.37**, **5.29**, **7.21**, 3.10, **7.05**, **7.08**, 1.27, 2.12, 4.84, **7.62**
	10	99.67	0.19	94.27	2.80	**5.61**, 1.08, 2.04, **5.20**, **7.05**, 2.49, 1.27, **6.49**, 4.00, 1.50, **7.62**, **8.07**, **7.12**, 2.10, **5.10**
JRip-EA	5	99.67	0.57	92.09	2.92	1.64, 3.63, **6.46**, 2.04, **6.64**, **6.72**, **6.79**, **8.07**, **6.25**, 0.88, **5.17**, 1.02, **6.16**, **9.18**, **6.82**
	10	99.90	0.21	92.77	2.82	2.12, 1.08, **5.29**, 1.86, **7.05**, **7.08**, **6.79**, **5.79**, 1.56, 0.81, 2.76, **6.46**, 2.04, 1.60, **8.07**

Contrarily to PLS-DA and RF supervised learning algorithms, the wrapper algorithms selected features that spread over all the ^{1}H-NMR spectral regions, but a predominance of meaningful resonances associated to the aromatic ring moiety of metabolites i.e., 5.50-8.25 ppm, could be detected, typically suggesting an important effect of, e.g., (poly)phenolic compounds in the classification models. Furthermore, it is also possible to notice some redundant information given by both wrapper methods.

The J48/JRip-EA wrapper methods showed to complement RF and PLS-DA since important features addressing the occurrence of phenolic compounds were identified, even suggesting the occurrence of phenolic acids (gallic – 7.05 ppm, *singlet*; t-cinnamic – 6.49, *duplet*; hydrocinnamic – 2.50 ppm, *triplet* and 7.12 ppm, *multiplet*; and caffeic – 8.07 ppm, *singlet*, 7.08 ppm, *double duplet*, 6.82 ppm, *singlet*, 6.79 ppm, *duplet*), as well as the tentatively assigned flavone apigenin (6.16 ppm-*duplet*, 6.46 ppm-*duplet*, and 6.72 ppm-*singlet*) [15] in the studied propolis. In fact, in this regard the application of the wrapper algorithms to the propolis metabolomics dataset expanded the possibilities of detecting relevant metabolite signatures typically found in that complex matrix. Such findings were further confirmed by reverse-phase high performance liquid chromatography coupled to a UV-visible detector (data not shown).

Besides, similarly to PLS-DA and RF analysis, among the significant top fifteen features identified by J48/JRip-EA wrapper methods a series of resonances (0.88, 1.27, 1.60, 2.04, 2.12, 2.76, and 5.29 ppm) associated to metabolite signatures of the, e.g., alkane moiety (C-CH$_3$, 0.88-1.02 ppm; C-CH$_2$-C, 1.27-1.30 ppm) and acetyl group (COCH$_3$, 2.04-2.37 ppm) [12, 13, 14] of monosaturated or unsaturated fatty acids was found in propolis samples. Finally, the presence of the nucleoside uridine in the samples is inferred as meaningful for the classification model, since typical resonances at 5.61 ppm-*duplet*, 5.37 ppm-*duplet*, and 3.63 ppm-*double duplet* were identified by the J48/JRip-EA algorithms and further confirmed by 2D-NMR (TOCSY and HSQC experiments).

The wrapper models showed very high mean fitness (≥ 99%) and prediction accuracy (≥92%) on the cross-validation studies. The validation of each final solution was conducted by doing an independent validation procedure, performing a 10 times 5-fold cross-validation process, using the set of selected features coming from the EA's best solution. Such a finding is worth mentioning taking into account the effect of the EA as optimization engine in controlling overfitting in classification-tree models. It is quite interesting to notice that the performance of the classifiers is quite acceptable with only 5 features, showing the ability of the classifiers to provide high accuracy models with a very limited set of features.

Taken together, the several test domains performed by running the J48/JRip-EA interfaces showed to be effective for feature selection and to develop a classification model tree with high prediction accuracy and consistency.

4   Conclusions

The selected classification methods PLS-DA, RF and the wrapper methods J48/EA and JRip/EA based on machine learning and feature selection appear usable tools for building classification models for the Brazilian propolis metabolomics, with high prediction accuracy.

PLS-DA, RF and J48-EA/JRip-EA analyses of the NMR-based propolis metabolomics dataset showed to be complementary approaches by retrieving and expanding the set of class discriminating features and by adding relevant information for the identification of the class determining metabolites. This allowed further elucidation of the system under investigation in regards to the metabolite signature of important compounds, i.e., chemical fingerprint, and geographic origin of Brazilian propolis.

Acknowledgments. The authors are indebted to CAPES, CNPq, FAPESC, and National Laboratory of Bioscience.

The work is partially funded by ERDF - European Regional Development Fund through the COMPETE Programme (operational programme for competitiveness) and by National Funds through the FCT (Portuguese Foundation for Science and Technology) within projects ref. COMPETE FCOMP-01-0124-FEDER-015079 and PEst-OE/EEI/UI0752/2011. RC's work is funded by a PhD grant from the Portuguese FCT (ref. SFRH/BD/66201/2009).

References

1. Raamsdonk, L.M., Teusink, B., Broadhurst, D., Zhang, N., Hayes, A., Walsh, M.C., Berden, J.A., Brindle, K.M., Kell, D.B., Rowland, J.J., Westerhoff, H.V., van Dam, K., Oliver, S.G.: A functional genomics strategy that uses metabolome data to reveal the phenotype of silent mutations. Nature Biotechnology 19, 45–50 (2001), doi:10.1038/83496
2. van den Berg, R.A., Hoefsloot, H.C., Westerhuis, J.A., Smilde, A.K., van der Werf, M.J.: Centering, scaling and transformations: improving the biological information content of metabolomics data. BMC Genomics 7, 142–147 (2006), doi:10.1186/1471-2164-7-142
3. Weljie, A., Newton, J., Mercier, P., Carlson, E., Slupsky, C.: Targeted profiling: quantitative analysis of 1HNMR metabolomics data. Analytical Chemistry 78, 4430–4442 (2006), doi:10.1021/ac060209g
4. Witten, I.H., Frank, E., Hall, M.A.: Data Mining: Practical Machine Learning Tools and Techniques, 3rd edn. Morgan-Kaufmann, Burlington (2011)
5. Watson, D.G., Peyfoon, E., Zheng, L., Lu, D., Seidel, V., Johnston, B., Parkinson, J.A., Fearnley, J.: Application of principal components analysis to ¹H-NMR data obtained from propolis samples of different geographical origin. Phytochemical Analysis 17, 323–331 (2006), doi: 10.1002.pca
6. Hackstadt, A.J., Hess, A.M.: Filtering for increased power for microarray data analysis. BMC Bioinformatics 10, 11–23 (2009), doi:10.1186/1471-2105-10-11
7. Brodsky, L., Moussaie, A., Shahaf, N., Aharoni, A., Rogachev, I.: Evaluation of peak picking quality in LC-MS metabolomics data. Analytical Chemistry 15, 9177–9187 (2010), doi:10.1021/ac101216e
8. Xia, J., Psychogios, N., Young, N., Wishart, D.S.: MetaboAnalyst: a web server for metabolomic data analysis and interpretation. Nucleic Acids Research 37(Web Server issue), W652–W660 (2009), doi:10.1093/nar/gkp356
9. Wehrens, R., Mevik, B.H.: Pls: partial least squares regression (PLSR) and principal component regression (PCR) (2007), R package version 2.1-0
10. Kuhn, M., Wing, J., Weston, S., Williams, A.: Caret: classification and regression training (2008), R package version 3.45
11. Liaw, A., Wiener, M.: Classification and regression by random Forest (2002), R News

12. Leyden, D.E., Cox, R.H.: Analytical Applications of NMR. John Wiley & Sons, New York (1977)
13. Waterman, P.G., Mole, S.: Analysis of Plant Metabolites. Blackwell Scientific Publications, London (1994)
14. Fan, T.W.M., Lane, A.N.: Structure-based profiling of metabolites and isotopomers by NMR. Progress in Nuclear Magnetic Resonance Spectroscopy 52, 69–117 (2008), doi:10.1016/j.pnmrs.2007.03.002
15. Bertelli, D., Papotti, G., Bortolotti, L., Marcazzanb, G.L., Plessia, M.: ^{1}H-NMR simultaneous identification of health-relevant compounds in propolis extracts. Phytochemical Analysis 23, 260–266 (2011), doi:10.1002/pca.1352

Principal Component Analysis for Bacterial Proteomic Analysis

Y-h. Taguchi[1,*] and Akira Okamoto[2]

[1] Department of Physics, Chuo University, 1-13-27 Kasuga, Bunkyo-ku,
Tokyo 112-8551, Japan
[2] Department of School Nursing and Health, Aichi University of Education,
1 Hirosawa, Igaya-cho, Kariya, Aichi 448-8542, Japan

Abstract. Proteomic analysis is a very useful procedure to understand
the bacterial behavioural responses to the external environmental factors. This is because bacterial genome information is mainly devoted to
code enzyme for the control of the cellular metabolic networks. In this
paper, we have performed proteomic analysis of *Streptococcus pyogenes*,
which is known to be flesh-eating bacteria and can cause several human
life-threatening diseases. Its proteome during growth phase is measured
for four time points under two different culture conditions; with or without shaking. Its purpose is to understand the adaptivity to oxidative
stresses. Principal component analysis is applied and turns out to be
useful to depict biologically important proteins for both supernatant and
cell components.

Keywords: *Streptococcus pyogenes*, proteomic analysis, principal component analysis.

1 Introduction

Streptococcus pyogenes is an important pathogen. The estimated annual number
of *Streptococcus pyogenes* infection cases are more than 700 million. There are
over 650,000 cases of severe, invasive infections that have a mortality rate of 25
%. Although *S. pyogenes* is a normal bacteria flora, occasionally *S. pyogenes* can
also cause life-threatening diseases. This means, it will be important to know
what triggers the diseases that *S. pyogenes* causes. There are a huge number of
researches [2] that investigate transcrptome responses to external environmental
factor, but there are very few researches on how its proteome changes in response
to external stimulations.

In this paper, we have systematically compared proteome of *S. pyogenes* during growing phases under two distinct culture conditions; with or without shaking. The latter condition was designed to be more oxidative stress condition. The
purpose of this research is to know the proteomic response to these two different growth conditions. Using the principal component analysis (PCA) [12], we
have selected representative proteins. Many of the representative proteins play
biological roles during the incubation.

---

* Corresponding author.

T. Shibuya et al. (Eds.): PRIB 2012, LNBI 7632, pp. 141–152, 2012.

2   Material and Methods

2.1   Proteome Analysis

In this study, *Streptococcus pyogenes* (serotype M1) SF370 of a clinical isolate was investigated. The sample was incubated at 37 °C for 4, 6, 14 and 20 hours ($OD_{660} = 0.40, 0.83, 0.92$, and 0.90, respectively).

Bacterial cultures were separated into the supernatant and the cellular fractions by centrifugation. The reason why the cellular fraction was not divided into soluble/insoluble fractions in contrast to the previous researches [9,14] was because these two did not differ from each other so much in the preliminary investigations (not shown here). Proteins contained in each fraction were partially purified by ethanol-chloroform purification. After reduced alkylation, they were digested by Lysyl Endopeptidase and Trypsin and were provided as samples for mass spectrometry. Detection of digested proteins was performed by LTQ-Orbitrap XL (Thermo Fisher Sceintific Inc.). Spectrums obtained by LTQ were identified by MASCOT program combined with Paradigm MS4 LC system (Michrom BioResources Inc.), based upon the in-house amino acid database which consists of coding-sequence predicted by genomic analysis [4] and re-evaluation of genome [10]. To be identified, at least two unique amino acid sequences for each protein were required. False discovery rate was estimated by decoy databases constructed by randomized amino acid sequences. Each of two fractions was measured three times for each of four time points separately under two distinct culture conditions. Analyzed quantity by PCA was %emPAI[5,13], which expresses the amount of proteins and %emPAI was its normalized value. %emPAI was normalized to have zero mean and unit variance before any analyses.

Hereafter, each sample was denoted by the tag ID in the form of XXXYY_Z, where XXX is either "sha" (the incubation under the shaking condition) or "sta" (the incubation under the static condition), YY denotes the duration time of the incubation (05, 07, 14, and 20 hours for the shaking incubation condition, and 04, 06, 14 and 20 hours for the static incubation condition), and Z is "wc" (the whole cellular fraction) or "snt" (the supernatant fraction), respectively.

2.2   Transcriptome Data

Transcriptome data set [1] with the accession number GSE5179 was downloaded from Gene Expression Omnibus (GEO). Raw data files GSM1167X.csv (X ranges from 67 to 79) were loaded into analysis program and column data named as F532.Median was used for further analyses. Each sample was normalized so as to have zero mean and unit variance. Then, six samples in the stationary phase were compared with six samples in the growth phase.

2.3   Statistical Methods

Application of Principal Component Analysis to Proteome Data. Suppose that we have proteome data x_{sp}, which is the normalized %emPAI of pth protein at sth sample $(s = 1, \ldots, S, p = 1, \ldots, P)$. This data can be understood as two ways, i.e.,

Category 1. In total, there are supposed to be S kinds of samples, each of which is characterized by the set of amounts of P kinds of proteins; a set of P dimensional vectors, the number of which is S.

Category 2. In total, there are supposed to be P kinds of proteins, each of which is characterized by the amount of its expression at S kinds of samples; a set of S dimensional vectors, the number of which is P.

Principal component analysis (PCA) can be applied to both of the two cases. If PCA is applied to the former (Category 1), the S kinds of samples are characterized with D_s principal component scores (PCSs) $y_s^i, (i = 1, \ldots, D_s)$, as

$$\mathbf{x}_s = (y_s^1, y_s^2, \ldots, y_s^{D_s})$$
$$y_s^i = \sum_p a_{ip} x_{sp}$$

instead of P kinds of proteins. Alternatively, if PCA is applied to the later (Category 2), the P kinds of proteins are characterized with D_p PCSs y_p^i, $(i = 1, \ldots, D_p)$, as

$$\mathbf{x}_p = (y_p^1, y_p^2, \ldots, y_p^{D_p})$$
$$y_p^i = \sum_s a_{js} x_{sp}$$

instead of S kinds of samples.

Selection of Representative Proteins. In some cases, PCA can be used to select representative $P'(< P)$ proteins[9,14] as follows. At first, each protein is embedded into $D_p'(< D_p)$ dimensional space (typically, D_p' is taken to be 2) by category 2 PCA. Then, the set S_p of top P' proteins which are far from origin are decided, i.e.,

$$S_p \equiv \left\{ p \mid \mathrm{rank}_p \left[\sum_{i=1}^{D_p'} (y_p^i)^2 \right] \leq P' \right\}$$

where $\mathrm{rank}_p[f_p]$ is the descent rank order of the element f_p. For example, when $f_2 < f_3 < f_1 < \cdots$, $\mathrm{rank}_p[f_1] = 3, \mathrm{rank}_p[f_2] = 1$, and $\mathrm{rank}_p[f_3] = 2$.

P' is decided to take a minimum number such that $y_s^i, (i = 1, \ldots, D_s' < D_s)$, where typically D_s' is taken to be 2, computed only with the selected P' proteins does not differ very much from the original y_s^i computed with all proteins.

This procedure is repeated after removing P' proteins, i.e., PCA is applied to the remaining $P - P'$ proteins. Then we get additional set $S_{p'}$ of $P''(< P - P')$ proteins to express new PCSs obtained by $P - P'$ proteins.

P-Values to Describe the Difference of Transcriptome between the Growth Phase and the Static Phase. Using the two sided t-test, we get P-values to check if gene expression in each phase differs from each other. Then, the obtained P-values are attributed to each gene. After that, 1643 genes have significant P-values ($P < 0.05$) even after the application of FDR correction based upon BH criterion, among 1798 genes to which Spy-IDs are attributed.

3   Results

3.1   Overview of Proteome with PCA Analysis

Figure 1A shows two dimensional embedding of samples using the category 1 PCA. Then $P' = 23$ proteins (Table 1) are selected based upon the two dimensional embedding (not shown here) of proteins obtained by category 2 PCA. Hereafter we call this as round one selection. After that, all of samples are re-embedded into two dimensional space (Fig. 1B) by category 1 PCA. Since Fig. 1B is almost identical with Fig. 1A, configuration seen in Fig. 1A turns out to be dependent upon the selected P' proteins only.

Table 1. Round one representative proteins. Ribosomal proteins are underlined. The proteins in italic letter are mentioned in the text.

SPy1489:hlpA	*SPy2039:speB*	SPy1073:rplL	SPy2005	*SPy2018:emm1*
SPy0059:rpmC	*SPy0611:tufA*	*SPy0274:plr*	SPy0062:rplX	*SPy2043:mf*
SPy0613:tpi	*SPy2079:AhpC*	SPy1831:rpsF	SPy2160:rpmG	SPy1373:ptsH
SPy0731:eno	*SPy1371:gapN*	*SPy1881:pgk*	*SPy0711:speC*	SPy0071:rpmD
SPy2070:groEL	SPy0019	*SPy0712:mf2*		

Above these procedures are repeated again for the remaining $P - P'$ proteins and we have successfully selected round two representative proteins $P'' = 30$. (Figure 2 and Table 2).

Table 2. Round two representative proteins. Notations are the same as in Table 1.

SPy0076:rpmJ	SPy1888:rpmB	SPy0063:rplE	SPy0717:rpmE	*SPy1429:gpmA*
SPy0822:rpmA	*SPy0273:fus*	SPy2092:rpsB	SPy0051:rplW	*SPy1282:pyk*
SPy0055:rplV	SPy1835:trx	*SPy1889:fba*	SPy1294	SPy1544:arcB
SPy0857:mur1.2	SPy0460:rplK	SPy0069:rpsE	SPy0272:rpsG	SPy1932:rplM
SPy1261	SPy1547:sagP	SPy1801:isp2	SPy1262	*SPy1436:mf3*
SPy1234:rpsT	SPy0052:rplB	*SPy2072:groES*	SPy0913	SPy1613

A)

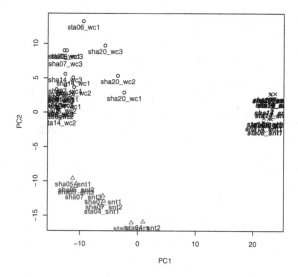

B)

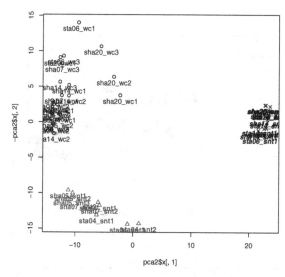

Fig. 1. A) Two dimensional embeddings of samples by Category 1 PCA. Black (○, normal): the whole cellular experiments (wc experiments), Red (△, bold): the early phase extracellular proteomes (sha05_snt, sha07_snt, and sta04_snt experimets), and Blue (×, bold italic): the late phase extracellular proteomes (sha14_snt, sha20_snt, sta06_snt, sta14_snt, and sta20_snt experiments) B) The same as A) but using only the selected $P' = 23$ proteins shown in Table 1. Cumurative contribution upto the second PC of the category 2 PCA is 82 %.

A)

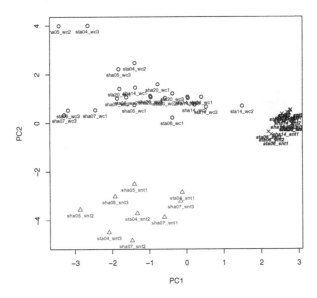

B)

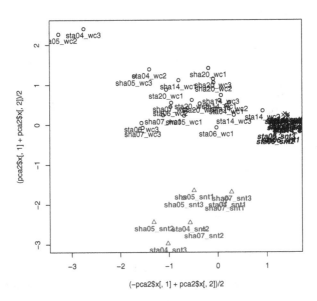

Fig. 2. A) Two dimensional embeddings of samples by Category 1 PCA, after the exclusion of P' proteins in Table 1. B) The same as A) but using only the selected $P'' = 30$ proteins shown in Table 2. Cumulative contribution up to the second PC of the category 2 PCA is 67 %.

The proteomes of *S. pyogenes* SF370, that grew under shaking or static culture condition, were clustered into three groups (Figures 1 and 2): the whole cellular proteome (all whole cellular experiments in Figures 1 and 2), the early phase extracellular proteome (sha05_snt, sha07_snt, and sta04_snt experiments in Figures 1 and 2), and the late phase extracellular proteome (sha14_snt, sha20_snt, sta06_snt, sta14_snt, and sta20_snt experiments in Figure 1 and 2), respectively. These results indicate that the proteomic phenotypes of *S. pyogenes* were divided into the two growth stages, the early growth phase that consists of the states at 5 and 7 hours under the shaking condition and the state at 4 hours under the static condition, and the late growth phase that consists of the states at the 14 and 20 hours under the shaking condition and the states at the 6, 14, and 20 hours under the static condition. It is suggested that the proteomic phenotype that grows under the static condition might rapidly grow from the early growth stage to the late growth stage compared with the shaking culture condition. Since the cell density (OD_{660}) at 5 hour under the shaking condition and the cell density at 4 hour under the static condition are the same value ($OD_{660} = 0.4$) and the cell density at 7 hour under the shaking condition and the cell density at 6 hour under the static condition are the same value ($OD_{660} = 0.8$), the proteome is dependent upon the cellular fraction (whole cell or extracellular) or the time development rather than the culture condition.

3.2 Biological Meanings of Representative Proteins

In Tables 1 and 2, we have shown representative proteins for rounds one and two. Figures 3 and 4 show expressions of the below mentioned proteins among those.

In this study, there are four designed experimental groups characterized by the combination of two criteria: two fractions (the whole cellular component or the supernatant component) and two culture conditions (incubation with or without shaking). Several proteins are group-specific and are picked up by PCA. For example, peroxiredoxin reductase (SPy2079:AhpC), which is estimated to be involved in oxygen metabolism and hydrogen peroxide decomposition, is found in shaking culture condition rather than static condition. It seems reasonable that the amount of AhpC increases in shaking condition because the shaking condition induces the higher oxygen stress. On the other hand, twenty out of the fifty-three representative proteins picked up with PCA are ribosomal subunit proteins (the proteins underlined in Tables 1 and 2). This number is as many as a half of ribosomal proteins identified in this study, while a total number of ribosomal proteins annotated in SF370 genome is fifty-three. These twenty ribosomal proteins were picked up with PCA due to the abundance in the cellular fraction (not shown here). The reason why several ribosomal proteins were also found in extracellular fraction (as a typical example, see SPy0055:rplV in Fig. 3) is possibly because of the leakage during cell division (see below).

Besides, many virulence associated proteins, pyogenic exotoxin B (SpeB; SPy2039), pyogenic exotoxin C (SpeC; SPy0711), mitogenic factors (Mf; SPy2043, Mf2; SPy0712, and Mf3; SPy1436), and M protein (Emm; SPy2018),

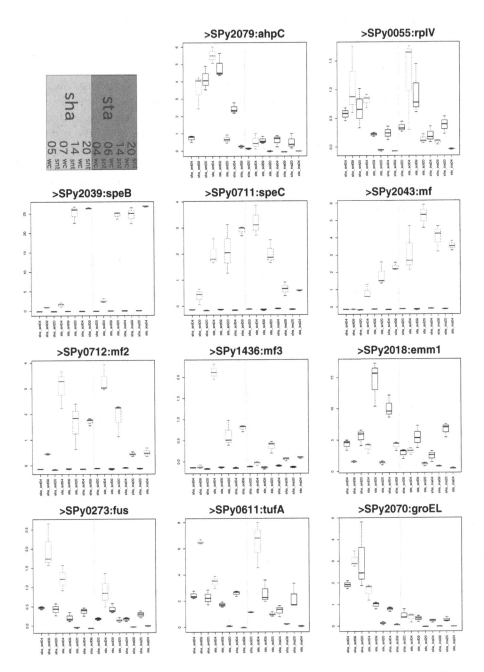

Fig. 3. Expression of representative proteins mentioned in the text. Colors and line types (black solid lines, red broken lines, and blue dotted broken lines) correspond to the colors in Figs. 1 and 2. The top-left panel: Schematic explanation of each panel.

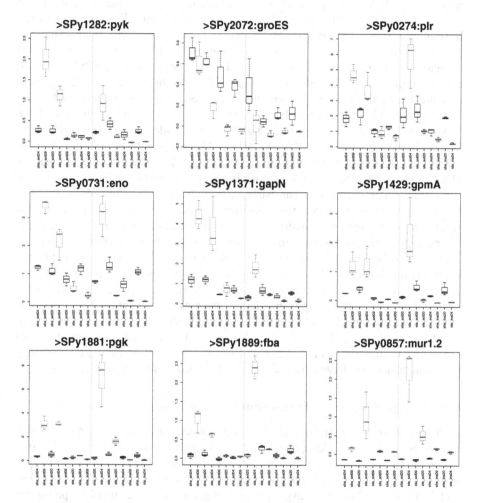

Fig. 4. Expression of representative proteins menthioned in the text. Notations are the same as Fig. 3.

are picked up by PCA analysis. These virulence-associated proteins have their own combination of the spatial and temporal distributions. SpeB increases monotonically in time, in both shaking and static culture condition. On the other hand, both Mf2 and SpeC increase under the shaking condition, but decrease under the static condition. The amount of both M protein and Mf increase and that of Mf3 decrease in shaking condition, although their amount keeps the constant value under the static incubation condition. The common distribution patterns are shared by the several abundant enzymes concerning the protein biosynthesis: such as an elongation factor EF-2 (Fus, SPy0273), an elongation factor Tu (TufA, SPy0611), a chaperonin (GroEL, SPy2070), and a co-chaperonin (GroES, SPy2072). The other common fashion of the protein distribution is also observed in enzymes involved in glycolysis: glyceraldehyde-3-phosphate dehydrogenase (Plr, SPy0274), phosphopyruvate hydratase (Eno, SPy0731), pyruvate kinase (Pyk, SPy1282), NADP-dependent glyceraldehyde-3-phosphate dehydrogenase (GapN, SPy1371), phosphoglyceromutase (GpmA, SPy1429), phosphoglycerate kinase (Pgk, SPy1881), and fructose-bisphosphate aldolase (Fba, SPy1889). Each protain is also observed by not small amount in the extracellular fraction at the early growth stage (sha05_snt, sha07_snt and sta04_snt, which are demonstrated by the red color in Fig. 3). They keep constant values throughout all sampling points in the whole cellular fraction. None of these proteins possessed signal sequence for secretion. Moreover, they are estimated to be intracellular enzymes such as the proteins involved in protein synthesis or glycolysis. It is confirmed the signal sequence-less proteins are always observed in the extracellular fraction of several bacterial species [6,7]. Most bacterial species that belong to firmicutes use autolytic enzymes, such as peptidoglycan hydrolase (Mur1.2, SPy0857), during the cell division processes[11,3,8]. Mur1.2 is also observed in early growth stage. It is supposed that these proteins are leaked from cytoplasm during cell division, especially in early growth stage.

In conclusion, we have successfully selected biologically important proteins.

3.3   Comparison with Transcriptome Analysis

Although there are no transcriptomic analyses performed to investigate the difference between the shaking or static incubation conditions, there is a research where the transcriptome is compared between the stationary phase and the exponential phase [1]. We also analysed these public domain data sets (see Materials and Methods) and tried to investigate if the gene coding the proteins picked up with PCA in this study show the significant difference between transcriptome between the static and exponential phases. In order to compare transcriptome between stationary and exponential phase, P-values, the rejection probability for the difference between the static and exponential phases, are attributed to transcruptome which corresponds to representative proteins. These P-values are compared with P-values for other proteins than representatives. Then P-values to depict the significant difference between two sets of P-values is obtained (Table 3). Both of P-values attributed to each of round one and two are mostly (21 out of 23 for round 1 and 23 out of 30 for round 2) less than 1×10^{-3} (Wilcoxon

Table 3. *P*-values (raw and BH corrected) attributed to representative proteins (Tables 1 and 2) obtained for transcriptome. Left:round one, midle and right: round two.

Round 1			Round 2					
Spy ID	P	FDR	Spy ID	P	FDR	Spy ID	P	FDR
SPy1489	9.71e-04	1.61e-03	SPy0076	6.32e-06	4.12e-05	SPy0857	4.73e-07	6.48e-06
SPy2039	3.70e-04	7.69e-04	SPy1888	3.66e-05	1.45e-04	SPy0460	5.15e-06	3.45e-05
SPy1073	1.91e-05	9.08e-05	SPy0063	1.25e-04	3.45e-04	SPy0069	1.05e-05	6.00e-05
SPy2005	8.54e-05	2.59e-04	SPy0717	3.70e-07	5.90e-06	SPy0272	2.83e-04	6.30e-04
SPy2018	7.19e-05	2.32e-04	SPy1429	4.31e-09	4.08e-07	SPy1932	1.58e-06	1.40e-05
SPy0059	9.17e-04	1.54e-03	SPy0822	6.10e-07	7.62e-06	SPy1261	1.37e-01	1.46e-01
SPy0611	7.09e-05	2.31e-04	SPy0273	9.05e-04	1.53e-03	SPy1547	1.20e-02	1.40e-02
SPy0274	1.98e-05	9.25e-05	SPy2092	6.83e-05	2.24e-04	SPy1801	1.60e-04	4.08e-04
SPy0062	4.11e-05	1.57e-04	SPy0051	8.19e-05	2.52e-04	SPy1262	4.40e-02	4.82e-02
SPy2043	1.49e-06	1.37e-05	SPy1282	8.51e-04	1.46e-03	SPy1436	2.31e-03	3.25e-03
SPy0613	1.34e-05	6.96e-05	SPy0055	1.32e-04	3.59e-04	SPy1234	2.20e-04	5.20e-04
SPy2079	1.03e-01	1.11e-01	SPy1835	5.42e-04	1.03e-03	SPy0052	2.77e-04	6.18e-04
SPy1831	2.09e-09	3.38e-07	SPy1889	2.00e-09	3.38e-07	SPy2072	6.36e-05	2.12e-04
SPy2160	5.43e-05	1.93e-04	SPy1294	1.83e-02	2.10e-02	SPy0913	4.05e-09	4.08e-07
SPy1373	2.01e-05	9.34e-05	SPy1544	7.16e-01	7.29e-01	SPy1613	2.71e-02	3.04e-02
SPy0731	6.13e-09	4.79e-07						
SPy1371	4.37e-06	3.10e-05						
SPy1881	5.01e-08	1.67e-06						
SPy0711	1.21e-03	1.91e-03						
SPy0071	4.39e-04	8.78e-04						
SPy2070	1.25e-04	3.45e-04						
SPy0019	3.79e-08	1.47e-06						
SPy0712	1.59e-04	4.06e-04						

test). We have also computed FDR corrected *P*-values, they are still mostly (22 out of 23 for round 1 and 28 out of 30 for round 2) highly significant, i.e., less than 5×10^{-2}. This means, proteins whose expression differs between two culture conditions are also significantly different with each other in transcriptome levels between exponential-phase and stationary-phase. Since the difference between two culture conditions is supposed to be the difference of time scale as mentioned above, our selection of representative proteins based upon proteome data turns out to be coincident with transcriptome analysis.

4   Conclusions

In this paper, we have performed proteome analysis of *Streptococcus pyogenes*, under two distinct culture conditions; with or withour shaking. Representative proteins are selected by iterative applications of PCA in two ways. These proteins turn out to be biologically informative and their trasctiptome expression also differs significantly between early or late stages.

Acknowledgement. This work was supported by KAKENHI (23300357).

References

1. Barnett, T.C., Bugrysheva, J.V., Scott, J.R.: Role of mRNA stability in growth phase regulation of gene expression in the group A streptococcus. J. Bacteriol. 189, 1866–1873 (2007)
2. Beyer-Sehlmeyer, G., Kreikemeyer, B., Hörster, A., Podbielski, A.: Analysis of the growth phase-associated transcriptome of streptococcus pyogenes. International Journal of Medical Microbiology 295(3), 161–177 (2005), http://www.sciencedirect.com/science/article/pii/S1438422105000421
3. Blackman, S.A., Smith, T.J., Foster, S.J.: The role of autolysins during vegetative growth of Bacillus subtilis 168. Microbiology (Reading, Engl.) 144 (pt. 1), 73–82 (1998)
4. Ferretti, J.J., McShan, W.M., Ajdic, D., Savic, D.J., Savic, G., Lyon, K., Primeaux, C., Sezate, S., Suvorov, A.N., Kenton, S., Lai, H.S., Lin, S.P., Qian, Y., Jia, H.G., Najar, F.Z., Ren, Q., Zhu, H., Song, L., White, J., Yuan, X., Clifton, S.W., Roe, B.A., McLaughlin, R.: Complete genome sequence of an M1 strain of Streptococcus pyogenes. Proc. Natl. Acad. Sci. U.S.A. 98, 4658–4663 (2001)
5. Ishihama, Y., Oda, Y., Tabata, T., Sato, T., Nagasu, T., Rappsilber, J., Mann, M.: Exponentially modified protein abundance index (emPAI) for estimation of absolute protein amount in proteomics by the number of sequenced peptides per protein. Mol. Cell Proteomics 4, 1265–1272 (2005)
6. Lei, B., Mackie, S., Lukomski, S., Musser, J.M.: Identification and immunogenicity of group A Streptococcus culture supernatant proteins. Infect. Immun. 68, 6807–6818 (2000)
7. Len, A.C., Cordwell, S.J., Harty, D.W., Jacques, N.A.: Cellular and extracellular proteome analysis of Streptococcus mutans grown in a chemostat. Proteomics 3, 627–646 (2003)
8. Mercier, C., Durrieu, C., Briandet, R., Domakova, E., Tremblay, J., Buist, G., Kulakauskas, S.: Positive role of peptidoglycan breaks in lactococcal biofilm formation. Mol. Microbiol. 46, 235–243 (2002)
9. Okamoto, A., Taguchi, Y.H.: Principal component analysis for bacterial proteomic analysis. IPSJ SIG Technical Report 2011-BIO-26, 1–6 (2011)
10. Okamoto, A., Yamada, K.: Proteome driven re-evaluation and functional annotation of the Streptococcus pyogenes SF370 genome. BMC Microbiol. 11, 249 (2011)
11. Oshida, T., Sugai, M., Komatsuzawa, H., Hong, Y.M., Suginaka, H., Tomasz, A.: A Staphylococcus aureus autolysin that has an N-acetylmuramoyl-L-alanine amidase domain and an endo-beta-N-acetylglucosaminidase domain: cloning, sequence analysis, and characterization. Proc. Natl. Acad. Sci. U.S.A. 92, 285–289 (1995)
12. Rao, P.K., Li, Q.: Principal Component Analysis of Proteome Dynamics in Iron-starved Mycobacterium Tuberculosis. J. Proteomics Bioinform. 2, 19–31 (2009)
13. Shinoda, K., Tomita, M., Ishihama, Y.: emPAI Calc–for the estimation of protein abundance from large-scale identification data by liquid chromatography-tandem mass spectrometry. Bioinformatics 26, 576–577 (2010)
14. Taguchi, Y.H., Okamoto, A.: Principal component analysis for bacterial proteomic analysis. In: 2011 IEEE International Conference on Bioinformatics and Biomedicine Workshops, pp. 961–963 (2011)

Application of the Multi-modal Relevance Vector Machine to the Problem of Protein Secondary Structure Prediction

Nikolay Razin[1], Dmitry Sungurov[1], Vadim Mottl[2], Ivan Torshin[2], Valentina Sulimova[3], Oleg Seredin[3], and David Windridge[4]

[1] Moscow Institute of Physics and Technology, Moscow, Russia
[2] Computing Center of the Russian Academy of Sciences, Moscow, Russia
[3] Tula State University, Tula, Russia
[4] University of Surrey, Guildford, UK

Abstract. The aim of the paper is to experimentally examine the plausibility of Relevance Vector Machines (RVM) for protein secondary structure prediction. We restrict our attention to detecting strands which represent an especially problematic element of the secondary structure. The commonly adopted local principle of secondary structure prediction is applied, which implies comparison of a sliding window in the given polypeptide chain with a number of reference amino-acid sequences cut out of the training proteins as benchmarks representing the classes of secondary structure. As distinct from the classical RVM, the novel version applied in this paper allows for selective combination of several tentative window comparison modalities. Experiments on the RS126 data set have shown its ability to essentially decrease the number of reference fragments in the resulting decision rule and to select a subset of the most appropriate comparison modalities within the given set of the tentative ones.

Keywords: Protein secondary structure prediction, machine learning, multi-modal relational pattern recognition, Relevance Vector Machine, controlled selectivity of reference objects and object-comparison modalities.

1 Introduction

Within the currently dominant paradigm of the protein science, the primary structure of a protein uniquely determines its spatial structure, which in turn determines the biological roles of the protein. Consequently, one of the main tasks of theoretical protein biology and bioinformatics is the establishment of the laws that govern the relationship between the primary and the spatial protein structure.

The secondary structure represents a projection of the local geometry of the spatial (tertiary) protein structure into a sequence of letters in a certain alphabet, most commonly, H – helix, S – strand, C – coil. The secondary structure prediction is increasingly becoming the work horse for numerous methods aimed at solving the much more challenging problem of predicting the spatial structure [1,2].

T. Shibuya et al. (Eds.): PRIB 2012, LNBI 7632, pp. 153–165, 2012.

The problem of protein secondary structure prediction was first conceived in the early 1960s, when a number of protein structures were determined by X-ray crystallography. A significant increase in the prediction accuracy was achieved once machine learning approaches were applied for solving the problem [3]. Despite an increase in the average accuracy, there is an evident lack of progress in this area in recent decades. For example, experiments within the framework of the conference-tournament CASP (Critical Assessment of the Protein Structure Prediction) [4], which have been carried out since the early 1990s, clearly show the absence of any significant positive trend in the accuracy of protein secondary structure prediction for at least 10 years from 1992 to 2002. It is perhaps for this reason that the problem of protein secondary structure prediction was even removed from the list of problems studied within the CASP framework (see publications of CASP-5 [5]).

The absence of any measurable progress is likely to be the result of numerous auxiliary assumptions of biological sort that underlie the prediction scenarios. It appears appropriate to develop and test algorithms, which should be based on the minimum possible number of additional assumptions drawn from biology and include adequate procedures for the selection of features representing amino acid sequences [6], as well as incorporate adequate training procedures for inferring relationship between the primary and the secondary structure from sufficiently large sets of proteins with the known spatial structure.

The commonly adopted principle of predicting the secondary structure at a position t in the polypeptide chain is its estimation from the local context, i.e., an amino acid window of a fixed length symmetric in relation to the target location t [3]. Given a training set of proteins whose known secondary structures are represented by strings on the three-letter alphabet $\{h, s, c\}$, the problem of inferring the prediction rule is that of pattern recognition.

The Support Vector Machine (SVM) is the most popular method of machine learning in pattern recognition learning [7]. As applied to secondary structure prediction [8], one of its advantages is that it yields a decision rule of classifying amino acid windows in new proteins on the basis of their comparison with a relatively small number of so-called support fragments inferred from the training set as result of training. However, the pay-off for this advantage is the onerous restriction that the comparison function must be a kernel, i.e., must possess the mathematical properties of the inner product in some hypothetical linear space into which the kernel embeds any set of objects. Elements of a linear space are usually called vectors, and this has led to the name of Support Vector Machine.

This paper is motivated by two intents – first, to remove any restrictions on the manner of comparison between amino acid fragments in contrast to excessively exacting kernels, and, second, to essentially decrease the number of reference fragments in the resulting decision rule. With this purpose, we rest here not on Vapnik's traditional SVM, but on the SVM-based Relevance Vector Machine (RVM) by Bishop and Tipping [9]. Two main advantages of the RVM technique are, first, just tolerance to any kinds of object comparison and, second, usage in the decision rule, instead of relatively few support vectors yielded by SVM, a still smaller number of so-called relevance vectors. In the problem of secondary structure prediction, this means that the structure states at subsequent points in the polypeptide chain of a new protein will be predicted by comparison of the respective windows with only a few reference sequences cut out of the training proteins as some sort of benchmarking windows representing the classes of the secondary structure.

For the window-based prediction of the protein secondary structure, we apply the multi-modal modification of the Relevance Vector Machine described in [10], which, in addition, allows to select a subset of the most appropriate window comparison functions within the given set of tentative ones.

For verification of the proposed technique, we used the RS126 set of protein chains as the source of both training and test sets.

To test the ability of the multimodal RVM to select most relevant comparison functions, two kinds of comparison principles were examined jointly – position-dependence of amino acids in fragments corresponding to the same local secondary structure in a protein [11,12] and a newly developed principle based on Fourier representation of both sequences as functions along the polypeptide axis.

We restrict here our attention to detecting strands in the secondary structure of proteins, which, as practice shows, represent an especially problematic element of the secondary structure. The aim of the paper is rather to explore the performance of the Relevance Vector Machine in the problem of widow-based secondary structure prediction than achieving some record-breaking results. Nevertheless, experiments on the RS126 data set have shown the accuracy of about 75% in detecting strands as especially problematic element of the secondary structure.

2   The Local Machine-Learning Approach to Secondary Structure Prediction – Pattern Recognition in a Sliding Amino Acid Window

Let $\omega = (\alpha_t, t = 1, ..., M)$ be the finite amino acid sequence which represents the primary structure of a protein of individual length $M = M_\omega$, where $\alpha_t \in A = \{\alpha^1, ..., \alpha^m\}$, $m = 20$ are symbols corresponding to the alphabet of amino acids. The protein's hidden secondary structure will be completely represented by a symbolic sequence $\mathbf{y} = (y_t, t = 1, ..., M)$ of the same length $M = M_\omega$, whose elements $y_t \in Y = \{h, s, c\}$ are associated with three classes of structure: h – helix, s – sheet, c – unspecified structure usually referred to as coil.

Let, further, the observer be submitted a training set of proteins whose amino acid sequences are labeled by the "correct" assignments of secondary structure:

$$\left\{ (\omega_l, \mathbf{y}_l), l = 1, ..., N^0 \right\}, \quad \omega_l = (\alpha_{lt}, t = 1, ..., M_l), \quad \mathbf{y}_l = (y_{lt}, t = 1, ..., M_l) \qquad (1)$$

Given a new amino acid sequence $\omega = (\alpha_t, t = 1, ..., M_\omega)$ not represented in the training set, we are required to estimate the secondary structure of the respective protein $\hat{\mathbf{y}}(\omega) = (\hat{y}_t(\omega), t = 1, ..., M_\omega)$.

Following [13], in this paper we restrict our consideration to prediction based on the principle of a sliding amino acid window. This means that the decision on the class of secondary structure at position t is made from the symmetric interval

$\omega_t = (\alpha_\tau, t-T \le \tau \le t+T)$ of the entire amino acid chain $\boldsymbol{\omega} = (\alpha_t, t=1,...,N)$. The odd width $T = 2T+1$ of the sliding window is thus defined by its half-width T as a parameter to be preset. Estimation of the secondary structure of a protein thus takes place only within its amino acid sequence truncated at both sides by the window's half-width $\hat{\mathbf{y}}(\boldsymbol{\omega}) = (\hat{y}_t(\boldsymbol{\omega}), T+1 \le t \le M-T) = (\hat{y}_t(\omega_t), \ T+1 \le t \le M-T)$.

Thus, the original problem of predicting the entire secondary structure of a protein $\hat{\mathbf{y}}(\boldsymbol{\omega})$ is reduced to the series of independent problems $\hat{y}_t(\omega_t) = \hat{y}_t(\alpha_{t-T},...,\alpha_t,...,\alpha_{t+T})$ of estimating the class of secondary structure $\hat{y}_t \in \{h,s,c\}$ for the central amino acid α_t in the respective window.

The window-based approach implies treating the training set as an unordered assembly of all continuous amino acid fragments $\{(\omega_j, y_j), j=1,...,N\}$ cut out of the given set of indexed amino acid sequences $\omega_j = (\alpha_{j\tau}, t-T \le \tau \le t+T)$, $y_j \in \{h,s,c\}$ (1). As a simplification resulting from our restricting the problem to distinguishing between strands and other elements of the secondary structure, we shall train a two-class classifier: $y_j \in \{1,-1\} = \{s, \bar{s}\} = \{s, \{h,c\}\}$.

3   The Multi-modal Relevance Vector Machine

The mathematical and algorithmic technique we use for window-based prediction of protein secondary structure is that of the multi-modal Relevance Vector Machine outlined in [10] which rests on three well-established principles of pattern-recognition learning.

First of all, we proceed from the featureless approach proposed by Duin et al. [14] under the name of Relational Discriminant analysis, which consists in the idea of representing the pattern recognition objects ω, not by individual feature vectors $\mathbf{x}(\omega) \in \mathbb{R}^k$, but by an arbitrary real-valued measure of pair-wise relation between them. In terms of window-based secondary structure prediction, the idea is to treat the values of this function between an arbitrary amino acid fragment ω and those of the training set $\{(\omega_j, y_j), j=1,...,N\}$ as the vector of secondary features $(x_j(\omega) = S(\omega_j,\omega), j=1,...,N)$. Then, the standard convex SVM training technique will yield the parameters $(a_1,...,a_N,b)$ of a discriminant hyperplane in the linear space of secondary features $\mathbb{R}^N$

$$d(\omega) = \sum_{j=1}^N a_j S(\omega_j,\omega) + b \gtrless 0,$$ (2)

which can be applied it to any new amino acid fragment

$$\omega = (\alpha_\tau, -T \le \tau \le T).$$ (3)

In order to weaken the demand of storing very large numbers of reference amino acid fragments $\{\omega_j, j=1,...,N\}$, we apply Bishop and Tipping's Relevance Vector

Machine (RVM) [9], underpinned by the notion of selecting only a small number of most informative Relevance Objects in the training set:

$$d(\omega) = \sum_{j \in \hat{J}} a_j S(\omega_j, \omega) + b \gtrless 0, \quad \hat{J} \subset \{1, ..., N\}. \tag{4}$$

However, the Bayesian principle of selecting secondary features implied by the original RVM results in a non-convex training problem.

The novel aspect of [10] which is immediately applicable in this paper is the assumption that several comparison modalities for pair-wise object representation are available $S_i(\omega', \omega'')$, $i = 1, ..., n$. The presence of several object-comparison functions expands the number nN of secondary features for any object $\big(x_{ij}(\omega) = S_i(\omega_j, \omega),$ $i = 1, ..., n, j = 1, ..., N\big)$. A straightforward generalization of the doubly-regularized SVM [15] has led in [10] to the *multimodal* convex training criterion which we call the multi-modal Relevance Vector Machine and which we shall apply in this paper to training sets of amino acid fragments $\{(\omega_j, y_j), j = 1, ..., N\}$:

$$\begin{cases} \sum_{i=1}^{n} \sum_{l=1}^{N} \left[(1-\mu) a_{il}^2 + \mu |a_{il}| \right] + C \sum_{j=1}^{N} \delta_j \to \min(a_{il}, b, \delta_j), \\ y_j \left(\sum_{i=1}^{n} \sum_{l=1}^{N} a_{il} S_i(\omega_l, \omega_j) + b \right) \geq 1 - \delta_j, \delta_j \geq 0, j = 1, ..., N. \end{cases} \tag{5}$$

This training criterion differs from the usual SVM by a more complicated regularization term which is a mix of L_2 and L_1 norms of the direction vector with an additional weighting parameter $0 \leq \mu < 1$ instead of the pure L_2 norm in the classical case.

We shall use the following notations for sets of, respectively, object-comparison modalities, training objects and all secondary features:

$$I = \{1, ..., n\}, J = \{1, ..., N\}, F = \{ij, i = 1, ..., n, j = 1, ..., N\} = I \times J.$$

The training criterion (5) is both modality-selective and reference-object-selective, therefore, we refer to it as the modality-selective Relevance Vector Machine. The subset of relevant secondary features $\hat{F} = \{ij: \hat{a}_{ij} \neq 0\} \subseteq F$ determines the subsets of relevant modalities $\hat{I}$ and relevant objects $\hat{J}$:

$$\hat{F} = \{ij: \hat{a}_{ij} \neq 0\} \subseteq F: \Rightarrow \hat{I} = \{i: \exists j (a_{ij} \neq 0)\} \subseteq I = \{1, ..., n\}, \quad \hat{J} = \{j: \exists i (a_{ij} \neq 0)\} \subseteq J = \{1, ..., N\}. \tag{6}$$

As a result, the optimal discriminant hyperplane, being a generalized analog of (4), takes into account only the relevance modalities of any new object, and is completely determined by the relevance objects of the training set:

$$d(\omega) = \sum_{ij \in \hat{F}} a_{ij} S_i(\omega_j, \omega) + b \gtrless 0, \quad \hat{F} \subseteq F. \tag{7}$$

If $\mu = 0$, the method equates to the classical SVM retaining all the secondary features $x_{ij}(\omega) = S_i(\omega_j, \omega)$, namely, the entire training set as the set of reference objects (2) and all the object-comparison modalities expressed by functions $S_i(\omega_j, \omega)$. As the structural parameter grows $0 \to \mu \to 1$, the subset of relevance features $\hat{F}$ diminishes, and both subsets of relevance objects $\hat{J}$ and relevance comparison modalities $\hat{I}$ shrink along with it. If $\mu \to 1$, the criterion becomes extremely selective. Experiments have shown [10] that in the latter case it becomes practically equivalent to the original RVM [9] except for having the favourable feature of being convex.

4   Modalities of Pair-Wise Amino Acid Fragment Comparison for Protein Secondary Structure Prediction

In this paper, we experimentally apply the outlined multimodal RVM technique to protein secondary structure prediction by utilizing several different modalities of amino acid sequence comparison. Two kinds of comparison principles are jointly examined – similarity measures exploiting the position-dependence of amino acids in fragments corresponding to the same local secondary structure in a protein [12], and a newly developed class of similarity measures implied by Fourier representation of both sequences as functions along the polypeptide axis.

In accordance with (3), each comparison function $S_i(\omega', \omega'')$ must be applicable to any two amino acid fragments $\omega' = (\alpha'_\tau, -T \le \tau \le T)$ and $\omega'' = (\alpha''_\tau, -T \le \tau \le T)$ of length $2T + 1$ defined by the half-width parameter of the window T. In our experiments, we examined two different half-width parameters:

– $T = 6$, i.e., the window length $2T + 1 = 13$, for comparison from the viewpoint of amino acid positions,

– $T = 17$, i.e., the window length $2T + 1 = 35$, for Fourier-based comparison; this window length fulfills the goal of exploring long-range dependencies of protein secondary structure on the amino acid sequence.

On the basis of each of these two comparison principles, we constructed three different comparison functions. So, we consider all in all $n = 6$ functions of pair-wise amino acid fragment comparison.

4.1   Amino-Acid-Position-Based Comparison

This form of comparison implements and generalizes the method of [12]. Let $\mathbb{A} = \{\alpha^1, ..., \alpha^{20}\}$ be the alphabet of amino acids. For each position $-T \le \tau \le T$ in the window $\omega = (\alpha_\tau, -T \le \tau \le T)$ and each of 20 amino acids, a binary feature is defined $z_{\tau k}(\omega) = 1$ if $\alpha_\tau = \alpha^k$ and $z_{\tau k}(\omega) = 0$ if $\alpha_\tau \ne \alpha^k$. All the features jointly make the binary $20(2T+1)$-dimensional feature vector $\mathbf{z}(\omega) = (z_{\tau k}(\omega), -T \le \tau \le T, k = 1, ..., 20)$. We examined three fragment comparison functions based on such features:

$$S_1(\omega', \omega'') = \sum_{\tau=-T}^{T} \sum_{k=1}^{20} z_{\tau k}(\omega') z_{\tau k}(\omega''),$$

$$S_2(\omega', \omega'') = \exp\left\{-\gamma \left[z_{\tau k}(\omega') - z_{\tau k}(\omega'') \right]^2 \right\}, \tag{8}$$

$$S_3(\omega', \omega'') = \sum_{\tau=-T}^{T} \sum_{k=1}^{20} \left| z_{\tau k}(\omega') - z_{\tau k}(\omega'') \right|.$$

Two former comparison functions were examined separately in [12]; both of them are kernels on the set of amino acid fragments, but this fact is out of significance in our approach.

It is shown in [12] that the amino-acid-position-based principle of comparison is more adequate to relatively short windows, therefore, we use it with the recommended window length $2T+1 = 13$.

4.2   Fourier-Transform-Based Comparison

This method is proposed here for the first time. It rests on the fact that both PAM and BLOSUM amino acid substitution matrices result from the same PAM evolutionary model [16], namely, an assumed ergodic and reversible Markov chain, and the main difference between them lies in the different initial data for estimating unknown transition probabilities [17]. Moreover, it is shown in [17] that that all PAM and BLOSUM substitution matrices express probabilities of the existence of a common ancestor for each pair of amino acids and are, by their nature, positive semidefinite matrices. This innate positive semi-definiteness is absent in published matrices only because of traditional logarithmic representation and rounding down to whole numbers.

The initial positive definite PAM matrices for any evolutionary distance can be easily computed from the estimated transition probabilities PAM1 available in [16] via the algorithm outlined in [17].

For the Fourier representation of amino acid fragments $\omega = (\alpha_\tau, -T \leq \tau \leq T)$, we use the positive definite PAM250 matrix, which we denote as $\mathbf{M} = \big(\mu(\alpha^k, \alpha^l),$ $k, l = 1, ..., 20 \big)$. Its positive eigenvalues $\eta^q > 0$ and eigenvectors $\mathbf{M} \mathbf{h}^q = \eta^q \mathbf{h}^q$, $\mathbf{h}^q = (h_1^q \cdots h_{20}^q) \in \mathbb{R}^{20}$, $q = 1, ..., 20$, satisfy the equality $\mathbf{M} = \sum_{q=1}^{20} \eta^q \mathbf{h}^q (\mathbf{h}^q)^T$, i.e., $\mu(\alpha^k, \alpha^l) = \sum_{q=1}^{20} \eta^q h_k^q h_l^q$. It follows from this equality that all the amino acids α^k may be represented by vectors $\mathbf{a}^k = (a_1^k \cdots a_{20}^k)^T = \big((\eta^1)^{1/2} h_1^k \cdots (\eta^{20})^{1/2} h_{20}^k \big)^T \in \mathbb{R}^{20}$, whose inner products completely coincide with elements of the substitution matrix $\mu(\alpha^k, \alpha^l) = (\mathbf{a}^k)^T \mathbf{a}^l$.

Thus, from the viewpoint of a specified substitution matrix, any initially discrete symbolic fragment of the amino acid chain $\omega = (\alpha_\tau \in A, -T \leq \tau \leq T)$ may be considered as a real-valued 20-dimensional signal $(\mathbf{a}_\tau = (a_{1\tau} \cdots a_{20\tau})^T \in \mathbb{R}^{20}, -T \leq \tau \leq T)$. The idea is then to represent each scalar component $(a_{k\tau} \in \mathbb{R}, -T \leq \tau \leq T)$ of this vector signal $i = 1, ..., 20$ in the form of the vector of Fourier coefficients with respect to the pairs of orthogonal basic harmonic signals, $\cos(i(\pi/T)\tau)$ and $\sin(i(\pi/T)\tau)$, of incrementing frequency $\{i(\pi/T), i=0,1,...,T\}$ in the interval $-T \leq \tau \leq T$.

Let $(a_\tau \in \mathbb{R}, -T \leq \tau \leq T)$ be a scalar signal. Its cosine and sine spectra are expressed by the following formulas:

$$\begin{cases} u_0 = \dfrac{1}{2T+1}\sum_{\tau=-T}^{T} a_\tau, i=0, \quad u_l = \dfrac{1}{2T+1}\sum_{\tau=-T}^{T} a_\tau \cos(l(\pi/T)\tau), l=1,...,T, \\ v_l = \dfrac{1}{2T+1}\sum_{\tau=-T}^{T} a_\tau \sin(l(\pi/T)\tau), l=1,...,T. \end{cases} \tag{9}$$

To partially dampen the dependence of the Fourier expansion on the shift of the sliding window along the polypeptide axis, we take into account only $T+1$ elements of the amplitude spectrum and ignore the phase of the Fourier transform:

$$f_0 = u_0, l=0, \quad f_l = (u_l^2 + v_l^2)^{1/2}, \quad l=1,...,T. \tag{10}$$

An amino acid fragment $\omega = (\alpha_\tau \in \mathcal{A}, -T \leq \tau \leq T)$ will yield a vector signal $(\mathbf{a}_\tau = (a_{1\tau} \cdots a_{20\tau})^T, -T \leq \tau \leq T)$ and, respectively, 20 spectra represented by the $T+1$ 20-dimensional vectors $l=0,1,...,T$ corresponding to the series of increasing frequencies in accordance with (9) and (10). In this work, we exploit four first harmonics along with the zero-frequency constant:

$$\mathbf{f}(\omega) = [\mathbf{f}_0(\omega), \mathbf{f}_l(\omega), l=1,...,4] \in \mathbb{R}^{20 \times 5} = \mathbb{R}^{100},$$
$$\mathbf{f}_0(\omega) = (f_{k0}(\omega), k=1,...,20), \mathbf{f}_l(\omega) = (f_{kl}(\omega), k=1,...,20). \tag{11}$$

The essence of the Fourier-transform-based comparison of amino acid fragments (ω', ω'') is thus exploitation of the feature vector (11) within a single comparison modality $S(\omega', \omega'') = S(\mathbf{f}(\omega'), \mathbf{f}(\omega''))$. We examine here three comparison functions numbered as continuation of (8):

$$S_4(\omega',\omega'') = \mathbf{f}^T(\omega')\mathbf{f}(\omega'') = \sum_{k=1}^{20} f_{k0}(\omega')f_{k0}(\omega'') + \sum_{k=1}^{20}\sum_{l=1}^{4} f_{kl}(\omega')f_{kl}(\omega''),$$
$$S_5(\omega',\omega'') = \exp\{-\gamma\|\mathbf{f}(\omega') - \mathbf{f}(\omega'')\|^2\} =$$
$$\exp\left\{-\gamma\left[\sum_{k=1}^{20}(f_{k0}(\omega') - f_{k0}(\omega''))^2 + \sum_{k=1}^{20}\sum_{l=1}^{4}(f_{kl}(\omega') - f_{kl}(\omega''))^2\right]\right\}, \tag{12}$$
$$S_6(\omega',\omega'') = \sum_{k=1}^{20}|f_{k0}(\omega') - f_{k0}(\omega'')| + \sum_{k=1}^{20}\sum_{l=1}^{4}|f_{kl}(\omega') - f_{kl}(\omega'')|.$$

This class of fragment comparison functions is meant to be appropriate for exploring long-range dependencies in protein secondary structure prediction. With this purpose, we use relatively large window length $2T+1=35$.

5   Experiments

To determine the performance of the multimodal Relevance Vector Machine in the context of protein secondary structure prediction at different levels of relevance-selection for amino acid fragments and fragment comparison functions, we used the RS126 data set that contains 126 proteins having less than 25% sequence identity for lengths greater than 80 amino acids.

All in all, the proteins in RS126 produce the set Ω of $|\Omega|=19075$ amino acid windows $\omega \in \Omega$ of length $2T+1=35$, each labelled by an index of the structural state at the center; $y=\pm 1$, i.e., strand/not-strand. We performed four experiments with this data set.

In each experiment, we independently partitioned the set of all amino acid windows into the training set $\Omega_{tr} \subset \Omega$ of size $N=|\Omega_{tr}|=1600$ randomly drawn from Ω, and the rest $\Omega_{test}=\Omega \backslash \Omega_{tr}$ of size $|\Omega_{test}|=17475$ which served as the source of test sets.

The set of six competing and concurrent fragment comparison functions remained the same, being those derived via functions (8) and (12), $n=6$. The Fourier-transform-based comparison of amino acid windows (12) utilizes the full length of the windows $2T+1=35$, whereas the amino-acid-position-based comparison (8), in accordance with the accepted strategy, is to be applied to shorter windows $2T+1=13$ obtained from the initial ones by ignoring the 11 amino acids at both ends.

Each of the four experiments consisted in training the multi-modal Relevance Vector Machine (5) seven times from the same training set Ω_{tr}, $N=1600$, with seven incrementing values of the selectivity parameter: $\mu_1=0$, $\mu_2=0.3$, $\mu_3=0.5$ $\mu_4=0.6$, $\mu_5=0.8$, $\mu_6=0.9999$, and $\mu_7=0.99999$ ($\mu \rightarrow 1$). Thus, the Relevance Vector Machine was run $4\times 7=28$ times.

The immediate result of each run of the training algorithm with a heuristic initial value of the selectivity parameter μ is the subset of relevant secondary features $\hat{F}(\mu)=\{ij: \hat{a}_{ij}(\mu)\neq 0\}\subseteq F$ and parameter values $\left(a_{ij}(\mu)\in \hat{F}, b(\mu)\right)$ of the discriminant hyperplane (7). Of particular importance are the resulting subsets of relevant objects (amino acid fragments of the training set), $\hat{J}(\mu)=\{j: \exists i(a_{ij}\neq 0)\}\subseteq \{1,...,N\}$, and relevant comparison modalities $\hat{I}(\mu)=\{i: \exists j(a_{ij}\neq 0)\}\subseteq I=\{1,...,n\}$. Their numbers are denoted, respectively, as $\hat{N}(\mu)=|\hat{J}(\mu)|\leq N$ and $\hat{n}(\mu)=|\hat{I}(\mu)|\leq n$.

We then randomly partitioned the remaining set $\Omega_{test}=\Omega \backslash \Omega_{tr}$ of 19075 amino acid windows into 10 test sets of approximately 1900 windows each, and computed the accuracy of recognition of secondary structure states $\{s,\bar{s}\}$, i.e., {strand} versus {not strand}, as the respective percentage values. The overall percentage accuracy in all the test sets for each selectivity μ_k, $k=1,...,7$, was assessed by the average value $Acc(\mu)$ and root-mean-square scatter $\sigma(\mu)$. Finally, the confidence interval was computed for each average percentage as $Acc(\mu)\pm 2\sigma(\mu)$.

Figure 1 visually displays the dependence of the accuracy percentage $Acc(\mu)$ and the number of relevant amino acid fragments participating in the final decision rule $\hat{N}(\mu)$ at selectivity level μ. All the results are represented in Table 1.

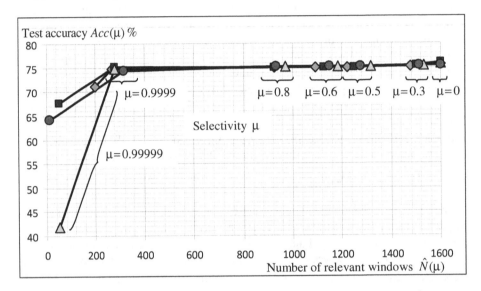

Fig. 1. Experimental dependence of the number of relevant amino acid fragments $\hat{N}$ and the test-set accuracy of detecting strands Acc on the level of secondary feature selectivity μ

It is evident from Table 1 and Figure 1 that, in all experiments, the best average accuracy of approximately 75.5% is achieved with zero selectivity $\mu=0$, when all the 1600 amino acid fragments constituting the training set and all the 6 comparison functions participate in the discriminant hyperplane (7) (which is thus defined in the $nN=9600$-dimensional space of secondary features of a single amino acid window $\omega=(\alpha_\tau\in A, -T\leq\tau\leq T)$). What is especially interesting is that no traces of overfitting are evident in the determination of the discriminant hyperplane in the linear space of secondary feature vectors, $\mathbf{x}(\omega)=\left(x_{ij}(\omega), i=1,...,6, j=1600\right)$, whose dimension, $\mathbf{x}(\omega)=\mathbb{R}^{9600}$, exceeds, by six times, the size of the training set.

The growth of μ thus diminishes both the number, $\hat{N}(\mu)$, of relevant training-set fragments and the number, $\hat{n}(\mu)$, of relevant fragment-comparison modalities forming the secondary features of current amino acid windows, and initially results in a minor decrease of the test-set accuracy. However, it is worth noting that the accuracy percentage remains practically the same in all independent experiments up to the selectivity level $\mu=0.9999$, when about 300 relevant amino acid fragments of the initial number of 1600 remain in the decision rule for strand detection, and only $\hat{n}=3$ comparison functions are required to classify new windows in the test set. The respective drop of accuracy relative to the absence of any selectivity $\mu=0$ does not exceed 1%.

Table 1. Results of four independent experiments (markers as in Figure 1)

			Accuracy of detecting strands $Acc(\mu)$	Number of relevant windows $\hat{N}(\mu)$	Number $\hat{n}(\mu)$ and list of relevant comparison functions
Experiment 1 ◆	Selectivity	μ	0	75.63 ± 1.78%	1600	$6, \hat{I}=\{1,2,3,4,5,6\}$
		0.3	75.04 ± 1.75%	1476	$5, \hat{I}=\{1,2,\cancel{3},4,5,6\}$
		0.5	74.95 ± 1.74%	1222	$5, \hat{I}=\{1,2,\cancel{3},4,5,6\}$
		0.6	74.96 ± 1.74%	1094	$5, \hat{I}=\{1,2,\cancel{3},4,5,6\}$
		0.8	74.96 ± 1.72%	924	$4, \hat{I}=\{1,2,\cancel{3},4,5,\cancel{6}\}$
		0.9999	74.63 ± 1.76%	267	$3, \hat{I}=\{1,2,\cancel{3},\cancel{4},5,\cancel{6}\}$
		0.99999	71.04 ± 1.69%	200	$2, \hat{I}=\{1,2,\cancel{3},\cancel{4},\cancel{5},\cancel{6}\}$
Experiment 2 ■	Selectivity	μ	0	75.85 ± 1.52%	1600	$6, \hat{I}=\{1,2,3,4,5,6\}$
		0.3	75.23 ± 1.72%	1501	$5, \hat{I}=\{1,2,\cancel{3},4,5,6\}$
		0.5	75.01 ± 1.63%	1247	$5, \hat{I}=\{1,2,\cancel{3},4,5,6\}$
		0.6	75.01 ± 1.65%	1127	$5, \hat{I}=\{1,2,\cancel{3},4,5,6\}$
		0.8	75.01 ± 1.65%	924	$5, \hat{I}=\{1,2,\cancel{3},4,5,6\}$
		0.9999	75.10 ± 1.73%	278	$3, \hat{I}=\{1,2,\cancel{3},\cancel{4},5,\cancel{6}\}$
		0.99999	67.60 ± 0.80%	49	$3, \hat{I}=\{1,2,\cancel{3},\cancel{4},5,\cancel{6}\}$
Experiment 3 ◀	Selectivity	μ	0	75.70 ± 1.22%	1600	$6, \hat{I}=\{1,2,3,4,5,6\}$
		0.3	75.30 ± 0.79%	1531	$5, \hat{I}=\{1,2,\cancel{3},4,5,6\}$
		0.5	75.10 ± 0.94%	1317	$5, \hat{I}=\{1,2,\cancel{3},4,5,6\}$
		0.6	75.08 ± 0.99%	1183	$5, \hat{I}=\{1,2,\cancel{3},4,5,6\}$
		0.8	75.08 ± 0.99%	971	$5, \hat{I}=\{1,2,\cancel{3},4,5,6\}$
		0.9999	74.74 ± 0.79%	280	$3, \hat{I}=\{1,2,\cancel{3},\cancel{4},5,\cancel{6}\}$
		0.99999	41.84 ± 2.33%	51	$3, \hat{I}=\{1,2,\cancel{3},\cancel{4},5,\cancel{6}\}$
Experiment 4 ●	Selectivity	μ	0	75.33 ± 0.99%	1600	$6, \hat{I}=\{1,2,3,4,5,6\}$
		0.3	75.30 ± 0.95%	1514	$5, \hat{I}=\{1,2,\cancel{3},4,5,6\}$
		0.5	75.07 ± 0.97%	1275	$5, \hat{I}=\{1,2,\cancel{3},4,5,6\}$
		0.6	75.03 ± 0.99%	1150	$5, \hat{I}=\{1,2,\cancel{3},4,5,6\}$
		0.8	75.03 ± 0.99%	933	$5, \hat{I}=\{1,2,\cancel{3},4,5,6\}$
		0.9999	74.27 ± 1.53%	318	$3, \hat{I}=\{1,2,\cancel{3},\cancel{4},5,\cancel{6}\}$
		0.99999	64.16 ± 1.81%	12	$3, \hat{I}=\{1,2,\cancel{3},\cancel{4},5,\cancel{6}\}$

Beyond this limit, a further increase of selectivity results in a drastic loss of both recognition accuracy and stability with respect to different training sets.

6   Conclusions

Application of the machine learning techniques to the problems of bioinformatics, in particular feature generation and selection in the space of amino acid sequences, represents a fruitful direction of research both in computer science and in computational biology. In this proof-of-principle study, we applied a method based on the Relevance Vector Machines (RVM) methodology to the problem of the protein secondary structure prediction. A unique characteristic of this method is that it permits automatic selection of the most appropriate features (modalities) from the total number of possible modalities.

In our study, the average accuracy of the strand prediction was approximately 75%, a comparable accuracy to the current state-of-the-art. However, the use of relevance vector principles means that this accuracy figure is achievable with only a small fraction (less than a quarter) of the totality of features,   representing a potentially significant advantage in terms of parsimony, robustness and interpretability of the resulting classifications.

Acknowledgments. We would like to acknowledge support from grants of the Russian Foundation for Basic Research 11-07-00409 and 11-07-00728, grant of the Ministry of Education and Science of the Russian Federation, contract 07.514.11.4001, and from UK EPSRC grant EP/F069626/1 (ACASVA).

References

1. Branden, C., Tooze, J.: Introduction to Protein Structure, 2nd edn., p. 410. Garland Publishing, Inc., New York (1999)
2. Rost, B.: Protein secondary structure prediction continues to rise. Journal of Structural Biology 134(2-3), 204–218 (2001)
3. Yoo, P., Zhou, B., Zomaya, A.: Machine learning techniques for protein secondary structure prediction: An overview and evaluation. Current Bioinformatics 3(2), 74–86 (2008)
4. Critical Assessment of the Protein Structure Prediction. Protein Structure Prediction Center. Sponsored by the US National Library of Medicine (NIH/NLM),
   http://predictioncenter.org/http://predictioncenter.org/index.cgi?page=proceedings
5. Aloy, P., Stark, A., Hadley, C., Russell, R.: Predictions without templates: new folds, secon-dary structure, and contacts in CASP5. Proteins 53(suppl. 6), 436–456 (2003)
6. Torshin, I.Y.: Bioinformatics in the Post-Genomic Era: The Role of Biophysics. Nova Biomedical Books, NY (2007) ISBN: 1-60021-048
7. Vapnik, V.: Statistical Learning Theory, p. 736. John-Wiley & Sons, Inc. (1998)
8. Ward, J., McGuffin, L., Buxton, B., Jones, D.: Secondary structure prediction with support vector machines. Bioinformatics 19(13), 1650–1655 (2003)

9. Bishop, C., Tipping, M.: Variational Relevance Vector Machines. In: Proceedings of the 16th Conference on Uncertainty in Artificial Intelligence, pp. 46–53. Morgan Kaufmann (2000)
10. Seredin, O., Mottl, V., Tatarchuk, A., Razin, N., Windridge, D.: Convex Support and Relevance Vector Machines for selective multimodal pattern recognition. In: The 21th International Conference on Pattern Recognition, Tsukuba Science City, Japan, November 11-15 (2012)
11. Engel, D., DeGrado, W.: Amino acid propensities are position-dependent throughout the length of α-helices. J. Mol. Biol. 337, 1195–1205 (2004)
12. Ni, Y., Niranjan, M.: Exploiting long-range dependencies in protein β-sheet secondary structure prediction. In: Proceedings of the 5th IAPR International Conference on Pattern Recognition in Bioinformatics, Nijmegen, The Netherlands, September 22-24, pp. 349–357 (2010)
13. Cole, C., Barber, J., Barton, G.: The Jpred 3 secondary structure prediction server. Nucl. Acids Res. 36 (suppl. 2), W197–W201 (2008)
14. Duin, R., Pekalska, E., de Ridder, D.: Relational discriminant analysis. Pattern Recognition Letters 20, 1175–1181 (1999)
15. Wang, L., Zhu, J., Zou, H.: The doubly regularized support vector machine. Statistica Sinica 16, 589–615 (2006)
16. Dayhoff, M., Schwarts, R., Orcutt, B.: A model of evolutionary change in proteins. Atlas of Protein Sequences and Structures 5(suppl. 3), 345–352 (1978)
17. Sulimova, V., Mottl, V., Kulikowski, C., Muchnik, I.: Probabilistic evolutionary model for substitution matrices of PAM and BLOSUM families. DIMACS Technical Report 2008-16, Rutgers University, p. 17 (2008)

Cascading Discriminant and Generative Models for Protein Secondary Structure Prediction

Fabienne Thomarat, Fabien Lauer, and Yann Guermeur

LORIA – CNRS, INRIA, Université de Lorraine
Campus Scientifique, BP 239
54506 Vandœuvre-lès-Nancy Cedex, France
{Fabienne.Thomarat,Fabien.Lauer,Yann.Guermeur}@loria.fr

Abstract. Most of the state-of-the-art methods for protein seconday structure prediction are complex combinations of discriminant models. They apply a local approach of the prediction which is known to induce a limit on the expected prediction accuracy. A priori, the use of generative models should make it possible to overcome this limitation. However, among the numerous hidden Markov models which have been dedicated to this task over more than two decades, none has come close to providing comparable performance. A major reason for this phenomenon is provided by the nature of the relevant information. Indeed, it is well known that irrespective of the model implemented, the prediction should benefit significantly from the availability of evolutionary information. Currently, this knowledge is embedded in position-specific scoring matrices which cannot be processed easily with hidden Markov models. With this observation at hand, the next significant advance should come from making the best of the two approaches, i.e., using a generative model on top of discriminant models. This article introduces the first hybrid architecture of this kind with state-of-the-art performance. The conjunction of the two levels of treatment makes it possible to optimize the recognition rate both at the residue level and at the segment level.

Keywords: protein secondary structure prediction, discriminant models, class membership probabilities, hidden Markov models.

1 Introduction

With the multiplication of genome sequencing projects, the number of known protein sequences is growing exponentially. Knowing their (three-dimensional/tertiary) structure is a key in understanding their detailed function. Unfortunately, the experimental methods available to determine the structure, x-ray crystallography and nuclear magnetic resonance (NMR), are highly labor-intensive and do not ensure the production of the desired result (e.g., some proteins simply do not crystallize). As a consequence, the gap between the number of known protein sequences and the number of known protein structures is widening rapidly. To bridge this gap, one must resort to empirical inference. The prediction of protein structure from amino acid sequence, i.e., *ab initio*, is thus

T. Shibuya et al. (Eds.): PRIB 2012, LNBI 7632, pp. 166–177, 2012.

a hot topic in molecular biology. Due to its intrinsic difficulty, it is ordinarily tackled through a divide and conquer approach in which a critical first step is the prediction of the secondary structure, the local, regular structure defined by hydrogen bonds. Considered from the point of view of pattern recognition, this prediction is a three-category discrimination task consisting in assigning a conformational state α-helix (H), β-strand (E) or aperiodic/coil (C), to each residue (amino acid) of a sequence.

For almost half a century, many methods have been developed for protein secondary structure prediction. Since the pioneering work of Qian and Sejnowski [1], state-of-the-art methods are machine learning ones [2–5]. Furthermore, a majority of them shares the original architecture implemented by Qian and Sejnowski. Two sets of classifiers are used in cascade. The classifiers of the first set, named sequence-to-structure, take in input the content of a window sliding on the sequence, or the coding of a multiple alignment, to produce an initial prediction. Those of the second set, named structure-to-structure, take in input the content of a second window sliding on the initial predictions. The structure-to-structure classifiers act both as ensemble methods (combiners) and filters of the initial predictions. The goal of filtering is to increase the biological plausibility of the prediction by making use of the fact that the conformational states of consecutive residues are correlated. Other specifications can be incorporated in the combiners, such as the requirement to output indices of confidence in the prediction or, even better, class posterior probability estimates. Currently, the recognition rate of the best cascades is roughly 80%, depending on the details of the experimental protocol (see [5] for a survey). However, it is commonly admitted that their prediction accuracy faces a strong limiting factor: the fact that local information is not enough to specify utterly the structure. This limitation is only partly overcome by using recurrent neural networks [2]. A natural alternative consists in using generative models. The first hidden Markov model (HMM) [6] dedicated to protein secondary structure prediction was presented in [7]. Since then, new models have regularly been introduced, with the focus being laid on the derivation of an appropriate topology [8, 9]. However, their recognition rate has never exceeded 75% so far [9]. The main reason that can be put forward to explain this disappointing behavior rests in the fact that they are not well-suited to exploit evolutionary information under its standard form, i.e., a profile of multiple alignment, or more precisely a position-specific scoring matrix (PSSM) produced by PSI-BLAST [10]. A generative model that appears more promising to process PSSMs is the dynamic Bayesian network (DBN) [11]. However, the assessment of its potential is still in its infancy [5].

All these observations suggest the assessment of hybrid architectures cascading discriminant and generative models so as to combine the advantages of both approaches. This idea was popularized twenty years ago in the field of speech processing (see for instance [12]), and introduced more recently in bioinformatics, precisely for protein secondary structure predicition [13–15]. In short, discriminant models are used to compute class posterior probability estimates from which the emission probabilities of HMMs are derived, by application of Bayes' formula. This approach widens the context used for the prediction, and makes it possible to incorporate some pieces of information provided by the biologist, such as syntactic rules. So far, the best prediction method based on this kind of hybrid architecture was YASPIN [15], whose recognition rate is only slightly superior to 77%. In this article, we introduce a new hybrid architecture for protein secondary structure prediction. It is obtained by post-processing the outputs of the prediction method we have developed during the last few years, MSVMpred2 [16, 17], with an "inhomogeneous HMM" (IHMM) [18]. It exhibits state-of-the-art prediction accuracy both at the residue level and at the segment level. The gain of roughly 5% in recognition rate compared to YASPIN is partly due to the recent availability of a very large data set of proteins with known structure and low sequence identity: CM4675.

The organization of the paper is as follows. Section 2 summarizes the main characteristics of MSVMpred2. Section 3 introduces the whole hybrid architecture, and focuses on the features of the upper part of the hierarchy, i.e., the specification and implementation of the IHMM. Experimental results are reported in Section 4. At last, we draw conclusions and outline our ongoing research in Section 5.

2   MSVMpred2

MSVMpred2, the lower part of the hierarchy of treatments, is a cascade of discriminant models implementing the architecture introduced by Qian and Sejnowski. Its main specificities can be summarized as follows. First, the sequence-to-structure prediction is performed by dedicated classifiers. Second, the combiners at the structure-to-structure level are chosen so as to satisfy two requirements: they must output class posterior probability estimates and cover a wide range in terms of capacity. Third, capacity control at this level is implemented through a convex combination of the combiners (with the consequence that the global outputs of the cascade are also class posterior probability estimates). The topology of MSVMpred2 is depicted in Figure 1.

2.1   Sequence-to-Structure Prediction

We first characterize the descriptions (vectors of predictors) $x \in \mathcal{X}$ processed at this initial level of the prediction. The predictors are derived from PSSMs produced by PSI-BLAST. The sliding window is centered on the residue of interest. The description x_i processed at this level to predict the conformational state of the ith residue in the data set is thus obtained by appending rows of the PSSM associated with the sequence to which it belongs. Let n_1 be the integer such that $2n_1 + 1$ is the size of the sliding window. Then, the indices of these rows range from $i' - n_1$ to $i' + n_1$, where i' is the index of the residue of interest in its sequence. Since a PSSM has 20 columns, one per amino acid, this corresponds to $20(2n_1 + 1)$ predictors. More precisely, $\mathcal{X} \subset \mathbb{Z}^{20(2n_1+1)}$.

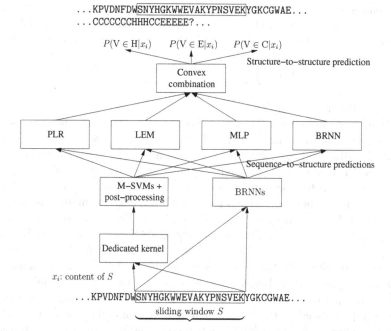

Fig. 1. Topology of MSVMpred2. The method computes estimates of the class posterior probabilities for the residue at the center of the sliding window S, here a valine (V).

Two kinds of classifiers are implemented at this level: multi-class support vector machines (M-SVMs) [19] and bidirectional recurrent neural networks (BRNNs) [20]. The kernel of the M-SVMs is an elliptic Gaussian kernel function applying a weighting on the predictors as a function of their position in the window. This weighting is learned by application of the principle of multi-class kernel target alignment [17]. The BRNNs are recurrent neural networks exploiting a context from both sides of the sequence processed. This makes them especially well-suited for the task at hand. Indeed, they obtain the highest prediction accuracy among all the neural networks assessed so far in protein secondary structure prediction [2, 14, 21]. Contrary to the BRNNs, the M-SVMs do not output class posterior probability estimates. In order to introduce homogeneity among the outputs of the different base classifiers, and more precisely ensure that they all belong to the probability simplex, the outputs of the M-SVMs are post-processed by the polytomous (multinomial) logistic regression (PLR) model [22].

2.2 Structure-to-Structure Prediction

We start with the characterization of the descriptions $z \in \mathcal{Z}$ processed at this level. Let N be the number of classifiers available to perform the sequence-to-structure prediction. The function computed by the jth of these classifiers (after the post-processing in the case of an M-SVM) is denoted $h^{(j)} = \left(h_k^{(j)} \right)_{1 \leqslant k \leqslant 3}$.

The second sliding window, of size $2n_2 + 1$, is also centered on the residue of interest. As a consequence, the description z_i processed by the combiners to estimate the probabilities associated with the ith residue in the data set is:

$$z_i = \left(h_k^{(j)} (x_{i+t}) \right)_{1\leqslant j\leqslant N, 1\leqslant k\leqslant 3, -n_2\leqslant t\leqslant n_2} \in U_2^{(2n_2+1)N},$$

where U_2 is the unit 2-simplex.

The four discriminant models used as structure-to-structure classifiers are the PLR, the linear ensemble method (LEM) [23], the multi-layer perceptron (MLP) [24] and the BRNN. They have been listed in order of increasing capacity [25]. Indeed, the PLR and the LEM are linear separators. An MLP using a softmax activation function for the output units and the cross-entropy loss (a sufficient condition for its outputs to be class posterior probability estimates) is an extension of the PLR obtained by adding a hidden layer. The boundaries it computes are nonlinear in its input space. At last, the BRNN can be seen roughly as an MLP operating on an extended description space. The availability of classifiers of different capacities for the second level of the cascade is an important feature of MSVMpred2. It makes it possible to cope with one of the main limiting factors to the performance of modular architectures: overfitting. The capacity control is implemented by the convex combination combining the four structure-to-structure classifiers. The behavior of this combination is predictable: it assigns high weights to the combiners of low complexity when the training set size is small (and the combiners of higher complexity tend to overfit the training set). On the contrary, due to the complexity of the problem, the latter combiners are favored when this size is large (see [17] for an illustration of the phenomenon).

3   Hybrid Prediction Method

In the field of biological sequence processing, the rationale for post-processing the outputs of discriminant models with generative models is two-fold: widening the context exploited for the prediction and incorporating high-level knowledge on the task of interest (mainly in the topology of the generative models). The generative model selected here to meet these goals is an IHMM with three states, one for each of the three conformational states. The advantage of this model compared to the standard HMM rests in the fact that its state transition probabilities are time dependent. This makes it possible to exploit a more suitable model of state durations, a necessary condition to get a high prediction accuracy at the conformational segment level. The global topology of the hierarchy is depicted in Figure 2.

For a given protein sequence, the final prediction is thus obtained by means of the dynamic programming algorithm computing the single best sequence of states (path), i.e., the variant of Viterbi's algorithm dedicated to the IHMM [18]. It must be borne in mind that this calls for a slight adaptation of the formulas, since MSVMpred2 provides estimates of the class posterior probabilities, rather

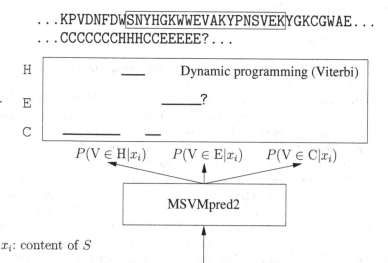

Fig. 2. Topology of the hybrid prediction method. The context available to perform the prediction exceeds that resulting from the combination of the two sliding windows of MSVMpred2.

than emission probabilities. Since the IHMM has exactly one state per conformational state, and the conformational state of each residue in the training set is known, the best state sequence is known for all the sequences of the training set. As a consequence, applying the maximum likelihood principle to derive the initial state distribution and the transition probabilities boils down to computing the corresponding frequencies on the training set.

Implementing a hybrid approach of the prediction is fully relevant only if the quality of the probability estimates computed by the discriminant models is high enough for the generative model to exploit them efficiently. Our hybrid architecture incorporates an optional treatment specifically introduced to address this issue: a basic post-processing of the outputs of MSVMpred2. This post-processing aims at constraining the final prediction so as to keep it in a vicinity of that of MSVMpred2. For each residue, the vector of probability estimates is replaced with a vector that is close to the binary coding of the predicted category. Precisely, given a small positive value ε, the highest of the class membership probability estimates is replaced with $1 - 2\varepsilon$, the two other estimates being replaced with ε. In this setting, the influence of the Viterbi algorithm on the path selected vanishes when ε goes to zero.

4   Experimental Results

4.1   Protein Data Sets

Our prediction method was assessed on two data sets. The first one is the well-known CB513 data set, fully described in [26], whose 513 sequences are made up of 84119 residues. The second one is the newly assembled CM4675 data set. It contains 4675 sequences, for a total of 851523 residues. The corresponding maximum pairwise percentage identity is 20%, i.e., it is low enough to meet the standard requirements of *ab initio* secondary structure prediction.

To generate the PSSMs, the version 2.2.25 of the BLAST package was used. Choosing BLAST in place of the more recent BLAST+ offers the facility to extract more precise PSSMs. Three iterations were performed against the NCBI nr database. The E-value inclusion threshold was set to 0.005 and the default scoring matrix (BLOSUM62) was used. The nr database, downloaded in May 2012, was filtered by pfilt [27] to remove low complexity regions, transmembrane spans and coiled coil regions. The initial secondary structure assignment was performed by the DSSP program [28], with the reduction from 8 to 3 conformational states following the CASP method, i.e., H+G $\rightarrow$ H (α-helix), E+B $\rightarrow$ E (β-strand), and all the other states in C (coil).

4.2   Experimental Protocol

The configuration chosen for MSVMpred2 includes the four main models of M-SVMs: the models of Weston and Watkins [29], Crammer and Singer [30], Lee, Lin, and Wahba [31], and the M-SVM^2 [32]. At the sequence-to-structure level, they are used in parallel with four BRNNs. The programs implementing the different M-SVMs are those of MSVMpack [33], while the 1D-BRNN package is used for the BRNNs. The sizes of the first and second sliding windows are respectively 13 and 15 ($n_1 = 6$ and $n_2 = 7$).

To assess the accuracy of our prediction method, we implemented a distinct experimental protocol for each of the data sets. The reason for this distinction was to take into account the difference in size of the two sets. For CB513, the protocol was basically the 7-fold cross-validation procedure already implemented in [16, 17] (with distinct training subsets for the sequence-to-structure level and the structure-to-structure level). At each step of the procedure, the values of the parameters of the IHMM that had to be inferred were derived using the whole training set. As for CM4675, it was simply split into the following independent subsets: a training set for the kernel of the M-SVMs (500 sequences, 98400 residues), a training set for the sequence-to-structure classifiers (2000 sequences, 369865 residues), a training set for the post-processing of the M-SVMs with a PLR (300 sequences, 52353 residues), a training set for the structure-to-structure classifiers (1000 sequences, 178244 residues), a training set for the convex combination (200 sequences, 34252 residues), and a test set (675 sequences, 118409 residues). Once more, the missing values of the parameters of the IHMM were derived using globally all the training subsets (4000 sequences, 733114 residues).

It can be inferred from the introduction that a secondary structure prediction method must fulfill different requirements in order to be useful for the biologist. Thus, several standard measures giving complementary indications must be used to assess the prediction accuracy [34]. We consider the three most popular ones: the recognition rate Q_3, Pearson-Matthews correlation coefficients $C_{\alpha/\beta/\text{coil}}$, and the segment overlap measure (Sov) in its most recent version (Sov'99).

4.3   Results

The experimental results obtained with MSVMpred2 and the two variants of the hybrid model (with and without post-processing of the outputs of MSVMpred2) are reported in Table 1.

Table 1. Prediction accuracy of MSVMpred2 and the hybrid model on CB513 and CM4675. Results in the last row were obtained with the optional treatment of the outputs of MSVMpred2 described in Section 3.

Method	CB513					CM4675				
	Q_3 (%)	Sov	C_α	C_β	C_{coil}	Q_3 (%)	Sov	C_α	C_β	C_{coil}
MSVMpred2	78.3	74.4	0.74	0.64	0.60	81.8	78.9	0.79	0.73	0.65
Hybrid	77.3	73.1	0.74	0.64	0.57	80.8	77.5	0.78	0.71	0.63
Hybrid ($\varepsilon = 0.01$)	78.3	**75.5**	0.74	0.64	0.60	81.8	**80.0**	0.79	0.73	0.65

In applying the two sample proportion test (the one for large samples), one can notice than even when using CB513, the superiority of MSVMpred2 over YASPIN appears statistically significant with confidence exceeding 0.95. Of course, such a statement is to be tempered since the figures available for YASPIN correspond to a different set of protein sequences. The recognition rate is always significantly above 80% when CM4675 is used. This highlights a fact already noticed in [5]: the complexity of the problem calls for the development of complex modular prediction methods such as ours. The feasibility of their implementation increases with the growth of the protein data sets available. The hybrid method is only superior to MSVMpred2 when it implements the post-processing described in Section 3. In that case, the gain is emphasized, as expected, by means of the Sov. This measure increases by the same amount (1.1 point) on both data sets.

The value of ε for which the results of the last row of Table 1 were obtained is a favorable one. We now present a short study of the prediction accuracy as a function of this parameter. Figures 3 and 4 illustrate the main phenomena observed.

The gain in Sov induced by the introduction of the IHMM can be obtained for ε varying is a relatively large interval (the precise boundaries depend on the data set chosen). In addition, to ensure that this gain is not balanced by a decrease of the Q_3, it suffices to choose a small enough value. This implies that the

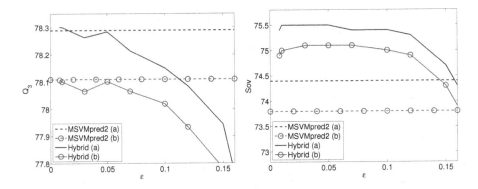

Fig. 3. Prediction accuracy on CB513 in terms of Q_3 (left) and Sov (right) as a function of the value of ε for MSVMpred2 (alone) and the hybrid model. In both cases, two variants of MSVMpred2 are considered: the one specified in Section 4.2 (a) and a simplified one including a single structure-to-structure classifier (b).

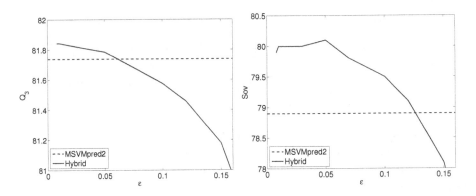

Fig. 4. Prediction accuracy on CM4675 in terms of Q_3 (left) and Sov (right) as a function of the value of ε for MSVMpred2 (alone) and the hybrid model

selection of an appropriate value for ε does not raise particular difficulties. Figure 3 also displays the results obtained with a simplified variant of MSVMpred2. By focusing on the difference between these results and those obtained with the standard MSVMpred2, we also see that the performance of the hybrid model is positively correlated with the prediction accuracy of the classifier providing the class membership probability estimates.

To sum up, these experiments support the thesis that hybrid models can be used to increase the performance of secondary structure prediction methods (at least at the level of the structural elements), while directly benefiting from improvements of the latters. The negative aspect is that we have not yet been able to make full use of the values of the class posterior probability estimates provided by MSVMpred2.

5 Conclusions and Ongoing Research

This article has introduced a new method for protein secondary structure prediction. This method, based on a hierarchical architecture obtained by cascading MSVMpred2 and an IHMM, is the first hybrid model exhibiting state-of-the-art performance. It takes benefit of the availability of a very large data set of proteins with known structure and low sequence identity: CM4675. So far, the main improvement resulting from introducing the generative model is an increase of the Sov'99 measure, i.e., an improvement of the prediction accuracy at the segment level. This should prove especially useful for the biologist using our method as an intermediate step of a tertiary structure prediction.

Obviously, there are many options one can think of to improve our method. A simple one consists in taking benefit of its flexibility to integrate knowledge sources and modules borrowed from the literature [4, 5]. The researcher in machine learning should be primarily interested in the following question: can we expect the class posterior probability estimates produced by MSVMpred2 to become accurate enough to be exploitable as is by the generative model? Answering this fundamental question is currently our main goal.

Acknowledgements. The authors would like to thank C. Magnan for providing them with the CM4675 data set and the 1D-BRNN package.

References

1. Qian, N., Sejnowski, T.J.: Predicting the secondary structure of globular proteins using neural network models. Journal of Molecular Biology 202, 865–884 (1988)
2. Pollastri, G., Przybylski, D., Rost, B., Baldi, P.: Improving the prediction of protein secondary structure in three and eight classes using recurrent neural networks and profiles. Proteins 47, 228–235 (2002)
3. Cole, C., Barber, J.D., Barton, G.J.: The Jpred 3 secondary structure prediction server. Nucleic Acids Research 36, W197–W201 (2008)
4. Kountouris, P., Hirst, J.D.: Prediction of backbone dihedral angles and protein secondary structure using support vector machines. BMC Bioinformatics 10, 437 (2009)
5. Aydin, Z., Singh, A., Bilmes, J., Noble, W.S.: Learning sparse models for a dynamic Bayesian network classifier of protein secondary structure. BMC Bioinformatics 12, 154 (2011)
6. Rabiner, L.R.: A tutorial on hidden Markov models and selected applications in speech recognition. Proceedings of the IEEE 77, 257–286 (1989)
7. Asai, K., Hayamizu, S., Handa, K.: Prediction of protein secondary structure by the hidden Markov model. CABIOS 9, 141–146 (1993)
8. Martin, J., Gibrat, J.-F., Rodolphe, F.: Analysis of an optimal hidden Markov model for secondary structure prediction. BMC Structural Biology 6, 25 (2006)
9. Won, K.-J., Hamelryck, T., Prügel-Bennett, A., Krogh, A.: An evolutionary method for learning HMM structure: prediction of protein secondary structure. BMC Bioinformatics 8, 357 (2007)

10. Altschul, S.F., Madden, T.L., Schäffer, A.A., Zhang, J., Zhang, Z., Miller, W., Lipman, D.J.: Gapped BLAST and PSI-BLAST: a new generation of protein database search programs. Nucleic Acids Research 25, 3389–3402 (1997)
11. Yao, X.-Q., Zhu, H., She, Z.-S.: A dynamic Bayesian network approach to protein secondary structure prediction. BMC Bioinformatics 9, 49 (2008)
12. Krogh, A., Riis, S.K.: Hidden neural networks. Neural Computation 11, 541–563 (1999)
13. Guermeur, Y.: Combining discriminant models with new multi-class SVMs. Pattern Analysis and Applications 5, 168–179 (2002)
14. Guermeur, Y., Pollastri, G., Elisseeff, A., Zelus, D., Paugam-Moisy, H., Baldi, P.: Combining protein secondary structure prediction models with ensemble methods of optimal complexity. Neurocomputing 56, 305–327 (2004)
15. Lin, K., Simossis, V.A., Taylor, W.R., Heringa, J.: A simple and fast secondary structure prediction method using hidden neural networks. Bioinformatics 21, 152–159 (2005)
16. Guermeur, Y., Thomarat, F.: Estimating the Class Posterior Probabilities in Protein Secondary Structure Prediction. In: Loog, M., Wessels, L., Reinders, M.J.T., de Ridder, D. (eds.) PRIB 2011. LNCS (LNBI), vol. 7036, pp. 260–271. Springer, Heidelberg (2011)
17. Bonidal, R., Thomarat, F., Guermeur, Y.: Estimating the class posterior probabilities in biological sequence segmentation. In: SMTDA 2012 (2012)
18. Ramesh, P., Wilpon, J.G.: Modeling state durations in hidden Markov models for automatic speech recognition. In: ICASSP 1992, pp. 381–384 (1992)
19. Guermeur, Y.: A generic model of multi-class support vector machine. International Journal of Intelligent Information and Database Systems (in press, 2012)
20. Baldi, P., Brunak, S., Frasconi, P., Soda, G., Pollastri, G.: Exploiting the past and the future in protein secondary structure prediction. Bioinformatics 15, 937–946 (1999)
21. Chen, J., Chaudhari, N.S.: Cascaded bidirectional recurrent neural networks for protein secondary structure prediction. IEEE/ACM Transactions on Computational Biology and Bioinfomatics 4, 572–582 (2007)
22. Hosmer, D.W., Lemeshow, S.: Applied Logistic Regression. Wiley, London (1989)
23. Guermeur, Y.: Combining multi-class SVMs with linear ensemble methods that estimate the class posterior probabilities. Communications in Statistics (submitted)
24. Anthony, M., Bartlett, P.L.: Neural Network Learning: Theoretical Foundations. Cambridge University Press, Cambridge (1999)
25. Guermeur, Y.: VC theory of large margin multi-category classifiers. Journal of Machine Learning Research 8, 2551–2594 (2007)
26. Cuff, J.A., Barton, G.J.: Evaluation and improvement of multiple sequence methods for protein secondary structure prediction. Proteins 34, 508–519 (1999)
27. Jones, D.T., Swindells, M.B.: Getting the most from PSI-BLAST. Trends in Biochemical Sciences 27, 161–164 (2002)
28. Kabsch, W., Sander, C.: Dictionary of protein secondary structure: pattern recognition of hydrogen-bonded and geometrical features. Biopolymers 22, 2577–2637 (1983)
29. Weston, J., Watkins, C.: Multi-class support vector machines. Technical Report CSD-TR-98-04, Royal Holloway, University of London, Department of Computer Science (1998)

30. Crammer, K., Singer, Y.: On the algorithmic implementation of multiclass kernel-based vector machines. Journal of Machine Learning Research 2, 265–292 (2001)
31. Lee, Y., Lin, Y., Wahba, G.: Multicategory support vector machines: Theory and application to the classification of microarray data and satellite radiance data. Journal of the American Statistical Association 99, 67–81 (2004)
32. Guermeur, Y., Monfrini, E.: A quadratic loss multi-class SVM for which a radius-margin bound applies. Informatica 22, 73–96 (2011)
33. Lauer, F., Guermeur, Y.: MSVMpack: a multi-class support vector machine package. Journal of Machine Learning Research 12, 2293–2296 (2011)
34. Baldi, P., Brunak, S., Chauvin, Y., Andersen, C.A.F., Nielsen, H.: Assessing the accuracy of prediction algorithms for classification: an overview. Bioinformatics 16, 412–424 (2000)

Improvement of the Protein–Protein Docking Prediction by Introducing a Simple Hydrophobic Interaction Model: An Application to Interaction Pathway Analysis

Masahito Ohue[1,2], Yuri Matsuzaki[1], Takashi Ishida[1], and Yutaka Akiyama[1]

[1] Graduate School of Information Science and Engineering,
Tokyo Institute of Technology, Tokyo, Japan
[2] Research Fellow of the Japan Society for the Promotion of Science
{ohue,y_matsuzaki,t.ishida}@bi.cs.titech.ac.jp,
akiyama@cs.titech.ac.jp

Abstract. We propose a new hydrophobic interaction model that applies atomic contact energy for our protein–protein docking software, MEGADOCK. Previously, this software used only two score terms, shape complementarity and electrostatic interaction. We develop a modified score function incorporating the hydrophobic interaction effect. Using the proposed score function, MEGADOCK can calculate three physicochemical effects with only one correlation function. We evaluate the proposed system against three other protein–protein docking score models, and we confirm that our method displays better performance than the original MEGADOCK system and is faster than both ZDOCK systems. Thus, we successfully improve accuracy without loosing speed.

Keywords: Protein–Protein Docking, MEGADOCK, Hydrophobic Interaction, Fast Fourier Transform, Protein–Protein Interaction.

1 Introduction

Proteins play a key role in virtually all biological events that take place within and between cells. Many proteins display their biological functions by binding to a specific partner protein at a specific site. Determining the structure of a given complex is one of the most important challenges in molecular biophysical research [1, 2]. In addition, the number of protein 3-D structures stored in the Protein Data Bank (PDB) [3] is currently increasing, allowing protein–protein interactions and complex structures to be connected using computational prediction methods, known as the 3-D interactome concept [4]. Against this background, there has been considerable research on protein–protein docking, which is the computational prediction of protein complex structures.

The goal of protein–protein docking is to determine the protein complex structure in atomic detail, starting from the coordinates of the unbound component molecules. Most current docking methods start with rigid-body docking,

T. Shibuya et al. (Eds.): PRIB 2012, LNBI 7632, pp. 178–187, 2012.

which generates a large number of docked conformations (called "decoys") with good surface complementarity. One of the major methods of simulating protein–protein docking is the Katchalski-Katzir algorithm [5], using a 3-D grid representation and fast Fourier transform (FFT) correlation approach. In the Katchalski-Katzir algorithm, the pseudo interaction energy score (called the docking score) between a receptor protein and a ligand protein is calculated by FFT and inverse FFT (IFFT) using a correlation of two discrete functions, as follows:

$$S(t) = \sum_{v \in \mathbb{N}^3} R(v)L(v + t) \tag{1}$$

$$= \mathrm{IFFT}[\mathrm{FFT}[R(v)]^* \mathrm{FFT}[L(v)]], \tag{2}$$

where R and L are the discrete score function of the Receptor and Ligand proteins, v is a coordinate in a 3-D grid space $\mathbb{N}^3$, and t is the parallel translation vector of the ligand protein. In order to find the best docking poses, possible ligand orientations are exhaustively examined at n_θ rotation angles for a given stepsize θ. For each rotation, the ligand protein is translated into $N \times N \times N$ patterns in the $\mathbb{N}^3$ grid space (where $N = |\mathbb{N}|$ is the grid size in each dimension). The decoy that yields the highest value of S for each rotation is recorded. In this manner, a total of $n_\theta \times N^3$ docking poses are evaluated for one protein pair. To directly execute the simple convolution sums in eq. (1), $\mathcal{O}(N^6)$ calculations are required; however, this is reduced to $\mathcal{O}(N^3 \log N)$ using the FFT in eq. (2).

There are a number of software packages using the Katchalski-Katzir algorithm [6–12]. Among them, ZDOCK [11, 12] is a widely used protein–protein docking software [13–15]. ZDOCK uses the original docking scores, which are accurate compared to other software. However, this requires two or more correlation function calculations, with a correspondingly large calculation time. Therefore, it is unrealistic to use ZDOCK in a situation where many docking calculations are needed, e.g., when aimed at predictions of a protein–protein interaction network [16–19] or an ensemble/cross-docking performing an all-to-all docking [20–22].

Our protein–protein docking software, MEGADOCK [23, 24], also uses the Katchalski-Katzir algorithm. By employing an original shape complementarity score function (called rPSC) and a general electrostatic interaction score model, MEGADOCK can calculate the docking score with only one correlation function, and thus exhibits quicker calculation times than ZDOCK. Accordingly, the docking prediction accuracy of MEGADOCK is lower than that of ZDOCK. ZDOCK calculates three physico-chemical effects: shape complementarity, electrostatics, and an empirical potential-based desolvation free energy as a hydrophobic effect, with two or more correlation functions. To improve the docking accuracy of MEGADOCK, we intend to incorporate a hydrophobic interaction effect to our scoring model. However, using the conventional score model employed by ZDOCK would cause an increase in the number of correlation functions to be calculated. Therefore, we need a new score model to make MEGADOCK suitable for varied applications.

In this study, we introduce a hydrophobic interaction effect to MEGADOCK. In particular, looking ahead to the application of an interaction network prediction, which is the final goal of MEGADOCK, we develop a simple hydrophobic interaction model that considers only the receptor protein. This increases the performance of the docking calculation without any detrimental effect on the speed.

2   Materials and Methods

2.1   Previous Score Model

In this subsection, we briefly explain our previously developed docking software, MEGADOCK version 2.5. MEGADOCK 2.5 uses a docking score function that combines two terms: the real Pairwise Shape Complementarity (rPSC) score term and the electrostatics (ELEC) score term, which is defined based on the FTDock force model [6] and the CHARMM19 atomic charge [25]. Each pair of proteins is first allocated a position on the 3-D grid space $\mathbb{N}^3$, which has a grid step size of 1.2 Å. Scores are then assigned to each voxel $v \in \mathbb{N}^3$ according to the location in the protein, such as surface or core.

The rPSC term is defined as follows:

$$\text{rPSC}(t) = \sum_{v \in \mathbb{N}^3} G_R(v) G_L(v + t),$$

$$G_R(v) = \begin{cases} \# \text{ of receptor atoms within } (3.6 \text{ Å} + r_{\text{vdW}}) & \text{(open space)} \\ -27 & \text{(inside of the receptor)}, \end{cases} \quad (3)$$

$$G_L(v) = \begin{cases} 1 & \text{(solvent excluding surface layer of the ligand)} \\ 2 & \text{(core of the ligand)}, \end{cases} \quad (4)$$

where G_R and G_L represent the rPSC grid value of the receptor/ligand proteins, r_{vdW} represents the van der Waals atomic radius, and t is the ligand translation vector. We omitted the zero value domain.

The ELEC term from FTDock potential is represented as the electric field $\varphi(i)$. $\varphi(i)$ is assigned to each voxel $i \in \mathbb{N}^3$ as follows:

$$\varphi(i) = \sum_{j \in \mathbb{N}^3} \frac{q(j)}{\varepsilon(r_{ij}) r_{ij}}, \quad \varepsilon(r) = \begin{cases} 4 & (r \leq 6 \text{ Å}) \\ 38r - 224 & (6 \text{ Å} < r < 8 \text{ Å}) \\ 80 & (8 \text{ Å} \leq r), \end{cases}$$

where $q(j)$ is the charge at grid point $j \in \mathbb{N}^3$, r_{ij} is the Euclid distance between grid points i and j, and $\varepsilon(r)$ is a distance-dependent dielectric function. ELEC term is defined as follows:

$$\text{ELEC}(t) = \sum_{v \in \mathbb{N}^3} E_R(v) E_L(v + t),$$

$$E_R(v) = \varphi(v) \quad \text{(open space)},$$

$$E_L(v) = q(v),$$

Table 1. Non-pairwise ACE scores. This table is reproduced from Table 1 of [26] in which Zhang, *et al.* defined the atom types and assigned ACE scores.

atom type	N	C^α	C	O	GC^α	C^β	KN^ζ	KC^δ	DO^δ
ACE score	−0.495	−0.553	−0.464	−0.079	0.008	−0.353	1.334	1.046	0.933
atom type	RN^η	NN^δ	RN^ε	SO^γ	HN^ε	YC^ζ	FC^ζ	LC^δ	CS^γ
ACE score	0.726	0.693	0.606	0.232	0.061	−0.289	−0.432	−0.987	−1.827

where E_R and E_L represent the ELEC grid values of receptor/ligand proteins, determined according to the charge of each voxel $q(v)$ in which atoms in the residues are assigned a charge according to CHARMM19.

Considering these two terms, the docking score $S(t)$ is represented as:

$$R(v) = G_R(v) + iE_R(v),$$
$$L(v) = G_L(v) + iw_eE_L(v),$$
$$S(t) = \Re\left[\sum_{v\in N^3} R(v)L(v+t)\right] = \text{rPSC}(t) - w_e\text{ELEC}(t),$$

where w_e is the weight parameter of ELEC term.

2.2 Proposed Method

In our proposed method, we used a non-pairwise-type atomic contact energy (ACE) score [26] to incorporate a hydrophobic interaction effect. For the current study, we introduce a simple model that considers only the receptor protein because, when both the receptor and ligand are taken into consideration, an increase in the number of correlation functions is unavoidable.

We modify the receptor rPSC value G_R in eq. (3) in order to introduce the ACE score. The new receptor value G'_R is defined as follows:

$$G'_R(v) = G_R(v) + w_hH_R(v),$$

$$H_R(v) = \begin{cases} \text{sum of ACE scores of receptor atoms} \\ \qquad\qquad\text{within } (3.6\text{ Å} + r_{\text{vdW}}) \text{ (open space)} \\ 0 \quad \text{(inside of the receptor)}, \end{cases}$$

where w_h is the weight parameter of H_R. Fig. 1 shows a pattern diagram of the proposed model. We use the ACE values given in Table 1.

This score model attains a value of $G_R(v) + w_hH_R(v)$ when the open space near the receptor surface is superposed on the ligand surface. The score of a ligand core of 2 depends on the penalty (−54) at the time of a core collision for enlargement. It is assumed that $2 \times \{G_R(v) + w_hH_R(v)\}$ will be obtained by the ligand core, depending on its position, under a situation where the core moves into a pocket that can obtain a high score, because a penalty (−27) is imposed on any collision between the ligand surface and a receptor. Therefore, we do not

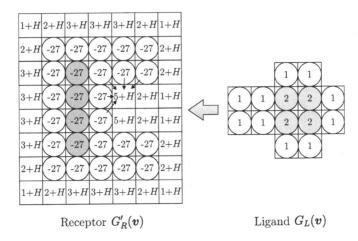

Receptor $G'_R(v)$ Ligand $G_L(v)$

Fig. 1. Proposed scoring model $G'_R(v)$ and $G_L(v)$. The model consists of 3-D grid, but here we show only two dimensions for simplicity. For clarity, grid points with a value of 0 have been omitted. Small arrows indicate the five atoms that are within the cutoff distance of a grid, and thus contribute to its score of $5 + H$, where H means $w_h H_R(v)$.

consider this situation to affect the good docking pose of the decoy. ZDOCK 2.3 [11] uses two correlation functions, and ZDOCK 3.0 [12] uses eight correlation functions to consider three effects—shape complementarity, electrostatics, and desolvation free energy—our score model can calculate docking scores under consideration of three effects with only one correlation function, while maintaining an advantage in terms of calculation speed.

2.3   Dataset

The protein complex structures used in this study were retrieved from a standard protein–protein docking benchmark set [27], containing 176 known 3-D structures of complex component proteins in both bound and unbound forms.

2.4   Evaluation of Docking Performance

To evaluate the docking pose prediction performance, we conducted a re-docking and unbound docking experiment using the benchmark dataset. We used the root mean square deviation (RMSD) of the ligand (L-RMSD), which is the RMSD of the predicted ligand position and that of the crystal complex structure calculated for all the atoms when the receptor positions are superimposed, in order to determine the accuracy of the docking predictions. The RMSDs of the unbound structures were only calculated for residues that were aligned by pairwise alignment of the amino acid sequences between the bound and unbound structures. We defined a "near-native decoy" as that for which L-RMSD was less than or equal to 5 Å. We compared the performance of the following docking methods: the proposed method, MEGADOCK 2.5, ZDOCK 2.3, and ZDOCK 3.0. For comparison with ZDOCK, we set parameters of 3,600 decoys per case and $\theta = 15°$ for the ligand rotation step. We compared the following widely used two values [1, 11, 12] to determine the docking performance:

- **Average Hit Count:** The average number of near-native decoys across the set of cases for a given number of top-ranked predictions per test case.
- **Success Rate:** The percentage of cases with near-native decoys for a given number of top-ranked predictions per test case.

3 Results and Discussions

3.1 Optimization of Weight Parameters

For determining parameter values w_e and w_h, we used only the bound dataset to avoid overfitting the unbound structures. We optimized the parameters for maximizing the Success Rate of 100 predictions. We searched the best combination of w_e and w_h, and tested w_e from 0.5 to 1.5 by 0.05 steps and w_h from 0.1 to 2.0 by 0.1 steps. As a result, we found the best values of $w_e = 1.15$ and $w_h = 0.6$.

3.2 Docking Prediction Accuracy

The Average Hit Count is shown in Fig. 2 since bound dataset was used for optimization of weight parameters, the results of unbound dataset are more important than bound dataset. We can see that our proposed method performed better than MEGADOCK 2.5 with both the bound and unbound sets. In addition, the proposed method displays an equivalent performance to ZDOCK 2.3 for the unbound set and is broadly similar for the bound set. However, our method is still less accurate than ZDOCK 3.0 for both sets. The performance of ZDOCK 3.0 is mainly due to its pairwise potential function, although this performance is obtained at the expense of calculation speed.

A similar trend is observed in the Success Rate of each method, as shown in Fig. 3. We see that the Success Rate of our proposed method is again better than that of MEGADOCK 2.5 for both sets. However, our proposed method is noticeably worse than ZDOCK 2.3 for the bound set. We think that G_R and H_R require further tuning using more complex structures in the PDB.

3.3 Calculation Time

Table 2 shows the average computation time for the benchmark dataset. All the calculations were conducted on the TSUBAME 2.0 supercomputing system, Tokyo Institute of Technology, Japan, which consists of two Intel Xeon 2.93 GHz (6 cores × 2) processors and 32 GB RAM, with operational nodes connected via an InfiniBand and Gigabit Ethernet. An average of 14.2 min was required for

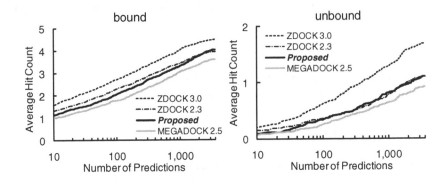

Fig. 2. Average Hit Count for all test cases of benchmark dataset. The Average Hit Count was defined as the average number of near-native decoys across the set of cases for a given number of top-ranked predictions per test case.

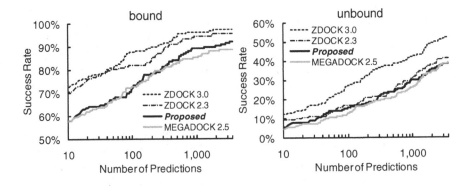

Fig. 3. Success Rate for all test cases of benchmark dataset. The Success Rate was defined as the percentage of cases with near-native decoys for a given number of top-ranked docking predictions per test case.

Table 2. Total time for 176 docking calculations using the benchmark dataset

	Proposed	MEGADOCK 2.5	ZDOCK 2.3	ZDOCK 3.0
time (hr)	41.7	41.6	157.3	365.6
speedup from ZDOCK 2.3	3.77	3.78	(1.0)	0.43
speedup from ZDOCK 3.0	8.77	8.79	2.32	(1.0)

each docking calculation using one CPU core. The proposed method obtained the almost same calculation speed as MEGADOCK 2.5 (only 0.7% of calculation time increase), some 3.8 times faster than ZDOCK 2.3 and 8.8 times faster than ZDOCK 3.0. Since FFT takes most of the execution time of MEGADOCK and the proposed method, if we increase the correlation function to 2 or 3 to get better performance of docking, calculation time will also increase 2- or 3-fold.

3.4 Application to Pathway Analysis

We also performed a case study using a biological interaction network by applying our proposed docking method to the protein–protein interaction prediction problem of bacterial chemotaxis pathways, which represents a typical target of signal transduction in the field of systems biology [28]. Docking and protein–protein interaction prediction were undertaken for $101 \times 101 = 10{,}201$ pairs corresponding to the constituent protein data of the 13 protein species present in the chemotaxis pathway [17].

We used the method of Matsuzaki *et al.* [17], with the improved MEGADOCK in place of ZDOCK 3.0. The docking score of 101×101 combinations was calculated for 101 protein structures and their affinity scores based on the literature [17]. We obtained an F-measure of 0.45 for this system, which is similar to that found in the previous study using ZDOCK 3.0 (F-measure of 0.49).

4   Conclusion

In this study, we added a hydrophobic interaction model to the protein docking software MEGADOCK. This additional component, which considers only the receptor protein, was combined with the considerations of shape complementarity and electrostatic interaction without increasing the calculation time. The proposed method succeeded in achieving the better level of accuracy as previous MEGADOCK. Although we need more better level of accuracy in bound cases, the proposed method achieved the same level of accuracy as ZDOCK 2.3 in unbound cases. It was also 3.8 times faster than ZDOCK 2.3 and 8.8 times faster than ZDOCK 3.0. However, to enhance the accuracy of the proposed model, further tuning of some system parameters is necessary in future. ACE was introduced only into the receptor side in the study because receptor term of rPSC was easy of introducing some atomic effects. We are attempting to develop a new score model with both receptor and ligand ACE term using only one correlation function. Additionally, we will apply our method to other large analyses, such as the interaction network prediction problem of other biological systems or the cross-docking of ensemble structures.

Acknowledgements. This work was supported in part by a Grant-in-Aid for JSPS Fellows (23·8750), a Grant-in-Aid for Research and Development of The Next-Generation Integrated Life Simulation Software, and Education Academy of Computational Life Sciences, all from the Ministry of Education, Culture, Sports, Science and Technology of Japan (MEXT).

References

1. Pons, C., Grosdidier, S., Solernou, A., Pérez-Cano, L., Fernández-Recio, J.: Present and future challenges and limitations in protein–protein docking. Proteins 78(1), 95–108 (2010)
2. Wass, M.N., David, A., Sternberg, M.J.E.: Challenges for the prediction of macromolecular interactions. Curr. Opin. Struct. Biol. 21(3), 382–390 (2011)
3. Berman, H.M., Westbrook, J., Feng, Z., Gilliland, G., Bhat, T.N., et al.: The Protein Data Bank. Nucleic Acids Res. 28(1), 235–242 (2000)
4. Stein, A., Mosca, R., Aloy, P.: Three-dimensional modeling of protein interactions and complexes is going 'omics. Curr. Opin. Struct. Biol. 21(2), 200–208 (2011)
5. Katchalski-Katzir, E., Shariv, I., Eisenstein, M., Friesem, A.A., Aflalo, C., et al.: Molecular surface recognition: Determination of geometric fit between proteins and their ligands by correlation techniques. Proc. Natl. Acad. Sci. USA 89, 2195–2199 (1992)
6. Gabb, H.A., Jackson, R.M., Sternberg, M.J.E.: Modelling protein docking using shape complementarity, electrostatics and biochemical information. J. Mol. Biol. 272(1), 106–120 (1997)
7. Vakser, I.A.: Evaluation of GRAMM low-resolution docking methodology on the hemagglutinin-antibody complex. Proteins (suppl. 1), 226–230 (1997)
8. Mandell, J.G., Roberts, V.A., Pique, M.E., Kotlovyi, V., Mitchell, J.C., et al.: Protein docking using continuum electrostatics and geometric fit. Protein Eng. 14(2), 105–113 (2001)
9. Cheng, T.M.-K., Blundell, T.L., Fernández-Recio, J.: pyDock: electrostatics and desolvation for effective scoring of rigid-body protein–protein docking. Proteins 68(2), 503–515 (2007)
10. Kozakov, D., Brenke, R., Comeau, S.R., Vajda, S.: PIPER: an FFT-based protein docking program with pairwise potentials. Proteins 65(2), 392–406 (2006)
11. Chen, R., Li, L., Weng, Z.: ZDOCK: an initial-stage protein-docking algorithm. Proteins 52(1), 80–87 (2003)
12. Mintseris, J., Pierce, B., Wiehe, K., Anderson, R., Chen, R., et al.: Integrating statistical pair potentials into protein complex prediction. Proteins 69(3), 511–520 (2007)
13. Hwang, H., Vreven, T., Pierce, B.G., Hung, J.-H., Weng, Z.: Performance of ZDOCK and ZRANK in CAPRI rounds 13–19. Proteins 78(15), 3104–3110 (2010)
14. Uchikoga, N., Hirokawa, T.: Analysis of protein–protein docking decoys using interaction fingerprints: application to the reconstruction of CaM-ligand complexes. BMC Bioinformatics 11(236) (2010)

15. Fleishman, S.J., Whitehead, T.A., Strauch, E.-M., Corn, J.E., Qin, S., et al.: Community-wide assessment of protein-interface modeling suggests improvements to design methodology. J. Mol. Biol. 414(2), 289–302 (2011)
16. Wass, M.N., Fuentes, G., Pons, C., Pazos, F., Valencia, A.: Towards the prediction of protein interaction partners using physical docking. Mol. Syst. Biol. 7(469) (2011)
17. Matsuzaki, Y., Matsuzaki, Y., Sato, T., Akiyama, Y.: In silico screening of protein–protein interactions with all-to-all rigid docking and clustering: an application to pathway analysis. J. Bioinform. Comput. Biol. 7(6), 991–1012 (2009)
18. Tsukamoto, K., Yoshikawa, T., Hourai, Y., Fukui, K., Akiyama, Y.: Development of an affinity evaluation and prediction system by using the shape complementarity characteristic between proteins. J. Bioinform. Comput. Biol. 6(6), 1133–1156 (2008)
19. Yoshikawa, T., Tsukamoto, K., Hourai, Y., Fukui, K.: Improving the accuracy of an affinity prediction method by using statistics on shape complementarity between proteins. J. Chem. Inf. Model. 49(3), 693–703 (2009)
20. Chaleil, R.A.G., Tournier, A.L., Bates, P.A., Kro, M.: Implicit flexibility in protein docking: Cross-docking and local refinement. Proteins 69(4), 750–757 (2007)
21. Dobbins, S.E., Lesk, V.I., Sternberg, M.J.E.: Insights into protein flexibility: The relationship between normal modes and conformational change upon protein–protein docking. Proc. Natl. Acad. Sci. USA 105(30), 10390–10395 (2008)
22. Venkatraman, V., Ritchie, D.W.: Flexible protein docking refinement using posedependent normal mode analysis. Proteins 80(9), 2262–2274 (2012)
23. Ohue, M., Matsuzaki, Y., Matsuzaki, Y., Sato, T., Akiyama, Y.: MEGADOCK: an all-to-all protein–protein interaction prediction system using tertiary structure data and its application to systems biology study. IPSJ TOM 3(3), 91–106 (2010) (in Japanese)
24. Ohue, M., Matsuzaki, Y., Akiyama, Y.: Docking-calculation-based method for predicting protein-RNA interactions. Genome Inform. 25(1), 25–39 (2011)
25. Reiher III, W.H.: Theoretical studies of hydrogen bonding. Ph.D. Thesis at Harvard University (1985)
26. Zhang, C., Vasmatzis, G., Cornette, J.L., DeLisi, C.: Determination of atomic desolvation energies from the structures of crystallized proteins. J. Mol. Biol. 267(3), 707–726 (1997)
27. Hwang, H., Vreven, T., Janin, J., Weng, Z.: Protein–protein docking benchmark version 4.0. Proteins 78(15), 3111–3114 (2010)
28. Baker, M.D., Wolanin, P.M., Stock, J.B.: Systems biology of bacterial chemotaxis. Curr. Opin. Microbiol. 9(2), 187–192 (2006)

Representation of Protein Secondary Structure Using Bond-Orientational Order Parameters

Cem Meydan and Osman Ugur Sezerman

Sabanci University, Biological Sciences & Bioengineering Dept., Istanbul, Turkey
{cemmeydan,ugur}@sabanciuniv.edu

Abstract. Structural studies of proteins for motif mining and other pattern recognition techniques require the abstraction of the structure into simpler elements for robust matching. In this study, we propose the use of bond-orientational order parameters, a well-established metric usually employed to compare atom packing in crystals and liquids. Creating a vector of orientational order parameters of residue centers in a sliding window fashion provides us with a descriptor of local structure and connectivity around each residue that is easy to calculate and compare. To test whether this representation is feasible and applicable to protein structures, we tried to predict the secondary structure of protein segments from those descriptors, resulting in 0.99 AUC (area under the ROC curve). Clustering those descriptors to 6 clusters also yield 0.93 AUC, showing that these descriptors can be used to capture and distinguish local structural information.

Keywords: bond-orientational order, secondary structure, machine learning, structural alphabet.

1 Introduction

In analysis protein structures, different models of representations on various levels of structural details are used. From coarse-grained to all-atom models, simplified lattice to continuous representations, each model can be used in different areas of research.

The need for abstraction in computational methods (such as structure search and comparison, fold matching, structural motif mining and other areas of pattern recognition) is especially high. The very high amount of data and precision in the 3D coordinates makes computational analysis very complex and very rigid in its applicability. Simplified models capture relevant information and hide unimportant details through abstraction, conferring the ability to group complex 3D information into manageable clusters that can be searched for, compared and "learned" by machine-learning algorithms in a flexible fashion.

The most common simplified representation of the protein states are the secondary structural assignments to the coordinates, which can be overlaid onto the sequence to create a 1D representation.

There have been other studies with aims to create local structural alphabets to represent the structure as a 1D sequence of structural blocks [1]. A structural alphabet

T. Shibuya et al. (Eds.): PRIB 2012, LNBI 7632, pp. 188–197, 2012.

is defined as a set of small prototypes that can approximate each part of the backbone. Creating such an alphabet requires the identification of a set of recurrent blocks that can identify all possible backbone conformations. A commonly used structural alphabet is PB [2], which uses the dihedral angles of the backbone structure in a sliding window to match the segment to one of 16 pre-defined blocks.

Another common approach for structure abstraction is to convert the protein structure into a graph from distance or contact maps. In this representation, each residue is coarse-grained into one center node that is connected to other nodes on the graph on the basis of distance (or other criteria). This allows each aminoacid to be represented with its contacts and the topology of the network around it. Representing the structure as a graph allows for sub-graph matching to find reoccurring common motifs in a data set [3], use of elastic network models for normal mode analysis [4] and other algorithms that can employ the graph theoretical properties.

The problem with different representation schemas is the amount of information lost to the abstraction. In case of secondary structure, representing the structure with two states (α-helix and β-sheet) causes the diversity of helices and sheets to be lost, as α-helices are frequently curved (58%) or kinked (17%) [5]. Use of local structural alphabets can capture this information; however as the name implies, the non-local neighbor information of the protein structure is missing. Graph based methods can capture both local and global information from the graph topology. However, since the 3D coordinates of the contacts are reduced to only edge weights, direction and the topology of the structure around each residue is lost.

To approximate both the local structural information with a relatively high degree of certainty and the non-backbone neighbor information and directionality of the contacts with a single model, we propose the use of bond-orientational order parameters. Bond-orientational order is a well-established metric that is used in analysis and comparison of the crystal structures packing of atoms [6]. Due to the use of spherical harmonics, they can capture the directional information around each residue, and since they are invariant of the rotations of the reference frame, matching two structures require only the comparison of numbers, instead of the more computationally costly and problem-prone structural alignment methods.

As a first step, we wanted to test whether the number and placement (angle and distance) of neighboring atoms around each residue show a repeating pattern in average protein structures. If there is such a pattern, we can use the protein descriptors to approximate the local structure around a center point. To test the feasibility of representing the protein structure with such orientational order descriptors, we tried to use those descriptors to capture and differentiate the secondary structural elements from each other. Recognizing and assigning secondary structures to atomic coordinates is a complex task [7] and require the ability to recognize both the local structure (for helices) and contact information (between β strands). If orientational order descriptors can predict secondary structural elements, it shows that they capture the necessary information and can be evaluated further for more complex motif discovery purposes.

2   Methods

2.1   Bond-Orientational Order

The bond-orientational order parameter is previously described by Steinhardt et al. [6] in the study of packed spheres. It has also been employed in the analysis of protein structures by means of local connectivity around each residue [8]. The bond-orientational order parameters are given as:

$$\bar{Q}_{lm}(i) = \frac{1}{N_b(i)} \sum_{n=1}^{N_b(i)} Y_{lm}[\theta(\vec{r}_n - \vec{r}_i), \phi(\vec{r}_n - \vec{r}_i)] \tag{1}$$

$$Q_l(i) = \left(\frac{4\pi}{2l+1} \sum_{m=-l}^{l} |\bar{Q}_{lm}(i)|^2 \right)^{1/2} \tag{2}$$

$$W_l(i) = \sum_{\substack{m_1, m_2, m_3, \\ m_1+m_2+m_3=0}} \begin{bmatrix} l & l & l \\ m_1 & m_2 & m_3 \end{bmatrix} \times \bar{Q}_{lm_1} \bar{Q}_{lm_2} \bar{Q}_{lm_3} \tag{3}$$

where l is the bond orientational parameter. $\vec{r}_n$ denotes the position vector of the nth residue. $(\vec{r}_n - \vec{r}_i)$ is the bond vector from residue i to n, and θ, ϕ are the polar angles of this bond, measured with respect to an arbitrary reference frame. $Y_{lm}[\theta(\vec{r}_n - \vec{r}_i), \phi(\vec{r}_n - \vec{r}_i)]$ are Laplace's spherical harmonic functions [9] for the given angles. $N_b(i)$ is the total number of contacts of i that are below a given cutoff distance. The coefficients shown as a matrix in Equation 3 are the Wigner-3-j symbols [10].

While the spherical harmonics of the bonds for a given l can change drastically by rotating the coordinate system, combining the Q_{lm} values into a quadratic invariant Q_l (Equation 2) and third-order invariant W_l (Equation 3) will result in a rotationally invariant parameter. These order parameters are invariant under reorientations of the external coordinate system. For l=2n, spherical harmonics are also invariant under inversion and therefore independent of reference frame.

In research of the crystal packing, most commonly used parameter is the Q_6 [6, 11, 12] as l=6 is the smallest value of l that can capture both cubic (simple, face centered, and body centered) and icosahedral orders (whereas Q_4 will miss icosahedral and Q_2 will miss both) [8].

2.2   Dataset

For experimentation, a total of 120 protein structures were collected from the Protein Data Bank [13]. Protein structures belonging to different SCOP [14] classes and folds were selected for more even representation of different folds in the dataset. On top of those, the benchmark set of non-homologous (<30% sequence identity) PDB proteins of Zhang et al. [15] were also added to the final dataset.

For each residue, Q_l (Equation 2) and W_l (Equation 3) values are calculated from the contacts of that residue, where contact is defined as residues with distance between the $C\alpha$ atoms that is less than a predefined cutoff threshold. During the calculation, different cutoff distances and l values were tried. Resulting Q_l and W_l values were merged in a feature vector by using sliding window on the backbone.

The secondary structure of each protein was calculated using STRIDE [16]. The secondary structure values of the windows were assigned as a class value on the basis of occurring in the majority of the segment (>60%) in a continuous fashion in the sliding window. The transition regions between different secondary structures that contain two or more different secondary structure classes in the protein segment were removed from the dataset since there is no clear secondary structure to be used in learning and prediction. After those removals, extracting the features from the 120 proteins (using a window size of 5) results in 15273 rows (protein segments) in the final dataset.

2.3   Secondary Structure Prediction

Secondary structures assigned to protein segments by STRIDE [16] are represented in a 3-class and 7-class fashion. The 7 classes are α helix, 3_{10} helix, π helix, β-sheet, coil, turn and bridge. Those 7 classes were simplified to 3 classes as "Helix", "Sheet" and "Loop". The final dataset was created using both 3-class and 7-class representations. However, in the resulting dataset the classes bridge, 3_{10} helix and π helix had only few copies, as either they are uncommon or are rarely found consecutively. Also, STRIDE is believed to underpredict π helices [17], possibly lowering their count even further. Due to very low sample size, 3_{10} helix and π helix classes were merged with the alpha helix class, and bridge regions were removed completely, resulting in a 4-class (helix, sheet, coil, turn) data.

In the feature vector, Q_l values always result in a value between 0 and 1, while W_l values can take arbitrary values. To overcome this, W_l values were normalized to the [0-1] range before the prediction.

Using the calculated Q_l and normalized W_l values from the sliding windows as the feature vector, and assigned secondary structure as the class value (for both 3-class and 4-class), a classification was performed using the SVM implementation libsvm [18] inside the Orange data mining software [19].

Optimization of the window size, l-values and the cutoff distance was carried out on a smaller independent set consisting of 15 proteins. The optimal results were obtained using a cutoff of 7 Å in conjunction with l=2 to 10, with a window size of 5.

Training and prediction was done on separate datasets, created from independent proteins (i.e. no protein segment was predicted with a classifier that was trained with a segment belonging to the same protein). The data was split in a 50-50% fashion (of the PDBs) to create the training and the testing sets.

3   Results

3.1   Prediction Results

The accuracy and the AUC (area under the receiver-operating-characteristic curve) of the predictions of the test set are given in Table 1. Accuracy of the prediction is 92.3% and the AUC is 0.993. AUC gives the probability that a randomly selected positive instance will score higher than a random negative instance, and is a more robust performance measure than accuracy itself [20].

Looking at the confidence table, helices (sensitivity of 0.99) can be represented exceptionally well by the bond-orientational order parameters, followed by sheet structures (sensitivity of 0.91). In 4-class representation, coils and turns have lower sensitivity (respectively 0.71 and 0.75). However, as can be expected, they are more likely to be mistaken as each other than a sheet or helix. In 3-class representation, assigning the class value of "loop-region" to coils and turns will result in a significantly higher sensitivity of 0.87.

Table 1. Area under the ROC curve, accuracy and confusion matrix of the test set predictions. In the confusion matrix, number of predicted instances and ratio of the correct predictions are given.The last row (C+T) represents Coil and Turns being classified as Loop-region in the 3-class prediction.

AUC	Accuracy
0.993	92.3%

		Predicted				
		Helix	**Sheet**	**Coil**	**Turn**	**Sensitivity**
Actual	**Helix**	3768 (98.9%)	12 (0.3%)	0 (0.0%)	26 (0.7%)	0.990
	Sheet	3 (0.2%)	1199 (90.70%)	12 (0.9%)	105 (7.9%)	0.907
	Coil	3 (0.7%)	27 (6.1%)	316 (71.0%)	99 (22.2%)	0.710
	Turn	71 (10.1%)	43 (6.1%)	48 (6.9%)	527 (75.3%)	0.753
	C+T	74 (6.5%)	70 (6.1%)	990 (87.3%)		0.873

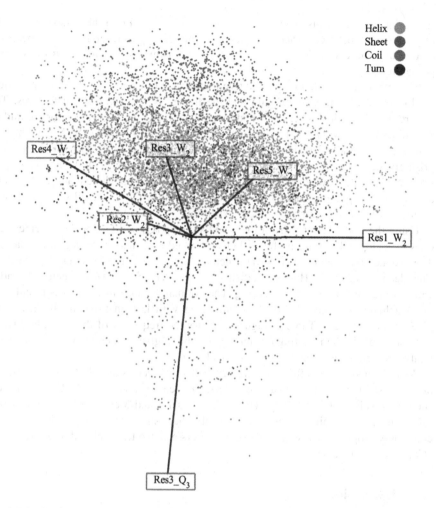

Fig. 1. Distribution of the class values with respect to different features in 2D linear projection. ResX_Y represents the feature Y of the residue X (out of 5) in the proteing segment.

3.2   Feature Analysis and Clustering

Due to the very high accuracy and AUC values, we investigated whether the high accuracy was because of the high predictive performance of SVM due to the use of non-linear kernels, or whether the accuracy could be replicated with a simple, human-understandable method.

We first investigated the effects of different Q_l and W_l features for each residue in the segment to the corresponding secondary structure. To see the importance of each feature and a visual representation of their relationship with samples, we created a 2D linear projection [21] of the data using 6 features, selected by running the VizRank heuristic [22] for 2000 generations on the training set. The rotation of the axes and the final projection was optimized using the FreeViz algorithm [23] to optimize separation of data points. The result is given in Figure 1. From the perspective of the $Q3$ and $W3$ parameters, sheets and coils form the opposing ends of the spectrum. Notice that the classes show a non-perfect but distinct separation even on a linear projection.

To further investigate the quantitative importance of each feature to the prediction, we looked at the information gain and the linear SVM weight of the features. The features that have the highest information gain are the Q_3 and Q_4 values for the middle 3 residues of the window of 5 aminoacids. When ranked by their SVM weights, Q_9 values of the middle 3 residues were also selected as well as the Q_4 values. Not surprisingly, the center portion of the window was ranked higher than the boundary portions. No W_1 values were selected as informative. We can conclude that Q_3, Q_4 and Q_9 are the most important features for classification, since they were all selected at least 3 times for that center portion without exception.

Using the top 6 features from the SVM weights (Q_4 and Q_9 for the 3 center residues of each window), we performed unsupervised k-means clustering on the dataset. The distance between each row was calculated as the distance between their vectors. Euclidean, Manhattan, Hamming distances and Pearson and Spearman correlation values were tried during the clustering. The optimal distance measure was found to be the Manhattan distance. Results for clustering with k=6 in k-means algorithm are given in Figure 2 and Table 2. Figure 2 shows the frequency of the secondary structural elements in the resulting clusters, and Table 2 gives the clustering accuracy and relative assignments of each class to each cluster.

As we can see, even after discretizing the feature vectors to only 6 clusters with an unsupervised method, the clustering has 84.6% accuracy and 0.932 AUC. The clusters show relatively high sensitivity. That is, clusters 1,2 and 3 can represent helix structures with high certainty, cluster 4 is mostly sheet structures and the cluster 5, 6 is commonly loop regions, with most of the errors are due to misclassifying "Turns" as "Coils" and vice versa.

4   Discussion

In our study, we tested the feasibility of using bond-orientational order parameters as descriptors of protein structure in predicting secondary structure from the coordinates Cα atoms. This resulted in 92.3% accuracy and 0.993 AUC. The helices can be predicted at ~99% sensitivity. Since helices are formed by local interactions that are established within the close vicinity of each amino acid, we can conclude that this structure can easily be captured by the orientational order parameters.

While helices can be predicted quite easily using backbone dihedral angles, this is not the case for sheet structures due to non-local, long range interactions. We show that orientational order parameters can capture the representation of β-sheets equally well (91% sensitivity) since strands stand parallel to each other to form the sheets. There is less information coming from the sequentially adjacent residues forming the sheet in comparison to helices (which makes it difficult to predict them in secondary structure prediction algorithms) but the orientational order descriptors can still capture

Table 2. Relative assigment of each class to the clusters. Cluster representations show which class is more likely to be in that cluster.

	# Helix	# Sheet	# Turn	# Coil	Representation
Cluster 1	98.3%	1.1%	0.5%	0.2%	Helix
Cluster 2	90.4%	4.1%	3.8%	1.7%	Helix
Cluster 3	90.1%	5.1%	3.1%	1.7%	Helix
Cluster 4	19.7%	68.9%	8.1%	3.3%	Sheet
Cluster 5	5.8%	20.9%	47.3%	26.0%	Loop region ~ Turn
Cluster 6	0.0%	0.7%	35.9%	63.4%	Loop region ~ Coil

Clustering Accuracy	84.6%
AUC	0.932

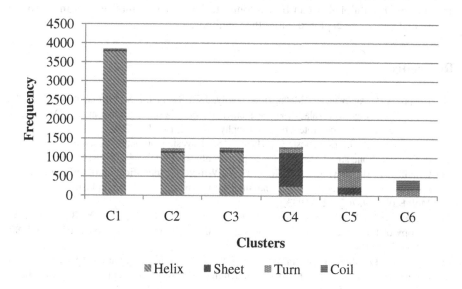

Fig. 2. The number of elements in each cluster by their secondary structural elements

the necessary local and neighbor information. Addition of orientational order parameters with higher cutoff distance values may help in this regard.

Turns and coils are more difficult to predict in comparison to helices and sheets, (75% and 71% sensitivity respectively). This is expected as they are short, can be found in different local environments (i.e. buried in the core or exposed to water) and lack a rigid structure. Turns are easier to predict than random coils since they are more structured and may have conserved hydrogen bonds between the backbone residues. Some coil structures can be mistakenly classified as turns (22.2%) but the rate of misclassification of turns as coils is not as high (6.9%).

While the continuous features are shown to be enough to capture secondary structure, we also investigated the applicability of comparing two orientational order feature vectors to evaluate structural similarity (i.e. whether a vector can be assigned to a class based on just a distance value and not by a complex rule learned by the SVM). By using an unsupervised clustering method with a simple Manhattan distance metric, we have obtained 6 clusters that correctly predict the secondary structure with 84.6% accuracy and 0.932 AUC, showing that similar structures definitely have similar vector characteristics, which is very important for use in structural alphabets. We can also see this effect in Figure 1; the classes have distinctive characteristics in their features that can be recognized even on a linear projection with few features.

We also looked at the relative importance of each feature in the descriptor vector. Q_3, Q_4 and Q_9 seem to be the most important features in prediction of the secondary structure elements, but a more through experimentation is needed.

We conclude that there is very strong potential application of orientational order parameters, especially in establishment of a new structural alphabet that takes local backbone structure as well as contact information from the neighboring regions into account. Such an alphabet can be exploited to identify structural motifs in a protein family that cannot be captured with other methods.

References

1. Joseph, A.P., Agarwal, G., Mahajan, S., Gelly, J.C., Swapna, L.S., Offmann, B., Cadet, F., Bornot, A., Tyagi, M., Valadie, H., Schneider, B., Etchebest, C., Srinivasan, N., De Brevern, A.G.: A short survey on protein blocks. Biophys. Rev. 2, 137–147 (2010)
2. de Brevern, A.G., Etchebest, C., Hazout, S.: Bayesian probabilistic approach for predicting backbone structures in terms of protein blocks. Proteins 41, 271–287 (2000)
3. Grindley, H.M., Artymiuk, P.J., Rice, D.W., Willett, P.: Identification of tertiary structure resemblance in proteins using a maximal common subgraph isomorphism algorithm. J. Mol. Biol. 229, 707–721 (1993)
4. Atilgan, A.R., Durell, S.R., Jernigan, R.L., Demirel, M.C., Keskin, O., Bahar, I.: Anisotropy of fluctuation dynamics of proteins with an elastic network model. Biophys J. 80, 505–515 (2001)
5. Martin, J., Letellier, G., Marin, A., Taly, J.F., de Brevern, A.G., Gibrat, J.F.: Protein secondary structure assignment revisited: a detailed analysis of different assignment methods. BMC Struct. Biol. 5, 17 (2005)
6. Steinhardt, P.J., Nelson, D.R., Ronchetti, M.: Bond-Orientational Order in Liquids and Glasses. Phys. Rev. B 28, 784–805 (1983)
7. Offmann, B., Tyagi, M., de Brevern, A.G.: Local protein structures. Curr. Bioinform. 2, 165–202 (2007)
8. Atilgan, C., Okan, O.B., Atilgan, A.R.: How orientational order governs collectivity of folded proteins. Proteins 78, 3363–3375 (2010)

9. Sternberg, W.J., Smith, T.L.: The theory of potential and spherical harmonics. Univ. of Toronto Press, Toronto (1946)
10. Landau, L.D., Lifshitz, E.M.: Quantum mechanics: non-relativistic theory. Pergamon Press; sole distributors in the U.S.A., Addison-Wesley Pub. Co., Reading, Mass., Oxford, New York (1965)
11. Truskett, T.M., Torquato, S., Debenedetti, P.G.: Towards a quantification of disorder in materials: distinguishing equilibrium and glassy sphere packings. Phys. Rev. E Stat. Phys. Plasmas Fluids Relat. Interdiscip. Topics 62, 993–1001 (2000)
12. Torquato, S.: Random heterogeneous materials: microstructure and macroscopic properties. Springer, New York (2002)
13. Berman, H.M., Westbrook, J., Feng, Z., Gilliland, G., Bhat, T.N., Weissig, H., Shindyalov, I.N., Bourne, P.E.: The Protein Data Bank. Nucleic Acids Res. 28, 235–242 (2000)
14. Murzin, A.G., Brenner, S.E., Hubbard, T., Chothia, C.: SCOP: a structural classification of proteins database for the investigation of sequences and structures. J. Mol. Biol. 247, 536–540 (1995)
15. Zhang, Y., Skolnick, J.: TM-align: a protein structure alignment algorithm based on the TM-score. Nucleic Acids Res. 33, 2302–2309 (2005)
16. Frishman, D., Argos, P.: Knowledge-based protein secondary structure assignment. Proteins 23, 566–579 (1995)
17. Fodje, M.N., Al-Karadaghi, S.: Occurrence, conformational features and amino acid propensities for the pi-helix. Protein Eng. 15, 353–358 (2002)
18. Chang, C.-C., Lin, C.-J.: LIBSVM: a library for support vector machines (2001)
19. Demšar, J., Zupan, B., Leban, G., Curk, T.: Orange: From Experimental Machine Learning to Interactive Data Mining. In: Boulicaut, J.-F., Esposito, F., Giannotti, F., Pedreschi, D. (eds.) PKDD 2004. LNCS (LNAI), vol. 3202, pp. 537–539. Springer, Heidelberg (2004)
20. Hanley, J.A., McNeil, B.J.: A method of comparing the areas under receiver operat-ing characteristic curves derived from the same cases. Radiology 148, 839–843 (1983)
21. Koren, Y., Carmel, L.: Visualization of labeled data using linear transformations. In: In-fovis 2002: IEEE Symposium on Information Visualization 2003, Proceedings, pp. 121–128, 248 (2003)
22. Leban, G., Zupan, B., Vidmar, G., Bratko, I.: VizRank: Data visualization guided by machine learning. Data Mining and Knowledge Discovery 13, 119–136 (2006)
23. Demsar, J., Leban, G., Zupan, B.: FreeViz–an intelligent multivariate visualization approach to explorative analysis of biomedical data. J. Biomed. Inform. 40, 661–671 (2007)

Diagnose the Premalignant Pancreatic Cancer Using High Dimensional Linear Machine

Yifeng Li and Alioune Ngom

School of Computer Sciences, 5115 Lambton Tower, University of Windsor,
401 Sunset Avenue, Windsor, Ontario, N9B 3P4, Canada
{li11112c,angom}@uwindsor.ca
http://cs.uwindsor.ca/uwinbio

Abstract. High throughput mass spectrometry technique has been extensively studied for the diagnosis of cancers. The detection of the pancreatic cancer at a very early stage is important to heal patients, but is very difficult due to biological and computational challenges. This paper proposes a simple classification approach which can be applied to the premalignant pancreatic cancer detection using mass spectrometry technique. Computational experiments show that our method outperforms the benchmark methods in accuracy and sensitivity without resorting to any biomarker selection, and the comparison with previous works shows that our method can obtain competitive performance.

Keywords: mass spectrometry, pancreatic cancer, classification, high dimensional linear machine.

1 Introduction

Proteomic mass spectrometry technique has great potential to be applied for clinical diagnosis and biomarker identification. The mass spectrometry data of a patient are obtained through measuring the ion intensities of tens of thousands of mass-to-charge (m/z) ratios of proteins and peptides. Analysis of such high throughput data is promising, but also difficult [1]. Some of the problems challenging the bioinformatics community include: 1) The data are quite noisy and subject to high variability. 2) Though the data are redundant, the amounts of useful and redundant information are not clear. 3) There are tens of thousands of features (m/z ratios), while there is only tens up to hundreds of samples, which is the well known large number of features versus small number of samples (LFSS) problem. This problem results in intolerable computational burden when using some prediction models, for example decision tree. Some models can not be applied on such data, because the number of their parameters grows exponentially as the number of dimensions increases, and therefore it is impossible to estimate these parameters using available training data. These problems are the notorious "curses of dimensionality" [2]. Due to the above problems, some models are easily subject to overfitting and hence have poor generalization. A lot of computational approaches dealing with the above problems have been proposed in two directions last decade. First of all, efforts of biomarker (and peak) identification and dimension reduction have been extensively taken for the clinical diagnosis, pathological, and computational purposes.

T. Shibuya et al. (Eds.): PRIB 2012, LNBI 7632, pp. 198–209, 2012.

As it is impossible to enumerate all works in this direction, we only give two highly cited examples in the following. Levner [3] tested many popular feature selection and feature extraction methods for biomarker identification coupled with nearest shrunken centroid classifier, and found that some state-of-art methods actually performs poorly on mass spectrometry data using consistent cross-validation. [4] is an excellent review on feature selection for mass spectrometry data. Second, kernel approaches have been invented [5]. These approaches are able to represent complex patterns and their optimization is dimension-free. Also, they are often robust to noise and redundant. Kernel approaches often have good capability generalization.

Patients with pancreatic cancer has a very high death rate. If the pancreatic cancer can be detected before the cancer develops, treated patients at the preinvasive stage can have a chance to survive. Unfortunately, there is no effective premalignant pancreatic cancer detection method by now [6]. In the precancerous stage, the proteins may have been developed differential signals. Proteomic mass spectrometry technique provides an insight into patient's protein profile, and therefore is quite promising to be applied to this area. Ge *et al.* [7] presented a framework of using ensemble of decision trees coupled with feature selection methods. Decision tree is very slow when learning on high dimensional data. Thus, in order to use decision tree as classifier, three feature selection methods (Student t-test, Wilcoxon rank sum test, and genetic algorithm) are used to reduce the dimension before classification. The performances of decision tree and its different ensembles were investigated in [7]. It was claimed that classifier ensembles generally have better prediction accuracy than single decision tree. However, most of the methods used in [7] still have low accuracies and low sensitivity. Another issue is that the candidate biomarkers selected by different methods are not consistent.

As we mentioned above, the curses of dimensionality actually imply, to a great extent, the difficulty of model selection due to LFSS in practice. Also the statement that"only very few features of mass spectrometry data are informative and the rest are redundant" is only an assumption in many circumstances. On another hand, the large number of features provides us with huge amount of information. Although the target informative knowledge hides in the data, we should have a chance of taking advantage of this using some data mining techniques. We can call this as one of the less-known "blessings of dimensionality" [8]. Although this principle has not yet been well understood theoretically, some studies based on this principle in computer vision have demonstrated prodigious results [9] [10]. In the high dimensional setting, real-world data points usually reside in manifolds. *Support vector machine* (SVM) [11] can be viewed as an example of taking advantage of high dimensionality. Its essential idea is that data points are mapped from the original low dimensional space to a very high (even infinite) dimensional space where the data points are likely to be linearly separable, and therefore a separating hyperplane could be implicitly learned through margin maximization.

In this paper, we shall prove that, in the case of LFSS, the mass spectrometry data are much likely to be linearly separable and we propose a high dimensional linear machine for such case to detect the premalignant pancreatic cancer at an early stage. The contributions of this study include

1. we bring the principle of blessings of dimensionality to the horizon of researchers in mass spectrometry data analysis;
2. we propose the high dimensional linear machine and show that it is a specific case of the general linear models for classification;
3. we propose a threshold adjustment method based on receiver operation curve.

The paper is organized as follows. In the next section, we first prove the linearity of the mass spectrometry in the case of LFSS, under some condition, and then describe our proposed method. The computational experiments and comparison results are then shown. After that related discussion are delivered. Finally, the paper is completed by some conclusions.

2   Methods

Suppose $D_{m \times n}$ is a training set with m features (m/z ratios) and n samples. These samples are from two groups: the premalignant pancreatic cancer group (denoted by +1) and the normal group (denoted by -1). The class labels of these n training samples are in the column vector c, and matrix $S_{m \times p}$ represents p unknown samples. Each of these p samples is either from premalignant pancreatic cancer class or normal class. The computational task is to predict the class labels of these p samples.

Linear models for classification, such as *linear Bayesian classifier*, *Fisher discriminative analysis* (FDA), and the state-of-art SVM, try to find a hyperplane between two groups of the training set. This hyperplane can be formulated as

$$g(x) = w^{\mathrm{T}} x = 0, \tag{1}$$

where $w, x \in \mathbb{R}^{m+1}$. w_0 is the bias and the corresponding $x_0 = 1$. w and x in this form are thus *augmented*.

In our case, the hyperplane should separate the two classes in D, that is

$$w^{\mathrm{T}}[1; D] = c^{\mathrm{T}}, \tag{2}$$

where the boldface 1 is a column vector accommodating n ones. $[1; D]$ uses MATLAB notation meaning the concatenation of 1 and D in row-wise direction. Using matrix transposition, we have

$$A^{\mathrm{T}} w = c, \tag{3}$$

where $A \in \mathbb{R}^{(m+1) \times n}$, $A = [1; D]$. Each column of A is an augmented training sample.

As $n < m$, this system of linear equations is underdetermined. The condition of existing a solution w is $rank(A^{\mathrm{T}}) = rank([A^{\mathrm{T}}, c])$. For rich high dimensional mass spectrometry data, this condition is not difficult to hold. In practice, due to biological complexity, it is much likely that the data is of full rank, that is $R(A^{\mathrm{T}}) = n$, in which case case, $rank(A^{\mathrm{T}}) = rank([A^{\mathrm{T}}, c])$ holds as $[A^{\mathrm{T}}, c]$ is also of full rank. Thus, we can state that it is much likely that mass spectrometry data are linearly separable. As long as the data are linearly separable, there are infinite solutions ws, and therefore there are infinite hyperplanes separating the two groups in A perfectly. That is we can obtain zero training error. The learning of a linear classifier should consider the trade-off between two efforts: minimizing the training error and maximizing the generalization capability. In this linear separable case, we need to focus on the second one. For any positive training sample, x^+, we have $g(x^+) = +1$, and for any negative training sample x^-, we have $g(x^-) = -1$. Since the distance of x^+ and x^- to the hyperplane $g(x) = 0$ is $d(x^+) = \frac{|g(x^+)|}{\|w\|_2} = \frac{1}{\|w\|_2}$ and $d(x^-) = \frac{|g(x^-)|}{\|w\|_2} = \frac{1}{\|w\|_2}$, the margin between the two classes is $m_\pm = d(x^+) + d(x^-) = \frac{2}{\|w\|_2}$. For the generalization purpose, this margin should be as wide as possible, that is the effort should be maximizing $\frac{2}{\|w\|_2}$ which is equivalent to minimizing $\|w\|_2$. Now let us summarize our task formally as below,

$$\min_{w} \frac{1}{2}\|w\|_2, \tag{4}$$
$$\text{s.t. } A^T w = c,$$

where the objective is to maximize the generalization capability and the constraint is to keep zero training error. We coin this method as *high dimensional linear machine* (HDLM). Equation 4 is the well-known least l_2-norm problem, and therefore has analytical optimal solution: $w^* = (A^T)^\dagger c$ where $(A^T)^\dagger = A(A^T A)^{-1}$ is the Moore-Penrose pseudoinverse [12]. Therefore, the hyperplane is

$$g(x) = w^{*T} x = (A^T A)^{-1} A^T x = 0. \tag{5}$$

Since $A^T A$ might be singular, its inverse can be computed by *singular value decomposition* (SVD) [13].

After obtaining w^*, the learning step is finished. The second step is the prediction step. Given a unknown sample, s (augmented), the class label of s is predicted through the relation of s and the hyperplane learned. That is the decision rule is defined as

$$d(s) = \begin{cases} +1 & g(s) > 0 \\ -1 & g(s) < 0, \\ rand\{-1, +1\} & g(s) = 0 \end{cases} \tag{6}$$

where $rand\{-1, +1\}$ returns either -1 or +1 with equal probabilities (suppose equal priors).

2.1 General Linear Models for Classification

HDLM looks similar with hard-margin SVM which is expressed as

$$\min_{w} \frac{1}{2}\|w\|_2, \tag{7}$$
$$\text{s.t. } A^T w \geq c.$$

In fact, both HDLM and SVM are the special cases of the following general linear model for classification:

$$\min \frac{1}{2}\|w\|_2 + \lambda l(A, c, w), \tag{8}$$

where the second term is a loss function of training, and parameter λ controls the trade-off between the capability of generalization and training precision. For SVM, $l(\boldsymbol{A}, \boldsymbol{c}, \boldsymbol{w}) = \sum_{i=1}^{n} \max(0, 1 - c_i \boldsymbol{a}_i^{\mathrm{T}} \boldsymbol{w})$, where $\boldsymbol{a}_i^{\mathrm{T}}$ is the i-th row of $\boldsymbol{A}$. This is the well-known hinge loss. The loss function of HDLM is essentially square loss which is expressed as $l(\boldsymbol{A}, \boldsymbol{c}, \boldsymbol{w}) = \sum_{i=1}^{n} (c_i - \boldsymbol{a}_i^{\mathrm{T}} \boldsymbol{w})^2 = \|\boldsymbol{c} - \boldsymbol{A}\boldsymbol{w}\|_2^2$. From this we can see that the optimization of HDLM is essentially rigid regression. The advantage of HDLM over SVM is that HDLM makes use of the specific assumption of linear separability of high-dimensional mass spectrometry data, and has analytical solution which is fast to compute.

2.2   Kernel HDLM

Most of the linear models can be kernelized due to the fact that their training and prediction step only require the inner products between samples. This is also indeed true for HDLM. From Equation 5, we can see that the prediction of a unknown sample $\boldsymbol{s}$ only needs the inner products $\boldsymbol{A}^{\mathrm{T}}\boldsymbol{A}$ and $\boldsymbol{A}^{\mathrm{T}}\boldsymbol{s}$. Therefore HDLM can be kernelized via replacing the inner products by kernel matrices $k(\boldsymbol{A}, \boldsymbol{A})$ and $k(\boldsymbol{A}, \boldsymbol{s})$. Because of this, we can find that the computation of HDLM is dimension-free, and the kernelization provides HLDM a flexible choice of representing complex patterns and dealing with noise and redundancy.

2.3   Increase the Performance

The classification performance can be measured by sensitivity ($sen. = \frac{TP}{TP+FN}$), specificity ($spec. = \frac{TN}{TN+FP}$), accuracy ($acc. = \frac{TP+TN}{TP+FN+TN+FP}$), and balanced accuracy ($BACC = \frac{sen.+spec.}{2}$), where TP, TN, FP, and FN are defined as the numbers of true positive, true negative, false positive, and false negative samples, respectively. Due to unbalanced group sizes and distributions of the groups, the sensitivity and specificity may be unbalanced, and therefore the accuracy may not reflect the true discriminative capability of the classifier. As a linear classifier, HDLM use the default threshold 0 in the decision rule (Equation 6). Thus, we need to adjust threshold, which is a variable that can be denoted by t. Our threshold learning method is described as below. As t increases from a reasonable value, the sensitivity increases to 1 while the specificity decreases to 0. Therefore, the sensitivity and specificity are functions with respect to t, respectively. The sensitivities and the corresponding specificities can be described by a *receiver operating characteristic* (ROC) curve [14]. An example of a ROC curve is shown in Figure 1. The far the ROC curve is away the line passing $(0, 0)$ and $(1, 1)$, the better a classifier is. The general quality of a classifier can be measured by area under the ROC curve (AUC). For application, we are also interested in choosing a threshold parameter of a specific classifier which leads to better performance than other thresholds. The distance between a point on the ROC curve to the straight line passing $(0, 0)$ and $(1, 1)$ is denoted by $d(t)$. We define the optimal pair of sensitivity and specificity as the one corresponding to the optimal $d(t)$, that is $d(t^*)$. t^* is the optimal threshold to learn. From Figure 1, we can easily find the relation between $d(t)$ and sensitivity and specificity: $d(t) = \frac{1}{\sqrt{2}}(Sen. - (1 - Spec.)) = \frac{1}{\sqrt{2}}(2BACC - 1)$. Practically, we can obtain t^* through measuring the mean BACC of k-fold CV of a binary linear classifier taking threshold t over the training set. The mean BACC can be denoted by function $MBACC(t, k, trainingset)$. Formally, $t^* = \arg_t \max MBACC(t, k, trainingset)$, where $t = -1 : 0.01 : 1$ (a MATLAB notation that generating a vector through increasing -1 to 1 by step 0.01). For narrative convenience, we coin this threshold adjusted HDLM as TA-HDLM.

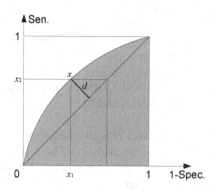

Fig. 1. ROC Curve

Other techniques such as feature selection [4], sample selection [15], classifier ensemble [16], and transductive learning [11, 17] are studied in machine learning to increase the performance of a classification approach. But for HDLM the most direct way is to tune the threshold in the decision rule. Since HDLM works in high dimensional setting, feature selection for the dimension reduction purpose is not applicable here. The linearity of samples in the high dimension discourages us to apply sample selection. Classifier ensemble is effective for weak classifiers, HDLM, however, does not fall into such class. Therefore we do not employ the bagging and boosting strategies. Transductive learning is a good choice when there are few labeled training samples, and many unlabeled training samples and testing samples. In our bioinformatics application of this study, it is unlikely to have a large number of unknown samples at once waiting for being diagnosed. For this reason we do not choose transductive learning.

2.4   Benchmark Methods

In order to be aware of the classification performance of the HDLM classifier, it is necessary to compare with other benchmark classification approaches. We includes two categories of classifiers as benchmark methods.

The first category is composed of the instance learning methods: *1-nearest neighbor* (1-NN), and two sparse representation methods. *Sparse representations* [18] [19] are novel and effective methods in the filed of pattern recognition. The fundamental idea is that an unknown sample can be represented by a linear combination of all the training samples for all of the classes. The sparse combination coefficients are obtained by minimizing the l_1-norm. Then the coefficients are partitioned according to classes.

The linear regression residual can be computed for each class using such corresponding coefficients. The unknown sample is assigned to the class which obtains the smallest regression residual. The implementation of *sparse representation classifier* (SRC) proposed in [18] works well in the case that the number of training samples is equal to or greater than the number of features, while it is difficult to control the regression error in the constraint. Therefore SRC is not applicable in the classification of mass spectrometry without any aid of feature selection. With the specific purpose to classify data of LFSS, the *non-negative least squares* (NNLS) classifier is recently proposed in [20]. The main idea of the NNLS classification method is that any new sample with unknown class label is assumed to be a sparse non-negative linear combination of the training samples. The combination coefficient is the non-negative least squares solution. And the training sample with the dominantly largest coefficient should reside in the same class as this new sample. Bootstrap NNLS (BNNLS) is also proposed in [20] to improve the prediction performance. NNLS and BNNLS are included in our benchmark methods in this study.

The second category consists of two SVMs and the recently proposed *extreme learning machine* (ELM) [21]. As state-of-art method, SVM is studied intensively and has been successfully applied in various fields, for example bioinformatics. The idea of SVM is to map the samples to a higher dimensional space where samples are likely to be linearly separable, and then to maximize the (hard or soft) margin between two groups. The kernel trick avoids the direct mapping and does optimization in the original space. Two kernel functions, *radial basis function* (rbf) and linear kernels, are utilized for SVM in the study. In fact, the linear kernel does not conduct any mapping. As we point out above, the mass spectrometry data are likely to be linearly separable in the original space. Therefore, the linear SVM may be enough instead of using any other kernel trick. ELM, as a variant of single layer feed-forward neural network, is claimed to be competitive with SVMs even outperform SVMs. As generalization of rbf neural network, ELM randomly assigns the weights, connecting the input to the hidden layer, instead of learning them. And then the weights connecting the hidden layer to the output are obtained as least squares or minimum norm optimizations.

3   Computational Experiments and Discussions

Our proposed methods are evaluated and compared with other benchmark approaches over a pancreatic cancer dataset: PanIN (human pancreatic intraepithelial neoplasia) [22]. This dataset was obtain from mice with premalignant pancreatic cancerous and normal statuses. The dataset is shortly described in Table 1. This dataset contains 101 normal samples and 80 premalignant pancreatic samples. 6771 m/z ratios compose the feature list. Readers are referred to [22] for more description about the collection of the data. The data is downloadable from [23]. Let matrix $A_{6772 \times 181}$ represent the data with each column is an augmented sample. The rank of A^{T} is estimated to be 181, which means this data is of full rank. Of course in such case, the rank of $[A^{\mathrm{T}}, c]$, which is a matrix concatenating A^{T} and the class labels (column vector c) in column-wise direction, is also 181. Any cancerous sample has the class label +1, and -1 for any normal sample. $Rank(A^{\mathrm{T}}) = Rank([A^{\mathrm{T}}, c])$ indicates that the data are linearly separable. Therefore a subset of A is also linearly separable.

Table 1. Datasets

Data	#Classes	#Features	#Samples	$Rank(\boldsymbol{A}^T)$	$Rank([\boldsymbol{A}^T, \boldsymbol{c}])$
PanIN [22]	2	6771	101+80=181	181	181

We used 10-fold cross-validation (CV) to split the data into training and test sets. The classifiers (NNLS, BNNLS, 1-NN, rbf-SVM, linear-SVM, ELM, and HDLM) learn on the training set, and predict the class labels of the test set. The classification performance was measured by sensitivity, specificity, accuracy, and balanced accuracy. The cancer samples are defined as positive samples, while normal samples negative. 10-fold CV reran for 20 times and the averaged result are shown in Table 2. We have the following observations. First, we can observe that the instance learning methods including the sparse representation methods do not perform well. The accuracies are just slightly better than random assignment. This may be because the data are very noisy and the distributions of the cancer and normal groups overlap largely. Second, the well known SVM with rbf kernel loses its power in the premalignant cancer diagnosis, sensitivity of 0 is obtained. But the linear-SVM classifier performs better in such high dimensional data because the data are linearly separable. Third, although it was claimed that ELM has similar even better performance than SVM [21], it performs poorly on this data. Forth, we can see that the performance of HDLM is significantly better than the benchmark approaches. It obtained a specificity of 0.758 and a sensitivity of 0.662. As we have stated above, the sensitivity is much crucial than specificity in disease diagnoses. Though the specificity and sensitivity are still unbalanced, this can be tackled through threshold adjustment. As we can be seen at the last row of Table 2, The sensitivity is increased to 0.710. Although this sacrifice some specificity, the accuracy and BACC do not degenerate dramatically.

The running time, including the training and test time for each pair of training and test sets, was recorded for each classifier. The averaged result is also listed in the last column of Table 2. We can observe that HDLM is much faster than SVMs and NNLSs. Although 1-NN and ELM are much efficient than HDLM, their accuracies are not competitive with HDLM. The fastness of HDLM is because it only needs to solve linear equations, while methods such as decision tree and neural network would be very intolerantly slow to learn over such high dimensional data. This is why feature selection or feature extraction have to be done when using decision tree and neural networks. Due to the threshold adjustment, TA-HDLM has the highest computational cost. This cost is still clinically acceptable to learn on about 163 training samples and predict about 18 unknown samples. unlike instance-based learning, once the learning of the HDLM model finishing, the prediction is actually very fast.

Next, we compared our methods with the performance reported in [7]. Readers should be aware that we find that the normal samples are incorrectly defined as positive samples while cancer samples negative in [7]. This can be proved as follows.

Table 2. Classification Performance

Method	Spec.(STD)	Sen.(STD)	Acc.(STD)	BACC(STD)	Time (CPU sec.)
NNLS	0.573(0.036)	0.491(0.039)	0.536(0.028)	0.532(0.028)	0.312
BNNLS	0.580(0.040)	0.484(0.033)	0.538(0.029)	0.532(0.028)	12.276
1-NN	0.583(0.030)	0.479(0.040)	0.537(0.020)	0.531(0.021)	0.056
rbf-SVM	1(0)	0(0)	0.558(0)	0.500(0)	1.393
linear-SVM	0.802(0.025)	0.424(0.040)	0.635(0.021)	0.613(0.022)	1.395
ELM	0.513(0.043)	0.488(0.056)	0.501(0.030)	0.500(0.031)	0.053
HDLM	0.758(0.020)	0.662(0.033)	**0.716**(0.021)	**0.710**(0.022)	0.193
TA-HDLM	0.704(0.031)	**0.710**(0.040)	**0.707**(0.021)	**0.707**(0.021)	75.633

Suppose $\frac{TP+FN}{TN+FP} = \alpha$. According to the definitions of sensitivity and specificity, we have $TP = Sen.(TP + FN) = Sen.\alpha(TN + FP)$ and $TN = Spec.(TN + FP)$. According to the definition of accuracy, we further have

$$
\begin{aligned}
Acc. &= \frac{TP + TN}{TP + FN + TN + FP} \\
&= \frac{Sen.\alpha(TN + FP) + Spec.(TN + FP)}{(1 + \alpha)(TN + FP)} \\
&= \frac{Sen.\alpha + Spec.}{1 + \alpha}.
\end{aligned}
\tag{9}
$$

Therefore we have $\alpha = \frac{Spec.-Acc.}{Acc.-Sen.}$. Take C4.5 in Table 4 in [7] for example, $\alpha = \frac{0.21-0.6444}{0.6444-0.99} = 1.2569$ which approximates to $\frac{101}{80} = 1.2625$ (the ratio of the number of normal samples to the number of cancer samples in the whole data) or $\frac{10}{8} = 1.25$ (the ratio of number of normal samples to the number of cancer samples in a test set). Readers can verify this using more results from [7]. Therefore we need to swap the sensitivity and specificity in the results of [7] and compare them with the results of our methods. Now back to our comparison. The comparison result is shown in Table 3. 10-fold CV was also used in [7]. The first 3 blocks in this table are top results from [7] with respect to accuracy. $+S$, $+W$, and $+G$ mean the combinations with Student t-test feature ranking, Wilcoxon rank test, and genetic feature selection, respectively. It can be seen that HDLM outperforms these methods, except Logistic+S, in [7], in accuracy and BACC. TA-HDLM obtained the highest sensitivity (0.71) among these methods. The Multiboost+W as one of the classifier ensemble methods only obtained a sensitivity of 0.660. Logistic+S and Neural Network+S achieved the sensitivity of 0.700 which is slightly lower than TA-HDLM.

It has to be noted that we did not conduct any preprocessing for our proposed methods and benchmark methods applied in this study, except that, for the cases of SVM and ELM, the ion intensities of each m/z ratio in the training set are normalized to have mean 0 and standard deviation 1. The normalization parameters estimated from the training set are used to normalize the test set. Our normalization is different from the one in the preprocessing stage of [7] where the whole dataset are normalized and

Table 3. Classification Performance

Method	Spec.	Sen.	Acc.	BACC
Logistic+S	0.790	**0.700**	0.750	0.745
Neural Network+S	0.700	**0.700**	0.700	0.700
Random Forest+W	0.790	0.590	0.700	0.690
Multiboost+W	0.730	0.660	0.700	0.695
SVM+G	0.720	0.530	0.633	0.625
Logitboost+G	0.680	0.540	0.617	0.610
Adaboost+G	0.670	0.550	0.617	0.610
linear-SVM	0.802	0.424	0.635	0.613
HDLM	0.758	0.662	0.716	0.710
TA-HDLM	0.704	**0.710**	0.707	0.707

scaled. Care has to be taken in this and other preprocessing in [7], because the test samples should keep intact before the test stage of inductive learning. If the preprocessing of the training set is influenced by the test set, the predicting ability of a classifier (and a feature selection method) is inflated, because the more or less information in the test set has been divulge in the learning stage. It is not to say that the information in the test set should not be used in the learning stage. We have to discuss this in two aspects. For inductive learning of a feature selection and a classifier, the test set should never be touched during learning in order to have a fair evaluation of the capability of feature selection and classifier. One common mistake is that feature selection, that is biomarker or peak identification in the study of mass spectrometry data analysis, is done over the whole data (training set and test set), after that a classifier learns over the training set, and the prediction accuracy of the test set is reported as the evaluation of quality of the feature selection. This actually overestimates the capability of the feature selection. However, if the purpose is not to evaluate a feature selection or a classifier, but is the prediction accuracy, the information of the unlabeled testing samples can be used during learning. This falls into the category of transductive learning [11] and semi-supervised learning [17]. Actually the prediction accuracy obtained using transductive learning is often higher than that using inductive learning. Since all samples are utilized during the preprocessing including baseline correction, sample scaling, and smoothing in [7], though the prediction accuracy is acceptable for the purpose of classification, the performances of feature selection and classifiers are more or less overestimated, and the biomarkers reported more or less overfit the whole data as well. Overfitting can lead poor capability of generalization. A suggested way of biomarker identification is that the performances of feature selection and classifier are evaluated over k-fold CV without using test information during preprocessing, and once such confidence is obtained about the feature selection and classifier (no biomarker is reported as they vary from fold to fold), the biomarkers are selected over the whole data and reported (because the confidence of such feature selection method and the quality of the selected features has already established before). All in all, purposes must be clear when design computational experiments. And special care has to be taken that the class labels of a test set should never be used in any model for any purpose.

4   Conclusion and Future Work

It is crucial to diagnose premalignant pancreatic cancer in a very early stage in order to increase the survival rate of patients. However, it is clinically and computationally difficult. This paper propose to apply fast HDLM as computational model to predict the cancer samples obtained through high resolution mass spectrometry. HDLM can avoid overfitting through maximizing margin and kernelization. Its computation is dimension-free. Experiments show that our HDLM methods achieve competitive performance. Comparison with reported performance shows that our approaches significantly outperform most of the benchmark and proposed approaches. Due to high performance and simplicity of implementation, it will be beneficial to use our methods to the diagnosis of premalignant pancreatic cancer which is suffering low accuracy and sensitivity. And our approaches, combining with the mass spectrometry protein profiling technique, can also be applied to the prediction of other premalignant cancers at an early stage. As future work, our methods will be tested on more protein mass spectrometry data. The performance of different loss functions on high-dimensional mass spectrometry data is still unknown. We will statistically and experimentally compare the performance of HDLM with other liner models of different loss functions on more data. It is also worth investigating suitable kernels for mass spectrometry data.

Acknowledgments. This research has been supported by IEEE CIS Walter Karplus Summer Research Grant 2010, Ontario Graduate Scholarship 2011-2012, and Canadian NSERC Grants #RGPIN228117-2011.

References

1. Ma, B.: Challenges in Computational Analysis of Mass Spectrometry Data for Proteomics. Journal of Computer Science and Technology 25(1), 107–123 (2010)
2. Bishop, C.M.: Pattern Recognition and Machine Learning. Springer, New York (2006)
3. Levner, I.: Feature Selection and Nearest Centroid Classification for Protein Mass Spectrometry. BMC Bioinformatics 6, e68 (2005)
4. Saeys, Y., Inza, I., Larrañaga, P.: A Review of Feature Selection Techniques in Bioinformatics. Bioinformatics 23(19), 2507–2517 (2007)
5. Shawe-Taylor, J., Cristianini, N.: Pattern Recognition and Machine Learning. Cambridge University Press, Cambridge (2004)
6. Pawa, N., Wright, J.M., Arulampalam, T.H.A.: Mass Spectrometry Based Proteomic Profiling for Pancreatic Cancer. JOP. J Pancreas. 11(5), 423–426 (2010)
7. Ge, G., Wong, G.W.: Classification of Premalignant Pancreatic Cancer Mass-Spectrometry Data Using Decision Tree Ensembles. BMC Bioinformatics 9, 275 (2008)
8. Donoho, D.L.: High-Dimensional Data Analysis: The Curses and Blessings of Dimensionality. Lecture in Math Challenges of the 21st Century, pp. 1–32 (2000)
9. Knies, R.: Yi Ma and the Blessing of Dimensionality. Microsoft Research Featured Story (May 28, 2010), http://research.microsoft.com/en-us/news/features/dimensionality-052810.aspx
10. Kroeker, K.L.: Face Recognition Breakthrough. Communications of the ACM 52(8), 18–19 (2010)

11. Vapnik, V.: Statistical Learning Theory, pp. 339–371. Wiley, New York (1998)
12. Chong, E.K.P., Żak, S.H.: An Introduction to Optimization, 3rd edn., pp. 211–246. Wiley, New York (2008)
13. Golub, G.H., Van Loan, C.F.: Matrix Computations, 3rd edn., pp. 206–274. Johns Hopkins, Baltimore (1996)
14. Fawcett, T.: An Introduction to ROC Analysis. Pattern Recognition Letters 27, 861–874 (2006)
15. Mundra, P.A., Rajapakse, J.C.: Gene and Sample Selection for Cancer Classification with Support Vectors Based t-statistic. Neurocomputing 73(13-15), 2353–2362 (2010)
16. Rokach, L.: Ensemble-Based Classifiers. Artificial Intelligence Review 33(1-2), 1–39 (2010)
17. Chapelle, O., Schölkopf, B., Zien, A.: Semi-Supervised Learning, pp. 453–472. MIT Press, Cambridge (2006)
18. Wright, J., Yang, A.Y., Ganesh, A., Sastry, S.S., Ma, Y.: Robust Face Recognition via Sparse Representation. IEEE Transactions on Pattern Analysis and Machine Intelligence 31(2), 210–227 (2010)
19. Wright, J., Ma, Y., Mairal, J., Sapiro, G., Hung, T.S., Yan, S.: Sparse Representation for Computer Vision and Pattern Recognition. Proceedings of The IEEE 98(6), 1031–1044 (2010)
20. Li, Y., Ngom, A.: Classification Approach Based on Non-Negative Least Squares. Technical Report. No. 12-010, School of Computer Science, University of Windsor (2012)
21. Zhang, R., Huang, G.B., Sundararajan, N., Saratchandran, P.: Multicategory Classification Using an Extreme Learning Machine for Microarray Gene Expression Cancer Diagnosis. IEEE/ACM Transactions on Computational Biology and Bioinformatics 4(3), 487–495 (2007)
22. Hingorani, S.R., et al.: Preinvasive and Invasive Ductal Pancreatic Cancer and Its Early Detection in The Mouse. Cancer Cell 4, 437–450 (2003)
23. http://home.ccr.cancer.gov/ncifdaproteomics/ppatterns.asp

Predicting V(D)J Recombination
Using Conditional Random Fields

Raunaq Malhotra, Shruthi Prabhakara, and Raj Acharya

Department of Computer Science Engineering, Pennsylvania State University,
University Park, PA, 16801, USA
{rom5161,sap263,acharya}@cse.psu.edu

Abstract. V(D)J gene segments undergo combinatorial recombination
in the T-cells and B-cells to provide humans and other vertebrates with
a large number of antibodies required for immunity. Each such recombination further undergoes mutations in their DNA sequences so that
they can recognize diverse antigens. Predicting the combination of gene
segments which formed a particular antibody is an essential task for
studying disease propagation and analysis. We propose a model based
on conditional random fields (CRFs) for predicting the boundary positions between V-D-J gene segments. We train the CRFs by generating
synthetic gene recombinations using all of the alleles of the V, D and J
gene segments. The alleles corresponding to a read can be determined
by mapping the segmented reads to the DNA sequences of the gene segments using softwares like BLAST and usearch. We test our method on
simulated dataset as well as real data of Stanford_S22 individual.

Keywords: Conditional Random Fields, VDJ recombination, Mapping
of DNA sequences.

1 Introduction

The immune system of an organism provides protection against a wide range
of antigens with the help of a large number of antibodies. These antibodies
are encoded from genes within the B-cells, and bind to different antigens in
order to protect organisms from diseases. The large number of genes that encode
these antibodies are primarily produced by combinatorial recombination of gene
segments within the B-cells. Identifying the gene segments which encode for
a particular antibody is important for understanding the immune response to
different types of antigens, and in the study of infections.

In B-cells, three types of gene segments or germline components, namely variable (V), diversity (D) and joining (J), combine together to form the variable
region of the immunoglobulin gene [10]. This combination of gene segments takes
place in a combinatorial fashion, in which one of the many alleles of D gene segment combines with an allele of J gene segment. This complex then combines
with one of the alleles of V gene segment to form a rearranged gene, which
has deleted segments between the joined regions. This process of combinatorial

T. Shibuya et al. (Eds.): PRIB 2012, LNBI 7632, pp. 210–221, 2012.

recombination is known as the VDJ recombination. These antibodies can undergo somatic mutations in their DNA sequences by a process known as somatic hyper-mutation [18].

Each antibody molecule consists of light and heavy chain protein molecules [17]. The heavy chain molecule is made up of a VDJ recombination while the light chain consists of recombinations of V and J gene segments only. In humans, there are 281 V, 84 D and 12 J heavy chain alleles [20], which can produce 283,248 possible heavy chain molecules. The number of known functional heavy chain alleles, however, are lesser (50 for V, 27 for D, and 6 for J giving 8100 possible heavy chain molecules [13]). Two types of light chains are also known, the κ [15] and λ [6]. Thus, only considering the combinatorial rearrangements, there can be millions of possible antibodies.

A host of methods have been proposed that align the sequences to the germline gene segments in order to determine the V(D)J configuration. IMGT/V-QUEST maps the DNA sequences of the antibody to an immunoglobulin and T-cell database to identify the V, D and J alleles [8]. JOINSOLVER, on the other hand, determines the gene segments by identifying the conserved motifs in the target gene [20]. SoDA implements a 3-D lattice alignment based on dynamic programming to traverse through all possible states of VDJ gene segments to determine the single highest scoring alignment [21]. The above methods do not provide a meaningful way of evaluating different rearrangements. Moreover, the large number of possible configurations makes sequence alignment equally time consuming and computationally intensive.

iHMMune-align is a probabilistic model that uses Hidden Markov Models (HMMs) for modeling the genes of an antibody to determine their constituent gene segments [7]. The software creates an HMM model for each of the V gene segment alleles connected to all the possible D and J gene segments. It also models the N-nucleotide additions and exonuclease action around the V-to-D or D-to-J gene segment boundaries. Soda2 is another HMMs based statistical model [17]. Although HMMs have been used efficiently for sequential data tasks, a HMM only models the dependencies between a base and its preceding context. It assumes the distribution to be independent of bases in subsequent positions, given the preceding context. Also the transition probability between two states in an HMM are independent of the bases observed in the two states. Such assumptions reduces the model complexity and makes the model tractable. However, in a typical gene segment, the distribution of bases is dependent throughout the length of the sequence, rendering such assumptions invalid.

In this paper, we propose a model based on CRFs that takes into account such dependencies without increasing the inference computation drastically. CRFs are a special type of Markov random fields where the unknown output variables are conditioned on the input variables [12]. For gene allele prediction, as each gene is a combinatorial recombination of the V, D, and J gene segments, the task at hand is to predict the boundary between the gene segments that make up an antibody. First, we predict the boundary between V and D, using a consensus of all V and D alleles in the database. Next, we infer the specific configuration of V and D allele by mapping the segment before the boundary to the known alleles of V and after the boundary to the alleles of D. An identical process is followed for inferring the boundary between D and J gene segments and the corresponding J allele.

The CRFs are trained on a dataset of rearranged VDJ gene segments, where the boundary positions between the gene segments are known. After training, when given a DNA sequence, the CRF predicts a label for each base in the DNA sequence. The label for each base indicates the gene segment (V, D or J) that generated the corresponding base. The alleles constituting the DNA sequence can be determined by mapping the segmented DNA sequence to the database of known alleles.

The paper is organized as follows. Section 2 describes the method based on CRFs for predicting the label sequence corresponding to the input DNA sequence. Section 3 explains the experimental setup and results obtained for simulated dataset. We conclude the paper with a summary and a discussion of future extensions of this work.

2   Methods

We are given a set $\mathbf{X} = \{X_1, X_2, ...X_N\}$ of N reads, each of which is sampled from the rearranged genes. Here each of read is of the form $X_i = \{x_{i1}x_{i2}...x_{in}\}$ where $x_i \in \{A, G, C, T\}$. The read length n may vary from read to read. Our objective is to associate each read X_i with a sequence of labels $Y_i = \{y_{i1}y_{i2}...y_{in}\}$, where y_{ik} denotes the gene segment set from which the base x_{ik} was generated. These sets of gene segments are denoted as $\mathbf{V} = \{V_1, V_2, .., V_K\}$, $\mathbf{D} = \{D_1, D_2, ...D_L\}$, and $\mathbf{J} = \{J_1, J_2, ..., J_M\}$, where (K, L, M) denote the number of alleles for corresponding gene segments. Here, V_i, D_i, J_i represent an allele of the corresponding gene segments.

We address the problem of determining the gene segments constituting a read in two steps. In the first step, we identify the bases x_{ik} and x_{il} at which a transition from V-to-D and D-to-J gene segment occurs. If the boundaries are present within the read, we label each of the bases $(x_{i1}x_{i2}...x_{ik})$ as $\mathbf{V}$, the ones between V-to-D and D-to-J boundaries $(x_{ik+1}x_{ik+2}...x_{il})$ as $\mathbf{D}$, and the rest $(x_{il+1}x_{il+2}...x_{in})$ as $\mathbf{J}$. In the second step, we determine the alleles for each gene segment (V_i, D_j, J_k) by mapping the segmented portions of the read labeled $\mathbf{V}$, $\mathbf{D}$ and $\mathbf{J}$ to the corresponding alleles in the immunoglobulin database.

2.1   Conditional Random Fields for Gene Segment Boundary Detection

For the first part, we propose a model based on conditional random fields (CRFs) for predicting the boundaries between the gene segment set that generates a read. Formally, each read $\mathbf{x} = \{x_1x_2...x_n\} \in \mathbf{X}$ is associated with a sequence of labels $\mathbf{y} = \{y_1y_2...y_n\}$ using CRFs. CRFs were originally proposed as probabilistic models for segmentation and sequential labeling[12]. Such methods have been applied in natural language processing, bioinformatics, image and video segmentation [16,14,1].

We use the linear-chain model of CRFs, where an input node x_i represents a base at a position i in read $\mathbf{x} \in \mathbf{X}$ and an output node y_i denotes the corresponding gene segment label. The conditional probability of the label sequence $\mathbf{y}$ given the observation $\mathbf{x}$ is proportional to

$$\sum_i \exp\left(\sum_j \lambda_j h_j(\mathbf{y}, \mathbf{x}, i)\right) \tag{1}$$

Here $h_j(\mathbf{y}, \mathbf{x}, i)$ is a feature function defined on a subset of the input and output variables that form a clique on the undirected graph and also on the current position i in the input sequence $\mathbf{x}$. The exponential (log-linear) terms in the probability expression are also known as potential functions. For the linear chain graph, where each output label y_i is connected to the preceding output label y_{i-1}, and the input gene sequence $\mathbf{x}$, the feature function is of the form $h_j(y_i, y_{i-1}; \mathbf{x}, i)$. Another popular choice of feature functions are $h_j(y_i; \mathbf{x}, i)$, where the dependence of the current label on the input sequence is captured. These two feature functions are commonly known as the *transition* and *state* feature functions.

The feature functions can be designed to capture various aspects of the given dataset, such as modeling the dependencies on the entire sequence $\mathbf{x}$, as opposed to just the preceding context. This is one of the properties that makes conditional random variables more powerful than Hidden Markov Models for sequential labeling. Each feature function is weighted by λ_j, which determines its contribution in predicting the label. The normalizing constant $Z(\mathbf{x})$ is defined as the sum over all the output labels of all the log-linear potential functions defined above.

$$Z(\mathbf{x}) = \sum_{\mathbf{y}} \sum_i \exp\left(\sum_j \lambda_j h_j(\mathbf{y}, \mathbf{x}, i)\right) \tag{2}$$

Thus, the probability of a label sequence $\mathbf{y}$ given the input sequence $\mathbf{x}$ is given by

$$P(\mathbf{y}|\mathbf{x}) = \frac{1}{Z(\mathbf{x})} \sum_i \exp\left(\sum_j \lambda_j h_j(\mathbf{y}, \mathbf{x}, i)\right) \tag{3}$$

where $\mathbf{\Lambda} = \{\lambda_j\}$ are the parameters of the model. Given a training dataset D, containing a set $\mathbf{X}$ of N sequences and their labels $\mathbf{Y}$ in a training set, we define a log-likelihood parameterized by Λ over all the training samples as

$$L(\mathbf{\Lambda}) = \sum_{(\mathbf{x},\mathbf{y}) \in \mathbf{D}} \log P(\mathbf{y}|\mathbf{x}) \tag{4}$$

The parameter values that maximize the above likelihood are chosen as the model parameter values. To determine the maximum, one can use gradient ascent methods such as Margin Infused Relaxed Algorithm (MIRA) [4], Limited memory BFGS [3].

The model parameters $\boldsymbol{\Lambda}$ that maximize the conditional likelihood are used for predicting the sequence of labels for test read $\mathbf{x}^*$ as follows:

$$\mathbf{y}^* = \arg\max_{\mathbf{y}} P(\mathbf{y}|\mathbf{x}^*) \tag{5}$$

The predicted sequence of labels $\mathbf{y}^*$ indicates the boundaries of the gene segments present in the test sequence.

Feature Functions. The log-linear nature of the feature functions provide the ability to capture complex dependencies on the input data without exponentially increasing the computational complexity for the inference. For applications in text processing such as named entity recognition (NER), the feature functions can be defined to incorporate the grammar of the language, for instance, the word capitalization. In another example, $h_j(\mathbf{y}, \mathbf{x}, i)$ could be defined to count the number of words starting with a capital letter in a sentence. Incorporating feature functions which capture such information increases the predictive power of the model.

In the current context, there is no prior knowledge about such grammar rules for VDJ recombination. In order to overcome such a challenge, we created a set of features which captures different dependencies in the neighborhood of a given base, and learns their weighting parameters from the training dataset. Ideally, the feature functions relevant for determining a V-to-D or a D-to-J junction should get higher weights as compared to the others. The features used are listed in Table 1.

Table 1. Feature functions used for predicting the V,D,J gene segments

Size of neighborhood	Relation to current base
1-base	x_{i-2}
	x_{i-1}
	x_i
	x_{i+1}
	x_{i+2}
2-base	$x_{i-1}x_i$
	$x_i x_{i+1}$
3-base	$x_{i-2}x_{i-1}x_i$
	$x_i x_{i+1}x_{i+2}$
4-base	$x_{i-3}x_{i-2}x_{i-1}x_i$
	$x_i x_{i+1}x_{i+2}x_{i+3}$
5-base	$x_{i-4}x_{i-3}x_{i-2}x_{i-1}x_i$
	$x_i x_{i+1}x_{i+2}x_{i+3}x_{i+4}$
	$x_{i-2}x_{i-1}x_i x_{i+1}x_{i+2}$

2.2   Boundary Detection and Determination of Gene Segment Alleles

Once we obtain the sequence of labels $\mathbf{y}$ for a given sequence $\mathbf{x}$ using CRFs, we can determine the boundary between V-to-D and D-to-J gene segments as given in $\mathbf{y}$. A base's predicted label is considered to be spurious, if all the neighboring bases within a distance of 4 have a identical labels that is different from that of the base under consideration. We correct for such spurious predictions in our method using mode filtering.

The alleles of gene segment present in a read are determined by mapping boundary segmented parts of reads to their corresponding gene segment set. For example, if a part of the read that is predicted to be generated from $\mathbf{V}$ gene segment, we map the read to the alleles in the V-gene segment set to determine the closest matching allele V_i. We use a program *usearch* for mapping the sequence on the allele and assign to it the label of the allele with the highest scoring alignment [5].

3   Experiments and Results

First, we evaluate the performance of CRFs in predicting boundaries between gene segments on simulated datasets. We synthetically generated all the combinatorial recombinations of the alleles of gene segments. The allele sequences for V, D and J gene segments in humans, are known. The combinatorial rearrangements of V, D and J alleles are generated by concatenating an allele of V with an allele of D, followed by an allele of J gene segment. In humans, there are 281 V gene segments, 84 D gene segments and 12 J gene segments, giving rise to a total of 283,248 possible recombinations [20]. The downloaded gene segments are from the Kabat database available on the JOINSOLVER website[20]. The statistics of the V, D, and J gene segments are given in Table 2.

Table 2. Statistics of the alleles present in the Kabat gene sequence database

Gene Segment	Total Number	Average Length	Maximum Length	Minimum Length
V gene segments	281	287	305	103
D gene segments	84	25	37	11
J gene segments	12	53	63	48

We randomly choose 60% of these combinatorial rearrangements for training the CRFs, and use the remaining 40% for testing. We repeated the experiment 5 times in which different 60% of the dataset was used for training, and the remaining 40% for testing. For training the CRFs, we used the software package CRF++ [11]. This implementation allows us to select a set of feature functions based on arbitrary combinations of neighboring nucleotides. Table 1 shows the feature functions that were used for training the linear CRF. We use a combination of bi-,tri-,tetra-, and penta-mers to train the CRF. We did not incorporate the state transition type feature functions as such prior knowledge is usually not available for a real dataset. For training, the default LBFGS training algorithm in CRF++ was used.

The test data for the boundary prediction by CRFs is generated as follows. We randomly choose 10 combinatorial rearrangements from the 40% of the data not used for training and sample reads using 454 sequencing technology. We used MetaSim to simulate 454 sequencing reads [19] with an average length of 200 bps and standard deviation 20 bps.We simulated reads using the default parameters for 454 sequencing technology provided in MetaSim.

For a given read $\mathbf{x}$, the CRFs model returns a label sequence $\mathbf{y}$ where each label represents the gene segment from which the corresponding base was generated.

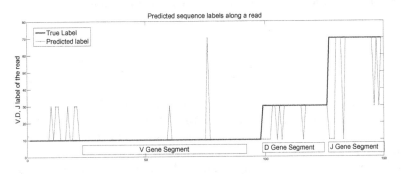

(a) Before the mode filtering

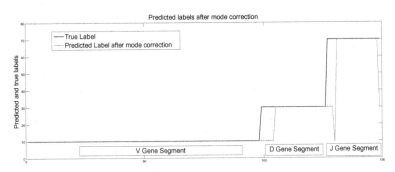

(b) After the mode filtering

Fig. 1. Predicted label sequence for one read

After obtaining the label sequence $\mathbf{y}$ for a given read $\mathbf{x}$, we need to predict the boundary positions between the gene segments. We predict a gene segment boundary at a base x_i, if all the bases after x_i are labeled by a different label as compared to the bases before x_i. A base x_i's label prediction y_i is considered to be spurious if it was surrounded by similarly labeled bases, that differ from the label y_i. For example, we observe that for most of the reads, there is one base labeled as D when all the other surrounding bases in the read are labeled as V. This is depicted in Figure 1, where we represent a read on the x-axis and the true and predicted labels for a read are shown on the y-axis. The V, D, and J labels are assigned levels of 10, 30, and 70 on y-axis for ease of representation.

Predicting a boundary at each position where we observe a change in labeling of the bases in the read, generates a large number of gene segment boundaries, which are not present in the read. We address this problem by first performing a 9-based wide mode filtering on the predicted label sequence. This technique relabels each of base to the mode label in a 9 base window centered on the current base. The window size of 9 was chosen heuristically. A boundary between V-to-D gene segments is called if there is a transition from V-to-D labels in the mode corrected label sequence. If there are multiple such transitions, then we call a boundary at a base having the minimum number of bases labeled as V after the transition. Also, in a given read, as a V-to-J transition is not a valid transition, and we ignore them. We also correct the labeling of all bases between the V-to-D transition and the D-to-J transition as D.

The time complexity for the overall method is same as the time complexity of the CRF method to predict the boundaries for a given set of reads. Once the models for V, D and J gene segments are trained, we can use them for prediction for any number of datasets. The boundary prediction correction, as described above, takes a linear time in terms of the number of reads, thus the time-intensive step being the training time for the CRF method.

Table 3. Precision, Recall, True Negative and Accuracy results for boundary detection of V-to-D and D-to-J gene segments

	V-to-D boundary	D-to-J boundary
Recall	95.7 ± 0.8%	64.1 ± 7.5
Precision	64.5 ± 3.2%	93.6 ± 3.9
True Negative	60.5 ± 6.7%	98.2 ± .8
Accuracy	75.6 ± 3.1%	88.9 ± 2.1

Table 3 reports the precision and recall rates for predicting a gene boundary averaged over the 5 test datasets. These values are calculated separately for the V-to-D and the D-to-J gene segment boundaries. The precision is defined as the number of reads in which a boundary is correctly detected divided by the total number of reads in which same boundary is detected. The recall rate is defined as the ratio of the number of reads in which the gene boundary is correctly detected to the number of reads which actually have that gene boundary. CRFs are more than 90% precise in detecting the boundary between the D-to-J gene segments and are more than 88% accurate for the same. However, the V-to-D boundary detection is not as precise. This can be attributed to the smaller lengths of the D gene segments, making it difficult to correctly predict a base as D.

For most cases, the gene segment boundary was predicted within 6 bases of the actual boundary. We compute the difference between the base position of a predicted gene segment boundary and the base position of a true gene segment boundary. The percentage of the reads in which the boundary was detected within a k base pairs from the true gene segment boundary is shown in Table 4 for $k = \{2, 3, 4, 5, 6\}$. We report the results separately for V-to-D and D-to-J boundaries. The algorithm predicts the boundary between gene segments within six base pairs with an average accuracy of 80%. One can segment the reads using the predicted boundary positions and map the segmented parts to the corresponding gene segments sets to determine the constituent allele within the read.

Table 4. Boundary prediction results as obtained after performing the mode filtering of the labeled sequences

Base pairs window	V-to-D	D-to-J
2 base pairs	$31.4 \pm 11.2\%$	$32.6 \pm 3.1\%$
3 base pairs	$48.2 \pm 8.7\%$	$46.2 \pm 4.8\%$
4 base pairs	$63.1 \pm 8.1\%$	$61.8 \pm 4.6\%$
5 base pairs	$71.9 \pm 7.5\%$	$69.6 \pm 2.9\%$
6 base pairs	$80.2 \pm 4.6\%$	$73.7 \pm 2.7\%$

Table 5 shows the 5-fold precision and recall values for the gene label prediction on a per base basis. The recall for V (D or J) gene segments is defined as the number of bases across all reads which were correctly identified as V (D, or J) divided by the total number of bases with true labels as V (D or J). The precision value is defined as the number of bases correctly labeled as V (D or J) gene segments divided by the total number of bases labeled as V (D or J) gene segments. We observe the highest precision and recall values for the longer V gene segments and lowest values for shorter D gene segments.

Table 5. Precision and recall values for the predicted gene segments on a per base basis

Gene Segment	Recall	Precision
V gene segments	$91.0 \pm 3.2\%$	$97.5 \pm 0.2\%$
D gene segments	$68.9 \pm 1.2\%$	$35.1 \pm 7.9\%$
J gene segments	$74.2 \pm 0.9\%$	$61.5 \pm 9.2\%$

For testing our models on real transcriptome dataset, we use the CRFs trained on all of the synthetic generated recombinations. As the transcriptome for S22 individual consists of rearranged V,D, J gene segments, and the CRFs are also trained on all the junctions obtained from human V,D and J genes, we believe that the usage of the CRFs trained above are a valid choice for the S22 individual.

The Stanford_S22 dataset consists of 13,153 reads from the rearranged VDJ genes for an individual. These reads were obtained from the DNA sequences derived from peripheral blood mononuclear cells [9]. The genotype of the individual is known through a previous study [2]. Thus, we can use the genotype to evaluate the predictions made by our model. We also compare our error rates with the iHMMune align method [7] mentioned in the benchmarking paper [9].

We used our model trained from all synthetically generated recombinations to predict the labels for each base in the reads of the Stanford_S22 dataset. The gene segment boundaries are determined in a read in a similar fashion to that used in the simulated dataset. Using all the predicted gene positions for a V-to-D (or a D-to-J) transition, we call a V-to-D (or a D-to-J) transition at a position which has the minimum number of V (or D and V) gene labeled bases after the gene position. If a D-to-J transition is absent in a read, we call a D-to-J boundary at a base position which is length of D base pairs after the V-to-D transition. This is easily obtained as the length of the D gene segments are known. We use similar corrections for incorrect prediction of a D-to-J transition before a V-to-D transition. Also, as before, a V-to-J transitions are ignored as they are incorrect.

To evaluate our method, we extract gene segments from each read based on the predicted boundaries. We map the predicted gene segments to the database of V, D and J genes using the software *usearch* [5]. An error in the mapping is counted if the mapped gene is not present in the genotype of the individual (given in the dataset). We compared these error results with that obtained for iHMMune align [7]. Table 6 summarizes our results. The error percentages reported for our method are comparable and even better than that for iHMMune-align. This can be explained on the basis that iHMMune align assumes an inherent Markov chain property where the prediction for a base is dependent on the previous bases only. In contrast a CRF uses potential functions dependent on all types of neighborhood relations between the bases. Also as all the genes of one type are modeled together, the general relationship between the genes of a type is captured in the CRF model. This helps in accurately predicting the boundaries between the gene segments. The relevant gene segments for a gene can be determined based on well established sequence searching algorithms (such as BLAST, usearch) once the boundaries are determined.

Table 6. Comparison of our method (CRF-based) to iHMMune Align. The numbers in the parenthesis are the number of errors for each gene type. The error was called for both using a similar technique.

Gene ID	Error % iHMMune Align	Error % CRF-based
V genes	(707) 5.3%	(136) 1.0%
D genes	(1008) 7.6%	(68) 0.5%
J genes	(10) 0.08%	(18) 0.13%

4   Conclusion and Future Work

We have applied the CRFs for identifying the junctions in VDJ recombination. The approach is very similar to Named Entity Recognition in the text domain. In the text domain, each word is labeled as a named entity or not, in a similar fashion, we label parts of the DNA sequences as belonging to the V, D, or J gene segments. The boundary predictions are within 6 base pairs difference of the actual transition in the simulated data. This is the approximately the number of bases that are deleted and inserted (N-nucleotide additions) when the recombination process happens. Thus our method is predicting the gene boundaries within the accepted accuracy. Our method also works well on the Stanford_S22 dataset, where the boundary predictions made lead to most of the gene segments mapping within the genotype of the individual. It is comparable and in some respects better than iHMMune align for predicting the gene segment boundaries. That being said, our method is a work in progress. We have not considered hyper-mutations of the VDJ recombinations, which often change the DNA sequences of these gene segments. These hyper-mutations introduce an additional challenge in predicting the boundaries between the gene segments. Nevertheless, boundary detection between the gene segments when combined with mapping of the detected sequences to the known DNA sequences will help in simplifying the prediction of individual alleles constituting a VDJ recombination.

Acknowledgements. The authors would like to thank Dr. Mary Poss for introducing us to the problem and her constant guidance and extreme patience in explaining the problem. We would also like to thank Bhargavi Panchangam and Dr. Daniel Elleder for insightful discussions.

References

1. Interactive Image Segmentation with Conditional Random Fields, vol. 2 (2008)
2. Boyd, S.D., Marshall, E.L., Merker, J.D., Maniar, J.M., Zhang, L.N., Sahaf, B., Jones, C.D., Simen, B.B., Hanczaruk, B., Nguyen, K.D., Nadeau, K.C., Egholm, M., Miklos, D.B., Zehnder, J.L., Fire, A.Z.: Measurement and clinical monitoring of human lymphocyte clonality by massively parallel v-d-j pyrosequencing. Science Translational Medicine 1(12), 12–23 (2009)
3. Byrd, R.H., Lu, P., Nocedal, J., Zhu, C.: A limited memory algorithm for bound constrained optimization. SIAM J. Sci. Comput. 16(5), 1190–1208 (1995)
4. Crammer, K., Singer, Y.: Ultraconservative online algorithms for multiclass problems. J. Mach. Learn. Res. 3, 951–991 (2003)
5. Edgar, R.C.: Search and clustering orders of magnitude faster than blast. Bioinformatics 26(19), 2460–2461 (2010)
6. Fippiat, J.-P., Williams, S.C., Tomlinson, L.M., Cook, G.P., Cherif, D., Le Paslier, D., Collins, J.E., Dunham, l., Winter, G., Lefranc, M.-P.: Organization of the human immunoglobulin lambda light-chain locus on chromosome 22q11.2. Human Molecular Genetics 4(6), 983–991 (1995)

7. Gata, B.A., Malming, H.R., Jackson, K.J.L., Bain, M.E., Wilson, P., Collins, A.M.: ihmmune-align: hidden markov model-based alignment and identification of germline genes in rearranged immunoglobulin gene sequences. Bioinformatics 23(13), 1580–1587 (2007)

8. Giudicelli, V., Chaume, D., Lefranc, M.-P.: IMGT/V-QUEST, an integrated software program for immunoglobulin and T cell receptor VJ and VD J rearrangement analysis. Nucleic Acids Research 32(suppl. 2), W435–W440 (2004)

9. Jackson, K.J.L., Boyd, S., Gaëta, B.A., Collins, A.M.: Benchmarking the performance of human antibody gene alignment utilities using a 454 sequence dataset. Bioinformatics 26(24), 3129–3130 (2010)

10. Jung, D., Giallourakis, C., Mostoslavsky, R., Alt, F.W.: Mechanism and control of v(d)j recombination at the immunoglobulin heavy chain locus. Annual Review of Immunology 24(1), 541–570 (2006)

11. Kudo, T.: Crf++: Yet another crf toolkit (2005)

12. Lafferty, J., Mccallum, A., Pereira, F.: Conditional Random Fields: Probabilistic Models for Segmenting and Labeling Sequence Data. In: Proc. 18th International Conf. on Machine Learning, pp. 282–289. Morgan Kaufmann, San Francisco (2001)

13. Lefranc, M.-P.: Imgt, the international immunogenetics database: a high-quality information system for comparative immunogenetics and immunology. Developmental &; Comparative Immunology 26(8), 697–705 (2002)

14. Li, M.-H., Lin, L., Wang, X.-L., Liu, T.: Protein protein interaction site prediction based on conditional random fields. Bioinformatics 23(5), 597–604 (2007)

15. Lorenz, W., Straubinger, B., Zachau, H.G.: Physical map of the human immunoglobulin k locus and its implications for the mechanisms of vkjk rearrangement. Nucleic Acids Research 15(23), 9667–9676 (1987)

16. Mccallum, A., Li, W.: Early results for named entity recognition with conditional random fields (2003)

17. Munshaw, S., Kepler, T.B.: SoDA2: a Hidden Markov Model approach for identification of immunoglobulin rearrangements. Bioinformatics 26(7), 867–872 (2010)

18. Neuberger, M.S.: Antibody diversification by somatic mutation: from burnet onwards. Immunolo. Cell Biol. 86, 124–132 (2008)

19. Richter, D.C., Ott, F., Auch, A.F., Schmid, R., Huson, D.H.: MetaSim A Sequencing Simulator for Genomics and Metagenomics. PLoS ONE 3(10), e3373+ (2008)

20. Souto-Carneiro, M.M., Longo, N.S., Russ, D.E., Sun, H.-W.W., Lipsky, P.E.: Characterization of the human Ig heavy chain antigen binding complementarity determining region 3 using a newly developed software algorithm, JOINSOLVER.. Journal of immunology (Baltimore, Md.: 1950) 172(11), 6790–6802 (2004)

21. Volpe, J.M., Cowell, L.G., Kepler, T.B.: Soda: implementation of a 3d alignment algorithm for inference of antigen receptor recombinations. Bioinformatics 22(4), 438–444 (2006)

A Simple Genetic Algorithm for Biomarker Mining

Dusan Popovic, Alejandro Sifrim, Georgios A. Pavlopoulos,
Yves Moreau, and Bart De Moor

ESAT-SCD / IBBT-KU Leuven Future Health Department, Katholieke Universiteit Leuven,
Kasteelpark Arenberg 10, box 2446, 3001, Leuven, Belgium
{Dusan.Popovic,Alejandro.Sifrim,Georgios.Pavlopoulos,
Yves.Moreau,Bart.DeMoor}@esat.kuleuven.be

Abstract. We present a method for prognostics biomarker mining based on a genetic algorithm with a novel fitness function and a bagging-like model averaging scheme. We demonstrate it on publicly available data sets of gene expressions in colon cancer tissue specimens and assess the relevance of the discovered biomarkers by means of a qualitative analysis. Furthermore, we test performance of the method on the cancer recurrence prediction task using two independent external validation sets. The obtained results correspond to the top published performances of gene signatures developed specially for the colon cancer case.

Keywords: genetic algorithm, feature selection, biomarker discovery, gene expressions, colon, cancer, gene signature, k-nearest neighbours, bagging.

1   Background

The recent advances in high-throughput technologies have opened a wide space of opportunities for studying complex diseases, such as cancer, at the molecular level. These led to the successful development of clinically approved diagnostic tests based on gene expression, such as the MammaPrint [1,2] for breast carcinoma. However, the complexity of resulting data from next generation sequencing or microarray experiments still poses a great analytical challenge. High dimensionality that characterizes high-throughput data, together with usually low number of available samples, renders classical statistical methodology nearly helpless when faced with data analysis tasks in this domain. This creates the increasing demand for data-driven modelling approaches capable of facilitating search for prognostics biomarkers. In this study we propose a methodology for mining cancer biomarkers from high-throughput data and demonstrate it on microarray samples in colon cancer.

Colorectal cancer is the third most common cancer type worldwide [3]. The disease starts as a benign polyp that develops to advanced adenoma and finally to invasive carcinoma. Although fairly curable if discovered on time (prior to stage III), a long term survival of initially successfully treated colorectal cancer patients critically depends on the stage of the disease at the time of diagnosis. As the current staging system does not always accurately reflect patient's individual risks [4], there is a

T. Shibuya et al. (Eds.): PRIB 2012, LNBI 7632, pp. 222–232, 2012.

growing need for patient-tailored diagnostics and prognostics tests. This resulted in increased efforts in the development of the gene signatures for this type of cancer [5-7].

The main objective of biomarker mining is to aid in the discovery of genes, proteins or other biological indicators that could be potentially associated with a particular clinical condition. By performing a part of this process in an automated fashion the costs of wet-lab analysis and the clinical trials could be sustainably reduced, which motivated a myriad of recent research initiatives in this direction. In general, one can distinguish between the two main types of tasks and the corresponding methods that fall within a category of biomarker mining. The first includes approaches for the identification of causative factors of disease development and progression, thus of potential therapeutic targets. The second consists of methods for searching biomarkers of which alternations are indicative with, but not necessarily directly involved, in disease onset. These are mostly used for diagnostics or prognostics purposes, which renders this task closely related to feature selection as known in the field of machine learning.

In this work we present a genetic algorithm-based method that facilitates the later approach to biomarker mining. It essentially searches through the space of possible gene combinations to optimize prediction accuracy, taking into account multivariate relations between genes. Also, in contrast to similar existing methods, it explicitly enforces short gene signatures through the fitness function with a constant shrinkage pressure. Furthermore, we employ an iterative randomized procedure similar to bootstrapping to enhance robustness of resulting gene signatures.

Genetic algorithms have been frequently used for feature selection as they scale well with increasing data dimensionality and do not rely on a particular decision surface form. This renders them suitable for solving multidimensional, non-differentiable, non-continuous and other types of problems of arbitrary complexity; such as in genetic biomarker discovery. Jourdan et al. [8] use GA for feature selection, taking into account spatial correlation between neighbouring genes on the chromosome. In [9] Jirapech-Umpai and Aitken proposed an evolutionary approach without cross-over for the same task and demonstrated it on two microarray data sets on cancer. They also compared it against a simple wrapper method based on genetic algorithm. However, both described approaches assume a fixed number of features. Ooi and Tan [10] partially address this problem in an implicit way - by introducing the gene that controls the size of a solution, but still within a predefined range.

This paper is organised as follows. The second section describes the method (2.1) and the datasets (2.2) used. Discussion on the method starts with an introduction to genetic algorithms, followed by a top-level view on the system, a detailed description of the fitness function, other particularities of our implementation and the experimental framework for the external evaluation. The sub-section on data sets (2.2) contains a description of the data together with the details on preprocessing. The third section discusses results in terms of qualitative biological analysis, followed by quantitative external evaluation. Finally, in the fourth section we present our conclusions.

2   Materials and Methods

2.1   Introduction to Genetic Algorithms

Genetic algorithms (GA) [11,12] are a class of search and optimization methods inspired by the "survival of the fittest" concept as known in evolutionary biology. They mimic the process of natural selection by repeatedly generating sets of solutions, called *populations*, from which the fittest *individuals* (sometimes also called *chromosomes*) are selected for producing the next generation. Here each and every individual represents one candidate solution of the optimization problem, usually by an array of binary values called *genes*. It is an iterative process that terminates when the given objective is achieved or when some stopping criteria is met.

The particular implementation of a genetic algorithm is completely characterized by its fitness function and the types of genetic operators used. The fitness function reflects the quality of a single individual (i.e. of a single solution) and thus affects the probability that it later would be kept in the next generation or selected for combining with other well adapted individuals. This function is essential for guiding the search process and therefore its form represents an important algorithm design choice.

The genetic operators play a crucial role in the diversification of the solution pool through chromosomal structure alterations. The two most important are the *crossover* and the *mutation*, while additional custom operators, such as a *random immigrant*, are also used sometimes. Crossover is a mechanism of exchanging genes between two individuals (*parents*) in a random manner to produce child solutions (Fig. 1). It could take various forms given the particular implementation of genetic algorithm, such as single-point, two points "cut and splice", half-uniform, uniform or other. The mutation operator affects one or more genes of a single chromosome in a way that is analogous to natural mutations. Usually, the value of a single bit of individual solution is flipped according to the predefined probability (Fig. 1).

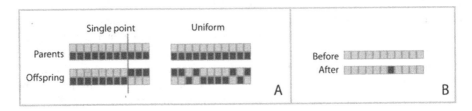

Fig. 1. Genetic operators: crossover (A) and mutation (B)

2.2   The Method

Our strategy for biomarker mining could be summarized by the following workflow (see Fig. 2). The core of our method is a genetic algorithm that optimizes a feature subset given the data and preferred classification performance metrics. This GA utilizes a customized fitness function based on supervised classification and the minimization of genetic signature length. The described optimization process is repeated iteratively, following a procedure similar to bagging [13] to facilitate robustness of the final result.

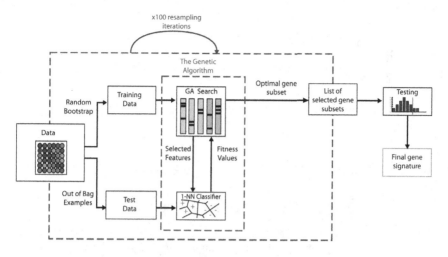

Fig. 2. The workflow of biomarker mining

We represent presence/absence of a biological gene in the signature by a value of a corresponding binary variable (*gene*) in a chromosome of the genetic algorithm. Thus a chromosome (candidate solution) works as a feature mask vector, having ones on the places of features (biological genes) to be selected and zeros elsewhere. A particular instance of potential predictive gene subset is then evaluated by the fitness function and discarded or retained for proliferation with chances proportional to its fitness. This is repeated for several chromosomes during many generations until GA reaches the execution limit, after which the most optimal genetic signature found is returned. We repeat described procedure one hundred times, saving these individual signatures from every iteration.

Each of GA optimization runs that we perform uses a different random sample from the whole training data for internal training of the classifier embedded in the fitness estimation procedure. For this we use Monte Carlo resampling with replacement, where the size of a resample is equal to that of the whole data set (bootstrap [14]). This leaves approximately 36.8% of total examples out, so that they can be utilized for the internal testing (*out-of-bag* examples). As we keep the counts on selected genes over all hundred runs, and use these for the final estimation of a particular gene importance, our procedure for model averaging emulates the bootstrap aggregation principle (bagging).

Counts across candidate genes approximately follow a negative binomial distribution which can be used for determining the threshold for selection. In general, the negative binomial distribution has relaxed assumptions compared to the Poisson distribution, which renders it appropriate for modelling a wider class of count data. Here we decide to include in the final signature genes that were selected more times than the 99% quantile of the estimated negative binomial distribution, which in this case corresponds to 17 or more (Fig 3). However, these counts could be also used as non-parametric ranks if one does not need to pose hard threshold for his/hers particular application, as is often the case in gene prioritization tasks.

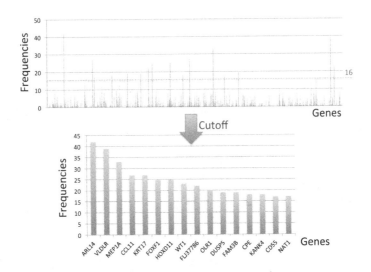

Fig. 3. Extracting the gene signature. The top figure shows how many times each gene has been selected in a signature (out of 100 independent GA runs). The figure on bottom shows these that suppressed the threshold, together with their names and frequencies (the final signature).

The Fitness Function. We use a fitness function that is based on the size of the individual solution and its performance on the independent test set. Firstly, we select genes based on a candidate solution and train one nearest neighbour (1-NN) classifier [15] on a bootstrap sample from the original data set. Then we measure performance of a trained classifier on the out-of-bag examples in terms of balanced accuracy :

$$B_{(j_c,g_c)} = \left(\frac{tp_{(j_c,g_c)}}{tp_{(j_c,g_c)} + fn_{(j_c,g_c)}} + \frac{tn_{(j_c,g_c)}}{tn_{(j_c,g_c)} + fp_{(j_c,g_c)}} \right) / 2 \tag{1}$$

where B stands for the balanced accuracy corresponding to a classifier based on a chromosome j_c from a generation g_c, and tp,tn,fp,fn for the obtained numbers of true positives, true negatives, false positives and false negatives, respectively.

We choose balanced instead of standard accuracy due to its robustness in presence of highly skewed class distributions, which is often a problem with the biomedical data sets in general. Furthermore, we choose 1-NN over more complex classification algorithms as it is very fast to evaluate and still able to capture non-linear relationships in data. When a new data point is presented to the trained algorithm, it simply assigns the outcome value of the closest (usually in terms of the Euclidean distance) example from training set to it. Thus, it also does not require any parameter tuning and, consequently, nested loops in algorithm. In addition, kNN asymptotically achieves Bayes error within a constant factor [16] and there is a body of empirical evidence suggesting that it could not be consistently outperformed by several more complex classification algorithms [17].

Furthermore, we penalize longer, and reward shorter solutions in terms of relative size gain or loss (S) when compared to average size of individuals from the initial generation:

$$S_{(j_c, g_c)} = \frac{Nc \sum_{i=1}^{Ng} f_{(i, j_c, g_c)}}{\sum_{j=1}^{Nc} \sum_{i=1}^{Ng} f_{(i, j, 0)}} \tag{2}$$

where Nc and Ng stand for number of chromosomes in a single generation and number of gene positions per chromosome, respectively; f is a binary variable that equals to one if a gene has been selected given the position (i), chromosome (j) and generation (g); g stands for a generation number (here zero and current generation – g_c). In this way, in addition to maximizing the performance measure we force algorithm to converge toward smaller solutions, hoping that this would lead to more robust and general feature subsets. Finally, given (1) and (2), the fitness function (F) takes the following form (3):

$$F_{(j_c, g_c)} = B_{(j_c, g_c)} - S_{(j_c, g_c)} + 1 \tag{3}$$

where the same weight is given to size and accuracy, while the constant 1 is added to assure that every possible fitness function value remains positive.

Implementation Details of the Proposed GA. We build up our code basing it on the SpeedyGA.m 1.2 Matlab script [18] that implements a simple genetic algorithm as described in [19]. Our initial population counts 200 randomly generated individuals with chance of 0.2 for each feature to be present in one. The probability of mutation per bit of individual chromosome has been set to 0.5 divided by the maximal length of solution (1000). We use uniform crossover, with the probability of reproduction without it set to zero. Selection is preformed proportionally to the sigma-scaled value [19] of the fitness function using the stochastic universal sampling [20]. We restrict the maximal number of generations to 500 and keep track on the best solution over all generations.

External Evaluation. To estimate the generalisation ability of the method we fit a simple linear regression to the selected biomarkers using all samples from the training data set and apply it to two independent test sets. In addition, we compare our algorithm against another frequently used multivariate feature selection method that utilizes bagging and supervised classification performance - namely Random Forest (RF) feature selection [21] on the same data sets. It estimates the importance of the single variable by comparing accuracy of each and every tree in the trained ensemble on corresponding out-of-bag examples against accuracy that is obtained when the values of former are randomly shuffled. To avoid influence of a solution size to the unbiased assessment, we set the number of genes to be selected by the RF to that obtained with our method and number of trees to be generated to hundred.

2.3   Data Sets

We utilize three independent publicly available microarray data sets containing colon cancer samples from the Gene Expression Omnibus (GEO) [22]. The set under GEO accession number GSE17536 [23] has been used for deriving the gene signature where sets GSE17537 [23] and GSE5206 [24] have been considered for the external evaluation of our method. All three data sets were generated on the Affymetrix HG U133 Plus 2.0 microarray platform. Prior to a public release, the first two data sets were preprocessed using the MAS5.0 [25] and the third one by the RMA [26]. In addition, we discard probes that correspond to multiple genes from all three data sets and average values over multiple probes associated with a single gene.

We use samples from the Moffitt Cancer Center (GSE17536) as the training set. This data set contains 177 samples from which 145 with known relapse status (36 out of 145 patients relapsed). Prior to application of the method, we pre-filtered it with the Wilcoxon Rank Sum test by keeping thousand of the most significant genes. The p-value of this particular non-parametric test corresponds to the area under ROC curve, so we use it here due to its robustness. The data from the Vanderbilt Medical Center (GSE17537) are used as one of our external validation sets. Here, the relapse status is determined for all 55 patients with 19 of them having developed recurrent cancer within a five years period. The second validation set (GSE5206) contains samples from 105 patients. We exclude non-diseased subjects and cases where the location of major diagnosis was not the colon, resulting in 74 retained examples in total, from which 16 with recorded recurrence.

3   Results and Discussion

Our final signature consists of 16 genes, namely (in the order of relative importance) : ARL14, VLDLR, MEP1A, CCL11, KRT17, FOXF1, HOXD11, WT1, FLI37786, OLR1, DUSP5, FAM3B, CPE, KANK4, CD55, NAT1. Firstly, we performed functional analysis to estimate the biological relevance of this result. We used Ingenuity Pathway Analysis (IPA) to determine if the signature was significantly enriched for particular pathways or functions of interest. We also performed a transcription factor association analysis in IPA. Out of the 16 signature genes, 10 genes (CCL11,CD55,DUSP5,FOXF1,HOXD11, KRT17, MEP1A, NAT1, OLR1, WT1) were functionally associated with cancer (p=4.84E-04), of which 3 were associated with colorectal cancer in particular (DUSP5,FOXF1, MEP1A, p-value=4.75E-02).

Interestingly the NF-κB complex regulates 5 (DUSP5,FOXF1, CCL11, OLR1,KRT17) of the 16 genes, of which 2 are associated with colorectal cancer: DUSP5 and FOXF1. NF-κB plays a well-studied role in the immune response, cell proliferation and cell survival by inhibition of apoptosis. DUSP5 [27,28] is a kinase phosphatase which negatively regulates members of the mitogen-activated protein (MAP) kinase family, which are associated with cellular proliferation and differentiation. Forkhead box F1 (FOXF1) is a gene associated with multiple cancer types and plays a role as a putative tumor suppressor gene [29,30]; also its inactivation causes megacolon, colorectal muscle hypoplasia and agangliosis [31]. FOXF1 has also been involved in paracrine signalling in association with the WNT signalling pathway, known to be involved in colorectal cancer development [32]. We found FOXF1 to be downregulated in our dataset coinciding with the hypothesis of it being a tumor-suppressor gene. Although no strong evidence supports associating other signature genes with colon cancer, their performance in the signature is likely related to their coexpression with functionally relevant markers, as we can see with the NF-κB regulated genes.

To assess reliability of our approach we test the gene signature that we obtained and another one generated by RF feature selection on the two independent test sets in a way previously described in "external evaluation" sub-section with following results (Fig 4): the GA based feature selection produces an AUC of 0.7705 on GSE5206 data and an AUC of 0.7266 on GSE17537, while the corresponding values for the RF feature selection are 0.7188 and 0.6564. Here we use the area under the ROC curve (AUC) as our preferred metrics for comparing classifiers due to its independence from a biased choice for a decision threshold. On these figures one can notice that, comparing to the Random Forest feature selection, our method yields better results on both testing data sets. In addition, it produces a stable set of biomarkers on repeated runs which the RF does not do.

Furthermore, our results are comparable or better to those already reported in literature [6,7], [33,34]. In [6] the 30-genes signature gives prediction accuracy of 80 and 76,3%, depending on a cross-validation scheme used. Wang et al. [7] suggest a gene signature that includes 23 genes and has corresponding AUC of 0.741. Jiang et al. [33] proposes further refinement of this signature (7 genes) and achieves an AUC of 0.66 on an independent validation set. In a study by Lin et al. [34], the authors test different combinations of classifiers and gene signatures augmented with clinical data on two data sets, resulting in AUCs of 0.73 and 0.80. They do not report AUCs obtained on gene expression data only.

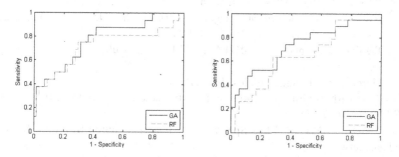

Fig. 4. ROC curves for linear regression based classification using the two feature selection methods on two test data sets (left - GSE5206, right - GSE17537)

However, most of these results are obtained using a single data set and some sort of internal validation. The predictive performance estimation in [6] and [7] relies on a training/validation split scheme (with addition of Monte Carlo crossvalidation in [6]), while [34] employs leave-one-out crossvalidation. We strongly believe that in order to prove robustness of a predictor and to avoid overestimation of its performance, one should test against external data set that originates from different cohort of patients.

Some of these studies [33,34] utilize additional or prior information, while some optimize choice on classifier to be used with biomarkers [34]. Finally, our gene signature is shorter than those reported in [6,7].

4   Conclusions

We present a simple genetic algorithm that is potentially applicable for a variety of biomarker discovery tasks and demonstrate it on the colon cancer recurrence prediction problem. The resulting gene signature displayed similar or better prediction performance than several of these proposed in the literature. Furthermore, in contrary to most of studies on the given problem, we utilize independent test sets for assessment of our method, which gave us indication of strong generalization properties of the resulting predictors. We also demonstrate biological relevance of particular biomarkers by means of a qualitative functional analysis.

In our future work we plan to improve the algorithm via finer tuning of its components and to introduce a dynamic version of the proposed fitness function to facilitate faster convergence. Furthermore, we will test it in conjunction with several popular classifiers to obtain fully optimized and complete classification system. In addition, we look forward to test the method on a wider class of biomarker mining problems and on data originating from various high-throughput platforms.

Acknowledgements. The authors would like to acknowledge support from:

- Research Council KUL: ProMeta, GOA MaNet, KUL PFV/10/016 SymBioSys , START 1, OT 09/052 Biomarker, several PhD/postdoc & fellow grants.
- Flemish Government:
    - o IOF: IOF/HB/10/039 Logic Insulin
    - o FWO: PhD/postdoc grants, projects: G.0871.12N (Neural circuits) research community MLDM; G.0733.09 (3UTR); G.0824.09 (EGFR)
    - o IWT: PhD Grants;   TBM-IOTA3, TBM-Logic Insulin
    - o FOD: Cancer plans
    - o Hercules Stichting: Hercules III PacBio RS
- EU-RTD: ERNSI: European Research Network on System Identification; FP7-HEALTH CHeartED
- COST: Action BM1104: Mass Spectrometry Imaging, Action BM1006: NGS Data analysis network

The scientific responsibility is assumed by its authors.

References

1. Van't Veer, L.J., Dai, H., van de Vijver, M.J., He, Y.D., Hart, A.A., Mao, M., Peterse, H.L., Van der Kooy, K., Marton, M.J., Witteveen, A.T., Schreiber, G.J., Kerkhoven, R.M., Roberts, C., Linsley, P.S., Bernards, R., Friend, S.H.: Gene expression profiling predicts clinical outcome of breast cancer. Nature 415, 530–536 (2002)

2. Glas, A.M., Floore, A., Delahaye, L.J., Witteveen, A.T., Pover, R.C., Bakx, N., Lahti-Domenici, J.S., Bruinsma, T.J., Warmoes, M.O., Bernards, R., Wessels, L.F., Van't Veer, L.J.: Converting a breast cancer microarray signature into a high-throughput diagnostic test. BMC Genomics 7, 278 (2006)
3. Jemal, A., Siegel, R., Ward, E., Hao, Y., Xu, J., Thun, M.J.: Cancer statistics, 2009. CA Cancer J. Clin. 59, 225–249 (2009)
4. O'Connell, J.B., Maggard, M.A., Ko, C.Y.: Colon cancer survival rates with the new American Joint Committee on Cancer sixth edition staging. J. Natl. Cancer. Inst. 96, 1420–1425 (2004)
5. Kerr D., Gray R., Quirke P., Watson D., Yothers G., Lavery I.C., Lee M., O'Connell M.J., Shak S., Wolmark N.: A quantitative multigene RT-PCR assay for prediction of recurrence in stage II colon cancer: Selection of the genes in four large studies and results of the independent, prospectively designed QUASAR validation study. J. Clin. Oncol. 27(suppl.), 169s, abstr 4000 (2009)
6. Barrier, A., Boelle, P.Y., Roser, F., Gregg, J., Tse, C., Brault, D., Lacaine, F., Houry, S., Huguier, M., Franc, B., Flahault, A., Lemoine, A., Dudoit, S.: Stage II colon cancer prognosis prediction by tumor gene expression profiling. J. Clin. Oncol. 24, 4685–4691 (2006)
7. Wang, Y., Jatkoe, T., Zhang, Y., Mutch, M.G., Talantov, D., Jiang, J., McLeod, H.L., Atkins, D.: Gene expression profiles and molecular markers to predict recurrence of Dukes' B colon cancer. J. Clin. Oncol. 22, 1564–1571 (2004)
8. Jourdan, L., Dhaenens, C., Talbi, E.-G.: A genetic algorithm for feature selection in datamining for genetics. In: Proceedings of the 4th Metaheuristics International Conference Porto (MIC 2001), Porto, Portugal, pp. 29–34 (2001)
9. Jirapech-Umpai, T., Aitken, S.: Feature selection and classification for microarray data analysis: evolutionary methods for identifying predictive genes. BMC Bioinformatics 6, 148 (2005)
10. Ooi, C.H., Tan, P.: Genetic algorithms applied to multi-class prediction for the analysis of gene expression data. Bioinformatics 19(1), 37–44 (2003)
11. Fraser, A.: Simulation of genetic systems by automatic digital computers. I. Introduction. Aust. J. Biol. Sci. 10, 484–491 (1957)
12. Holland, J.H.: Adaptation in natural and artificial systems: an introductory analysis with applications to biology, control, and artificial intelligence. University of Michigan Press (1975)
13. Breiman, L.: Bagging predictors. Machine Learning 24(2), 123–140 (1996)
14. Efron, B., Tibshirani, R.: An Introduction to the Bootstrap. Chapman & Hall/CRC, Boca Raton (1993)
15. Cover, T.M., Hart, P.E.: Nearest neighbor pattern classification. IEEE Transactions on Information Theory 13(1), 21–27 (1967)
16. Stone, C.J.: Consistent nonparametric regression. The Annals of Statistics 5(4), 595–620 (1977)
17. Stanfill, C., Waltz, D.: Toward memory-based reasoning. Commun. ACM 29(12), 1213–1228 (1986)
18. Keki, M.B.: Generative Fixation: A Unified Explanation for the Adaptive Capacity of Simple Recombinative Genetic Algorithms. Ph.D. Thesis, Brandeis University (2009)
19. Mitchell, M.: An Introduction to Genetic Algorithms. MIT Press (1996)
20. Baker, J.E.: Reducing Bias and Inefficiency in the Selection Algorithm. In: Proceedings of the Second International Conference on Genetic Algorithms and their Application, pp. 14–21. L. Erlbaum Associates, Hillsdale (1987)
21. Breiman, L.: Random Forests. Machine Learning 45(1), 5–32 (2001)

22. Edgar, R., Domrachev, M., Lash, A.E.: Gene Expression Omnibus: NCBI gene expression and hybridization array data repository. Nucleic Acids Res. 1:30(1), 207–210 (2002)

23. Smith J.J., Deane N.G., Wu F., Merchant N.B., Zhang B., Jiang A., Lu P., Johnson J.C., Schmidt C., Bailey C.E., Eschrich S., Kis C., Levy S., Washington M.K., Heslin M.J., Coffey R.J., Yeatman T.J., Shyr Y., Beauchamp R.D.: Experimentally derived metastasis gene expression profile predicts recurrence and death in patients with colon cancer. Gastroenterology 138(3), 958–968, PMID: 19914252 (2010)

24. Kaiser, S., Park, Y.K., Franklin, J.L., Halberg, R.B., Yu, M., Jessen, W.J., Freudenberg, J., Chen, X., Haigis, K., Jegga, A.G., Kong, S., Sakthivel, B., Xu, H., Reichling, T., Azhar, M., Boivin, G.P., Roberts, R.B., Bissahoyo, A.C., Gonzales, F., Bloom, G.C., Eschrich, S., Carter, S.L., Aronow, J.E., Kleimeyer, J., Kleimeyer, M., Ramaswamy, V., Settle, S.H., Boone, B., Levy, S., Graff, J.M., Doetschman, T., Groden, J., Dove, W.F., Threadgill, D.W., Yeatman, T.J., Coffey Jr., R.J., Aronow, B.J.: Transcriptional recapitulation and subversion of embryonic colon development by mouse colon tumor models and human colon cancer. Genome Biol. 8(7), R131, PMID: 17615082 (2007)

25. Hubbell, E., Liu, W.M., Mei, R.: Robust estimators for expression analysis. Bioinformatics 18(12), 1585–1592 (2002)

26. Irizarry, R.A., Hobbs, B., Collin, F., Beazer-Barclay, Y.D., Antonellis, K.J., Scherf, U., Speed, T.P.: Exploration, normalization, and summaries of high density oligonucleotide array probe level data. Biostatistics 4(2), 249–264 (2003)

27. Mandl, M., Slack, D.N., Keyse, S.M.: Specific inactivation and nuclear anchoring of extracellular signal-regulated kinase 2 by the inducible dual-specificity protein phosphatase DUSP5. Mol. Cell. Biol. 25(5), 1830–1845 (2005)

28. Ueda, K., Arakawa, H., Nakamura, Y.: Dual-specificity phosphatase 5 (DUSP5) as a direct transcriptional target of tumor sup-pressor p53. Oncogene 22(36), 5586–5591 (2003)

29. Watson, J.E., Doggett, N.A., Albertson, D.G., Andaya, A., Chinnaiyan, A., van Dekken, H., Ginzinger, D., Haqq, C., James, K., Kamkar, S., Kowbel, D., Pinkel, D., Schmitt, L., Simko, J.P., Volik, S., Weinberg, V.K., Paris, P.L., Collins, C.: Integration of high-resolution array com-parative genomic hybridization analysis of chromosome 16q with expression array data refines common regions of loss at 16q23-qter and identifies underlying candidate tumor suppressor genes in prostate cancer. Oncogene 23, 3487–3494 (2004)

30. Lo, P.K., Lee, J.S., Liang, X., Han, L., Mori, T., Fackler, M.J., Sadik, H., Argani, P., Pandita, T.K., Su-kumar, S.: Epigenetic inactivation of the potential tumor suppressor gene FOXF1 in breast cancer. Cancer Res. 70, 6047–6058 (2010)

31. Ormestad, M., Astorga, J., Landgren, H., Wang, T., Johansson, B.R., Miura, N., Carlsson, P.: Foxf1 and Foxf2 control murine gut development by limiting mesenchymal Wnt signaling and promoting extracellular matrix production. Development 133, 833–843 (2006)

32. Madison, B.B., McKenna, L.B., Dolson, D., Epstein, D.J., Kaestner, K.H.: FoxF1 and FoxL1 link hedgehog signaling and the control of epithelial proliferation in the developing stomach and intestine. J. Biol. Chem. 284, 5936–5944 (2009)

33. Jiang, Y., Casey, G., Lavery, I.C., Zhang, Y., Talantov, D., Martin-McGreevy, M., Skacel, M., Manilich, E., Mazumder, A., Atkins, D., Delaney, C.P., Wang, Y.: Development of a clinically feasible molecular assay to predict recurrence of stage II colon cancer. J. Mol. Diagn. 10, 346–354 (2008)

34. Lin, Y.H., Friederichs, J., Black, M.A., Mages, J., Rosenberg, R., Guilford, P.J., Phillips, V., Thompson-Fawcett, M., Kasabov, N., Toro, T., Merrie, A.E., van Rij, A., Yoon, H.S., McCall, J.L., Siewert, J.R., Holzmann, B., Reeve, A.E.: Multiple gene expression classifiers from different array platforms predict poor prognosis of colorectal cancer. Clin. Cancer. Res. 13, 498–507 (2007)

Finding Conserved Regions
in Protein Structures Using Support Vector
Machines and Structure Alignment

Tatsuya Akutsu, Morihiro Hayashida, and Takeyuki Tamura

Bioinformatics Center, Institute for Chemical Research, Kyoto University,
Gokasho, Uji, Kyoto 611-0011, Japan
{takutsu,morihiro,tamura}@kuicr.kyoto-u.ac.jp

Abstract. This paper proposes a novel method for finding conserved regions in three-dimensional protein structures. The method combines support vector machines (SVMs), feature selection and protein structure alignment. For that purpose, a new feature vector is developed based on structure alignment for fragments of protein backbone structures. The results of preliminary computational experiments suggest that the proposed method is useful to find common structural fragments in similar proteins.

1 Introduction

Analysis of protein structures is an important topic in bioinformatics and computational biology. In particular, classification of protein structures and identification of common structural patterns are very important. For that purpose, a lot of studies have been done and several databases have been developed such as SCOP [3] and CATH [13]. *Protein structure alignment* is a powerful approach to comparison of protein structures [1,9,16]. Furthermore, *multiple structure alignment* is useful to identify common patterns of multiple protein structures [12,17]. However, it is known that multiple structure alignment and identification of conserved regions are NP-hard if gaps (i.e., insertions and/or deletions of amino acid residues) are allowed [2]. Indeed, existing methods have some problems in computation time and/or accuracy and thus other approaches should also be studied.

Recently, *support vector machines* (SVMs) have been applied to classification of protein structures, where SVMs are a statistical method widely used in bioinformatics and other various areas [6,15]. In order to apply SVMs to protein structures, a *kernel function* or a *feature vector* for protein structure is required. Dobson and Doig developed a feature vector based on various information on proteins [7], which includes secondary-structure content, amino acid propensities, surface properties and ligands. Their feature vector was applied to classification of proteins into enzymes and non-enzymes. Borgwardt *et al.* developed kernel functions based on graph kernels [4], where each protein structure is represented

T. Shibuya et al. (Eds.): PRIB 2012, LNBI 7632, pp. 233–242, 2012.

as a graph using secondary structure information. In order to improve the prediction accuracy, they also used additional features similar to those used by Dobson and Doig. Qiu *et al.* proposed a kernel for protein structures using a structure alignment algorithm [14]. Though these methods are very useful for predictions, it is difficult to extract structural information or common regions of proteins from the results of SVM learning. Therefore, it is desirable to develop a method with which structural information and/or common regions can be extracted.

In this paper, we propose a simple feature vector for finding common regions of protein structures. The proposed feature vector is based on the concept of *spectrum kernel* for sequence data [11]. The spectrum kernel uses a feature vector based on the numbers of occurrences of substrings of fixed length, where the length is usually short (e.g., 2 or 3). Though it is very simple, this method or similar methods are effectively applied to various problems in bioinformatics. Instead of substrings, our proposed feature vector uses a set of *template fragments of protein backbone structures*. And then, occurrences of similar fragments are taken into account in the feature vector. Different from the spectrum kernel, we use longer fragments each of which consists of several tens of $C\alpha$ atoms. Moreover, similarities between fragments are measured by means of structural alignment because gaps cannot be ignored for such long fragments. For computing structural alignment, STRALIGN is employed, which was previously developed by one of the authors [1]. One of the important points of the proposed feature vector is that, different from existing methods [4,7], it uses structural information only and does not use any additional information such as secondary-structure content, amino acid propensities and so on.

The proposed feature vector is combined with SVMs in order to classify protein structures. Furthermore, it is combined with a feature selection method so as to find fragments conserved in multiple protein structures. To examine the proposed method, we performed computational experiments. The results suggest that the proposed method is useful to find common structural fragments in similar proteins.

2   Preliminaries

In this section, we briefly review SVM [6,15] and STRALIGN [1].

2.1   Support Vector Machine and Feature Vector

SVM is a kind of statistical learning method and is basically used for binary classification. Let POS and NEG be the sets of *positive examples* and *negative examples* in a training data set, where each example is represented as a point in d-dimensional Euclidean space (see Fig. 1). Then, an SVM finds a hyperplane h such that the distance between h and the closest point is the maximum (i.e., the margin is maximized) under the condition that all points in POS lie above h, and all points in NEG lie below h. Once this h is obtained, we can infer that a new test data is positive (resp. negative) if it lies above h (resp. below h).

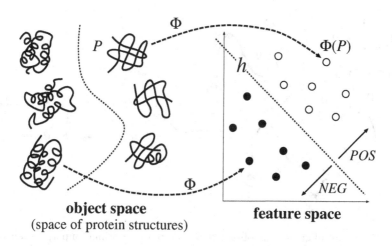

Fig. 1. Support vector classification (left) and feature map (right). In order to apply SVMs to analysis of protein structures, each structure should be mapped to a point (feature vector) in feature space.

If there does not exist h that completely separates POS from NEG, the SVM finds h which maximizes the *soft margin*, where we omit details of the soft margin [6,15].

In order to apply SVMs to real-world problems, it is important to design a *feature vector* or a *kernel function* suited to an application problem since objects to be classified are not usually points in Euclidean space. That is, we should find a feature mapping Φ from the object space $\mathcal{X}$ to the d-dimensional Euclidean space $\mathcal{R}^d$ (we can even consider infinite dimensional space). Then, $\Phi(x)$ is called a *feature vector* and $\mathcal{R}^d$ is called the *feature space*. That is, Φ transforms an object $x \in \mathcal{X}$ to a feature vector $\Phi(x)$:

$$x \in \mathcal{X} \Longrightarrow \Phi(x) \in R^d.$$

We also define a kernel K from $\mathcal{X} \times \mathcal{X}$ to $\mathcal{R}$ by

$$K(x, y) = \Phi(x) \cdot \Phi(y),$$

where $\Phi(x) \cdot \Phi(y)$ is the inner product between vectors $\Phi(x)$ and $\Phi(y)$. $K(x, y)$ is regarded as a measure of similarity between x and y.

2.2   STRALIGN

Protein structure alignment is a problem of finding amino acid pairs occupying spatially equivalent positions, given two 3D protein structures. Though the output of protein structure alignment is almost the same as that of pairwise sequence alignment, structural similarities are considered instead of similarities of

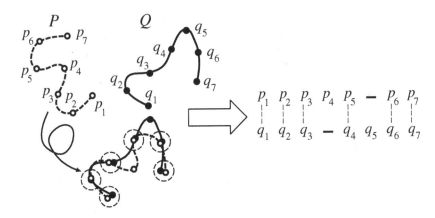

Fig. 2. Structure alignment is obtained by computing optimal superposition of two structures

amino acids (see Fig. 2). While many methods have been proposed for structure alignment [9,16], we use STRALIGN developed by one of the authors [1] because it is designed based on concrete theoretical foundation and is easy to modify. STRALIGN computes structure alignment in the following way.

STEP 1: A series of initial superpositions are computed from pairs of structural fragments (of length 10-20) using a standard technique to compute an optimal superposition without gaps (i.e., RMS (root mean squares) fitting).

STEP 2: For each of such superpositions, a rough alignment is first computed using a dynamic programming technique, and then is refined through an iterative improvement procedure which also uses dynamic programming.

STEP 3: Finally, the best alignment among those is selected as an output.

3   Method

In the proposed method, each protein structure P in training and test data sets is transformed into a feature vector $\Phi(P)$ and then SVM learning and classification are performed in a usual manner. Furthermore, feature selection is performed in order to extract common structural fragments. In the following, we describe outlines of computation of a feature vector and selection of important features.

3.1   Feature Vector Based on Similarity of Structural Fragments

In this work, each protein structure is represented by a sequence of positions of $C\alpha$ atoms. Let $P = (p_1, p_2, \ldots, p_n)$ be a sequence of positions of $C\alpha$ atoms. In the proposed method, a feature vector $\Phi(P)$ for protein structure P is defined as follows (see also Fig. 3).

Let L be the length of a structural fragment, where a fragment is a consecutive sequence of positions of Cα atoms, and $L = 40$ was employed in this work based on several trails. Let $\mathcal{T}$ be a set of template structures. Let $Q = (q_1, \ldots, q_m)$ be a template structure in $\mathcal{T}$. A set of fragments $frag(Q)$ from Q is defined by

$$frag(Q) = \{ (q_{i\Delta+1}, q_{i\Delta+2}, \ldots, q_{i\Delta+L}) \mid i = 0, 1, 2, \cdots \text{ and } i\Delta + L \le m \},$$

where $\Delta = 10$ was used in this work. Then, a set of template fragments $\mathcal{F}$ is defined as

$$\mathcal{F} = \bigcup_{Q \in \mathcal{T}} frag(Q).$$

That is, a set of template fragments contains several fragments from each template structure, where template structures are selected from positive and negative classes (but not included in training or test data set).

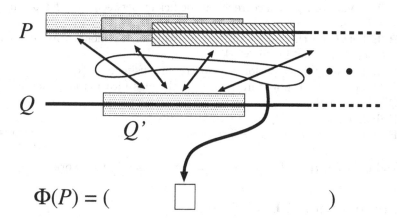

Fig. 3. Computation of a feature vector. Each coordinate in a feature vector corresponds to template fragment Q', where the coordinate value is defined by the sum of the scores for fragments in P against Q'.

For a structural fragment P' from a training or test protein structure P and a template fragment Q', we define the score $w(P', Q')$ by

$$w(P', Q') = \frac{\text{the number of superposed residue pairs}}{|P|},$$

where $|P|$ denotes the number of residues in P. We used this measure to evaluate the result of structural alignment between P' and Q' because STRALIGN tries to maximize the number of superposed residue pairs within some distance threshold. Then, the feature vector $\Phi(P)$ for a training or test protein structure P is defined by

$$\Phi(P) = \left(\sum_{P' \in frag(P)} w(P', Q') \right)_{Q' \in \mathcal{F}}.$$

That is, each coordinate value corresponding to a template fragment $Q' \in \mathcal{F}$ is defined by the sum of the scores for fragments of P against Q'.

3.2   Feature Selection

In order to find conserved structural fragments, we employ Recursive Feature Elimination (RFE) [8], which is a well-known feature selection method for SVMs. Different from the original RFE [8], we use the prediction accuracy (for the training data set) as a measure for eliminating features. Moreover, pre-processing based on Pearson correlation coefficient is introduced so as to eliminate redundant features efficiently. The following is an outline of our feature selection method:

STEP 1: Let $\mathcal{F}_0$ be a set of all template fragments.
STEP 2: Compute Pearson correlation coefficient between each $f \in \mathcal{F}$ and the class (i.e., positive or negative).
STEP 3: Let $\mathcal{F}$ be the subset of $\mathcal{F}_0$ consisting of fragments with H highest coefficients ($H = 30$ in this work).
STEP 4: For all $Q' \in \mathcal{F}$, perform SVM training using $\mathcal{F} - \{Q'\}$.
STEP 5: Let Q'' be the feature such that the classification accuracy for $\mathcal{F} - \{Q''\}$ is the highest.
STEP 6: Let $\mathcal{F} \leftarrow \mathcal{F} - \{Q''\}$.
STEP 7: Repeat STEPS 4-6 until reaching the specified number of features K.

It should be noted that $H = 30$ and $K = 3$ were used in this work.

4   Computational Experiments

We performed preliminary computational experiments in order to examine the potential power of the proposed method. We used protein structure data from ASTRAL [5] and SCOP [3] databases, where these two databases are closely related. We used the structure and classification data of ASTRAL SCOP 1.69 with less than 40% sequence identity. All experiments were performed on a PC cluster with AMD Opteron Model 280 (2.4GHz) CPUs, where each evaluation (i.e., combination of fold class and the feature selection method) took several minutes using one CPU in most cases.

First, we examined the accuracy of binary classification using SVM and feature vector $\Phi(P)$, where we employed SVM^{light} [10] for SVM learning and classification. We used the following eight SCOP fold classes each of which contains sufficient number of non-homologous proteins:

   a.24: Four-helical up-and-down bundle,
   a.118: Alpha-alpha superhelix,
   b.29: Concanavalin A-like lectins/glucanases
   b.40: OB-fold,
   c.1: TIM beta/alpha-barrel,

c.23: Flavodoxin-like,
d.15: Beta-Grasp (ubiquitin-like),
d.58: Ferredoxin-like.

For each fold class F, we examined binary classification (i.e., predict whether or not a given protein structure belongs to F). For that purpose, 40 protein structures were randomly selected from F as positive data and 40 protein structures were randomly selected from other fold classes as negative data. As for template structures, 4 protein structures were randomly selected from F and 4 protein structures were randomly selected from other fold classes, under the condition that these 8 structures were different from the above 80 protein structures. Then, $\Phi(P)$ was computed for each P of 80 protein structures. Using these feature vectors and SVM^{light}, 5-fold cross validation was performed. For each fold class, *sensitivity*, *specificity* and *overall accuracy* are shown in the left part (corresponding to "all features") of Table 1. It should be noted that these measures are defined by:

$$\text{sensitivity} = \frac{TP}{TP + FN},$$

$$\text{specificity} = \frac{TP}{TP + FP},$$

$$\text{accuracy} = \frac{TP + TN}{TP + FP + TN + FN},$$

where TP, FP, TN and FN denote the numbers of true positives (structures correctly classified to the target fold class), false positives, true negatives and false negatives, respectively. It is seen that overall accuracies are reasonably good though these may not be the best among existing methods. It should be noted that we do not pay much attention to optimization of classification accuracy in this paper. Instead, we are more interested in identification of conserved fragments.

Next, we applied the proposed feature selection method to the same data sets, where the target number of features (i.e., the number of structural fragments) was set to 3 (i.e., $K = 3$). For comparison, we examined a very simple method that selects features with 3 highest Pearson correlation coefficients. Then, 5-fold cross validation was performed for each case and the results are shown in the middle and right parts of Table 1. It is very interesting to note that feature selection is useful to increase the classification accuracy. The results suggest that protein structures are well-classified by using only a small number of fragments. It is also seen that the RFE-based selection method is better than or as good as the Pearson-based selection method in most cases. Thus, the proposed feature selection method is considered to be useful for selecting fragments for protein structure classification.

Finally, we measured the conservation ratio of the best fragment among 3 fragments selected by the proposed method. For each fold class and for each of positive and negative data sets, we calculated the ratio of the number of protein structures containing the fragment to the total number of protein structures

Table 1. Comparison of classification accuracy (%) for different sets of features. Bold numbers correspond to the best classification accuracies among the three methods.

Class	all features Sens.	Spec.	Acc.	REF-based selection Sens.	Spec.	Acc	Pearson-based selection Sens.	Spec.	Acc
a.24	90.0	90.0	90.0	92.5	90.2	**91.3**	92.5	90.2	**91.3**
a.118	100.0	93.0	96.4	100.0	97.6	**98.8**	97.5	100.0	**98.8**
b.29	72.5	100.0	86.3	87.5	89.7	**88.8**	80.0	86.5	83.8
b.40	90.0	81.8	85.0	70.0	96.6	**83.8**	65.0	92.9	80.0
c.1	87.5	83.3	85.0	87.5	87.5	87.5	95.0	84.4	**88.8**
c.23	72.5	85.3	80.0	87.5	89.7	**88.8**	80.0	86.5	83.8
d.15	77.5	93.9	86.3	75.0	93.8	**85.0**	67.5	96.4	82.5
d.58	52.5	80.8	70.0	60.0	85.7	**75.0**	57.5	88.5	**75.0**

Table 2. Conservation ratios (%) of selected fragments

	Fold Class a.24 a.118 b.29 b.40 c.1 c.23 d.15 d.58
Positive	85.0 92.5 91.4 42.5 80.0 70.0 62.5 72.5
Negative	30.0 10.0 28.6 0.0 10.0 20.0 7.5 40.0

(i.e., 40) in the data set, where protein structure P is regarded to contain fragment Q' if the number of superposed residue pairs is no less than $0.65 \cdot |Q'|$. It is to be noted that the ratio should be high for positive data whereas it should be low for negative data. The result is shown in Table 2, where the threshold of $0.8 \cdot |Q'|$ was used for the case of 'a.118' (since the ratios for positive/negative data sets were 100%/60% if the threshold of $0.65 \cdot |Q'|$ was used). It is observed that good conservation ratios were obtained for most cases. For the case of 'd.58', the ratios were not good. But, it is consistent with the classification result in Table 1. For the case of 'b.40', the ratio for positive data was low. However, the ratios were 65.0%/0% if the threshold of $0.5 \cdot |Q'|$ was used. In summary, the proposed feature selection method is considered to be useful for selecting conserved fragments. It should be noted that, though conservation of a single fragment was examined here, multiple fragments are required to obtain the results of Table 1 and thus selection of multiple features is still important.

5   Concluding Remarks

We proposed a method for finding conserved regions in similar proteins. The method is a combination of a new feature vector based on structure alignment for fragments with two techniques in statistical learning: support vector machines and feature selection. It should be noted that, different from a common approach to identify conserved regions, the proposed method does not use multiple structure alignment though it uses pairwise structure alignment for fragments. The results of preliminary computational experiments suggest that the proposed method is useful to identify important structural fragments.

One of important future work is to perform rigorous and larger scale computational experiments, which include (i) adjustment of parameters (e.g., L, Δ, K and H) used in the method, (ii) study of the sensitivity of these parameters, (iii) comparison with other kernels for protein structures (e.g., [4,7,14]), and (iv) examination of other feature selection methods. It is also important to study biological meaning and/or significance of the selected fragments.

In the proposed method, configurations between fragments are not taken into account. However, configurations between fragments may play an important role in protein functions. In particular, such information seems important if we would like to predict interactions between proteins and/or interactions between proteins and chemical compounds. Therefore, a feature vector and/or a kernel function reflecting such information should also be developed.

References

1. Akutsu, T.: Protein structure alignment using dynamic programming and iterative improvement. IEICE Trans. Inf. Syst. E79-D, 1629–1636 (1996)
2. Akutsu, T., Halldórsson, M.M.: On the approximation of largest common subtrees and largest common point sets. Theoret. Comp. Sci. 233, 33–50 (2000)
3. Andreeva, A., Howorth, D., Brenner, S.E., Hubbard, T.J.P., Chothia, C., Murzin, A.G.: SCOP database in 2004: refinements integrate structure and sequence family data. Nucleic Acids Res. 32, D226–D229 (2004)
4. Borgwardt, K.M., Ong, C.S., Schönauer, S., Vishwanathan, S.V.N., Smola, A.J., Kriegel, H-P.: Protein function prediction via graph kernels. Bioinformatics 21, i47–i56 (2005)
5. Chandonia, J.M., Hon, G., Walker, N.S., Lo Conte, L., Koehl, P., Levitt, M., Brenner, S.E.: The ASTRAL compendium in 2004. Nucleic Acids Res. 32, D189–D192 (2004)
6. Cortes, C., Vapnik, V.: Support-vector networks. Machine Learning 20, 273–297 (1995)
7. Dobson, P.D., Doig, A.J.: Distinguishing enzyme structures from non-enzymes without alignment. J. Mol. Biol. 330, 771–783 (2003)
8. Guyon, I., Weston, J., Barnhill, S., Vapnik, V.: Gene selection for cancer classification using support vector machines. Machine Learning 46, 389–422 (2002)
9. Holm, L., Sander, C.: Protein structure comparison by alignment of distance matrices. J. Mol. Biol. 233, 123–138 (1993)
10. Joachims, T.: Making large-scale SVM learning practical. In: Schölkopf, B., Burges, C., Smola, A. (eds.) Advances in Kernel Methods - Support Vector Learning, pp. 41–56. MIT Press (1999)
11. Leslie, C., Eskin, E., Noble, W.S.: The spectrum kernel: a string kernel for SVM protein classification. In: Proc. Pacific Symposium on Biocomputing, vol. 7, pp. 564–575 (2002)
12. Lupyan, D., Leo-Macias, A., Ortiz, A.R.: A new progressive-iterative algorithm for multiple structure alignment. Bioinformatics 21, 3255–3263 (2005)
13. Pearl, F.M., Bennett, C.F., Bray, J.E., Harrison, A.P., Martin, N., Shepherd, A., Sillitoe, I., Thornton, J., Orengo, C.A.: The CATH database: an extended protein family resource for structural and functional genomics. Nucleic Acids Res. 31, 452–455 (2003)

14. Qiu, J., Ben-Hur, A., Vert, J.-P., Noble, W.S.: A structural alignment kernel for protein structures. Bioinformatics 23, 1090–1098 (2007)
15. Shawe-Taylor, J., Cristianini, N.: Kernel Methods for Pattern Analysis. Cambridge Univ. Press (2004)
16. Shindyalov, I.N., Bourne, P.E.: Protein structure alignment by incremental combinatorial extension (CE) of the optimal path. Protein Eng. 11, 739–747 (1998)
17. Ye, Y., Godzik, A.: Multiple flexible structure alignment using partial order graphs. Bioinformatics 21, 2362–2369 (2005)

Aligning Discovered Patterns
from Protein Family Sequences

En-Shiun Annie Lee, Dennis Zhuang, and Andrew K.C. Wong

University of Waterloo,
Centre of Pattern Analysis and Machine Intelligence,
200 University Avenue West, Waterloo, Ontario N2L 3G1
http://www.pami.uwaterloo.ca

Abstract. A basic task in protein analysis is to discover a set of sequence patterns that characterizes the function of a protein family. To address this task, we introduce a synthesized pattern representation called Aligned Pattern (AP) Cluster to discover potential functional segments in protein sequences. We apply our algorithm to identify and display the binding segments for the Cytochrome C. and Ubiquitin protein families. The resulting AP Clusters correspond to protein binding segments that surround the binding residues. When compared to the results from the protein annotation databases, PROSITE and pFam, ours are more efficient in computation and comprehensive in quality. The significance of the AP Cluster is that it is able to capture subtle variations of the binding segments in protein families. It thus could help to reduce time-consuming simulations and experimentation in the protein analysis.

Keywords: Protein Analysis, Protein Function Identification, Pattern Discovery, Pattern Clustering, Hierarchical Clustering, Motif Finding, Local Alignment, Approximate String Matching.

1 Introduction

Proteins are involved in many biological processes of the organism, from enzyme catalysts to ligand binding. To rapidly and reliably find out from the primary sequence to which known protein family an uncharacterised protein belongs will help to understand its functions and roles in the cellular processes. A protein often assumes a specific function such as binding or enzymatic activity and thus its functionality constrains regions such as binding sites. In addition, domains are less subject to mutations, giving rise to certain discernible conserved segments in the primary sequence. In another word, proteins in the same family can be homologues or distantly related in their primary sequence but they might contain conserved segments (often called motifs, patterns, or fingerprints). It is therefore important to discover such conserved areas that characterize a protein family.

There are different approaches for identifying the similar conserved regions in the protein sequences that characterize a protein family. One approach is multiple sequence alignment, which takes a set of protein sequences aligned by

T. Shibuya et al. (Eds.): PRIB 2012, LNBI 7632, pp. 243–254, 2012.

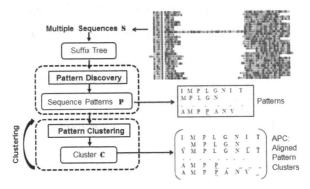

Fig. 1. The overview of the Pattern Alignment (PA) Process. Our method involves two steps: the Pattern Discovery Step and the Pattern Alignment Step. The results are the APC.

dynamic programming to come up with homologous regions, that may be of consequence of functional, structural, or evolutionary relationships. CLUSTAL W [1], T-Coffee [2], DIALIGN [3] and HMMER [4] are the representatives of such approaches. Concerning the computation complexity, it has been shown that finding the global optimal alignment is an NP-complete problem [5]. Even with heuristics, it is still not efficient enough to handle large scale dataset. Moreover, this approach is considered as more suitable for sequences that are globally homologous with have high level of similarity. The result would be unsatisfactory if the sequences are only distantly related or just share local similarities. Another approach is multiple local alignment, unlike the multiple sequence alignment [6] that aligns the whole sequence, attempts to locate and align locally similar subsequences and build up a probabilistic model for describing the conserved regions that represent the motif in the sequences. Hence it is also called the motif finding approach. A commonly used model is position weight matrix (PWM), which assumes independent position in the motif. However such assumption is not realistic in many cases. Furthermore, it is computationally expensive to obtain global optimum. Hence, heuristics such as Expectation Maximization and Gibbs Sampling are used to find the locally optimal model. Two well known methods are MEME [7] and GLAM [6]. This approach often returns one or more highest score solutions. It is likely to miss those motifs that are statistically significant. Furthermore, the reported motifs often have high false positive rate [8]. Another approach is to generate sequence patterns that repeat sufficient times precisely or approximately with variations in the sequences in an exhaustive fashion. YMF and Weeder are two examples for such approach. However, the common problem is that there are usually too many patterns discovered and each pattern often partially characterizes the functional regions in the sequences since even the functional sites may exhibit a certain degree of variability. To overcome this limitation, we present in this paper a new method that groups and aligns the similar patterns discovered by a sequence pattern discovery algorithm into aligned pattern clusters. The aligned pattern cluster is able to align significant patterns while capturing more variability.

Aligned Pattern (AP) Clusters are used to reveal and represent protein functional segments. For eachAPC obtained, We examine whether it corresponds to binding segments or other protein functional segments. When we applied our PA Process to the Cytochrome C. and Ubiquitin protein families, we did find such strong correspondence. Our PA Process is efficient. The results obtained are consistent with the motifs found in the two well known databases: pFam and PROSITE. This shows that the APCs obtained capture the functional regions of a protein family.

2 Methods

Our method (Fig. 1) takes the sequence patterns obtained by a previously developed method as input, and groups and aligns them into aligned pattern clusters. The resulting knowledge-rich representations is abbreviated as APC . We will briefly describe the pattern discovery process, but will focus mainly on the pattern aligning and clustering process.

2.1 Discover Sequence Pattern

Let Σ be an ALPHABET set containing the elements $\{\sigma_1, \sigma_2, \ldots, \sigma_{|\Sigma|-1}, \sigma_{|\Sigma|}\}$. Let $\mathbb{S} = \{S_1, S_2, \ldots, S_{|\mathbb{S}|-1}, S_{|\mathbb{S}|}\}$ be a set of MULTIPLE SEQUENCES. Each sequence is composed of consecutive elements taken from the alphabet Σ. For protein sequences, the alphabet can be the 20 amino acids. The pattern discovery method in [13] takes the multiple sequences and produces a list of sequence patterns $\mathbb{P} = \{P_1, P_2, \ldots, P_{|\mathbb{P}|-1}, P_{|\mathbb{P}|}\}$. Each pattern P is essentially a substring from the input sequences but passes three conditions. First, it is frequent, that is, it repeats itself sufficiently many times in the input sequences. Second, it is statistically significant, meaning that the pattern is not resulted by the random associations of elements given a random background model. Third, it is not redundant compared against the other patterns in the result set. The information provided by a non-redundant pattern cannot be accounted by other patterns. With these three conditions, a compact yet informative set of patterns are obtained. The running time of the pattern discovery process takes linear time to the input size and thus is efficient. The discovered patterns correspond to potential functional segments in the sequences. We devised a score to rank the patterns according to their interestingness. The score is $s = \frac{q_P}{N} \cdot z_P$, where q_P is the number of sequences where the pattern P appears, N is the number of sequences, and z_P is the statistical significance.

2.2 Aligning Similar Patterns

For the task of pattern alignment, we develop an algorithm which groups a set of similar patterns of different lengths obtained from the pattern discovery process and then align and cluster them into a set of APs of the same length by inserting gaps and wildcards. These APs are aligned into a matrix group where corresponding residues amongst the patterns are aligned on the same column, thus implying a common functionality among the APs [9]. An APC, C, is a group of similar patterns that have been aligned into a set of APs $\mathbb{P} = \{P_1, P_2, ..., P_m\}$ represented by C, which can be expressed as

$$C = \text{Align} \begin{pmatrix} P_1 \\ P_2 \\ \vdots \\ P_m \end{pmatrix} = \begin{pmatrix} s_1^1 & s_2^1 & \cdots & s_n^1 \\ s_1^2 & s_2^2 & \cdots & s_n^2 \\ \vdots & \vdots & \vdots & \vdots \\ s_1^m & s_1^m & \cdots & s_n^m \end{pmatrix}_{n \times m}, \tag{1}$$

where $s_j^i \in \Sigma \cup \{\_\}$ is an AP P_i with newly aligned column index j. Each of the m APs in the rows of C is of length n.

An ALIGNED PATTERN $P = s_1^P s_2^P ... s_{|P|}^P$ is a subsequence of order-preserving elements maximizing the similarity of the patterns within $\mathbb{P}$ with gaps and mismatches so that each $P \in \mathbb{P}$ is of length n. An ALIGNED COLUMN c_j in C represents the jth column of characters from the set of APs forming the current APC. Thus, $C = \begin{pmatrix} c_1 & c_2 & ... & c_n \end{pmatrix}$.

The Alignment Algorithm. The algorithm iteratively ALIGNs two APCs in a pairwise-manner based on their ALIGNMENT score and that they do not lie on the same sequences. The alignment algorithm combines two APC into one iteratively in the hierarchical manner. Two possible alignment algorithms are considered in this paper: the NeedlemanWunsch alignment algorithm, which is global, and the SmithWaterman alignment algorithm, which is local. The ALIGNMENT is essentially a dynamic programming algorithm that, first, recursively builds a score table from the optimal sub-scores by forward-scoring and, then, backtracks through the score table from the optimal score to arrive at the final solution. The runtime for computing the score table of two APCs, $\mathbb{P}_1$ and $\mathbb{P}_2$, in the dynamic programming algorithm is $O(|\mathbb{P}_1||\mathbb{P}_2|)$. Note that depending on the type of alignment score used, there may be an added linear time of complexity described in the next section.

The Alignment Score. Two major categories of ALIGNMENT scores are explored for computing the score of matching the combined aligned columns of two APCs: the sum-of-pair scores and the entropy-based scores. The sum-of-pair scores has the runtime of $O(m|\mathbb{P}_1|k|\mathbb{P}_2|)$ and the entropy-based scores has the runtime of $O((m + k)|\mathbb{P}_1||\mathbb{P}_2|)$.

Sum-of-Pair Scores. The sum-of-pair scores compare all pairs of residues from the two APCs' aligned columns and scores them using Hamming Distance. In addition to Hamming Distance, we also considered weighting the penalty of the Hamming Distance to prefer gaps or to prefer mismatches.

Entropy-Based Scores. The entropy-based scores constitute more variational information than the sum-of-pair scores. Instead, this category of scores uses the probability distribution of the existing character residues occurring at the combined aligned sites. The two different entropy-based scores considered are the Information Entropy Score and the Information Gain Score.

The Stopping Conditions. The STOPPING condition of the ALIGNMENT algorithms, like the ALIGNMENT scores chosen, also determines the quality of the final resulting APCs. The STOPPING conditions considered are the Number of Patterns per Cluster and the Final Number of Clusters.

3 Synthetic Results and Discussion

For demonstrating the runtime and quality of our method, we created nine sets of synthetic input data containing synthetic patterns of length 10, where each pattern occurs with a frequency of five and pattern has a 10% chance of mutation at a random position from the previous pattern. These nine datasets vary based on the number of synthetic patterns in each set in increments of five.

The Runtime Comparison. To compare the runtime of our PA Process against the combinatorial method, we plotted the experimental runtime of our PA Process. We measured the experimental runtime of our PA Process by counting the number of character comparisons and plot it against the number of synthetic patterns in the dataset. Five ALIGNMENT scores are plotted for the global alignment and for the local alignment resulting in ten combinations. As described in the methodology section, the pairwise-sum-of scores performed slower than the entropy scores due to a more complete pairwise comparisons (Fig. 2).

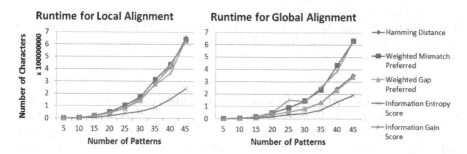

Fig. 2. The five ALIGNMENT scores are Hamming Distance, Weighted Mismatch Preferred, Weighted Gap Preferred, Information Entropy Score, and Information Gain Score. The first graph compare the runtime of five ALIGNMENT scores while executing local ALIGNMENT Algorithm while the second graph executes global ALIGNMENT Algorithm.

248   E.-S. Annie Lee, D. Zhuang, and A.K.C. Wong

The Alignment Algorithm and Score. To determine the parameters that yield the highest quality of APCs, we examined the combinations of the ALIGNMENT algorithms, and the ALIGNMENT scores. We compare the resulting quality of the APC by computing the Average Cluster Entropy of all the normalized entropy of the final clusters and their columns. The first set of tuning experiments identify the optimal combination of the ALIGNMENT algorithms with the ALIGNMENT scores (Fig. 3). Of the five ALIGNMENT scores compared, the sum-of-pairs scores performed better than entropy scores because they exhausively compare all pairs of amino acids from both aligned columns and take longer to execute. These observations indicate that sum-of-pair scores tend to perform better than entropy-based scores because these scores use the full residue and take longer to run. Global alignment performs better than local alignment because it aligns the full pattern rather than a sub-sequence of the pattern.

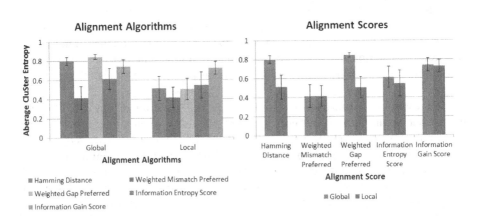

Fig. 3. The first graph divides the ten Average Pattern Quality into the two ALIGNMENT Algorithms. For global alignment, the Hamming Distance is the best ALIGNMENT Score; for local alignment, the Information Gain Score is the best. The second graph divides the ten Average Pattern Quality into the five ALIGNMENT Scores. Of the two ALIGNMENT algorithms compared, the global alignment results in a better APC than the local alignment.

The Stopping Conditions. To examine the properties of the STOPPING conditions, we fixed the ALIGNMENT algorithm to Global Alignment and the ALIGNMENT score to Hamming Distance. We measured the Average Cluster Quality and observed how it varies with the two STOPPING conditions (Fig. 4): 1) the Number of Patterns per Cluster, and 2) the Final Number of Clusters. The threshold is adjusted for each set of synthetic patterns. The first STOPPING condition, the Final Number of Clusters, results in an inverse exponential curve, since the threshold point occurs when the quality of the APCs decreases rapidly.

There is an ideal threshold point where the quality of the APC is close to the optimal value of one and increases slowly. The Second STOPPING condition by the number of clusters fits a logarithmic curve, because decreasing the number of clusters also increases the number of patterns which in turn increases the cluster entropy.

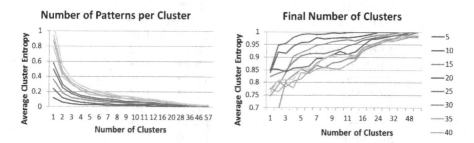

Fig. 4. Finally the two conditions considered for stopping the alignment are: 1) the Number of Patterns per Cluster, which fits a inverse exponential curve, and 2) the Final Number of Clusters, which fits a logarithmic curve

4 Biological Results and Discussion

4.1 Cytochrome C. Results and Discussion

To demonstrate that the binding segments of a protein family can be represented by APCs, we executed the pattern alignment method on a list of patterns that had resulted from the pattern discovery process. The downloaded input sequences from pFam are from the protein family Cytochrome C., which is uniquely identified by the pFam family identification number PF00034. The pFam seed sequences of the Cytochrome C. contains 238 essential sequences. The two binding residues in the Cytochrome C. protein that binds the heme ligand are (1) the proximal binding segment that binds the heme ligand from the proximal side of the protein and (2) the distal binding segment that binds the opposite side of the heme ligand from the distal side of the protein.

Table 1 shows the top ranked patterns. Most of them correspond to the proximal and distal binding segments for the Cytochrome C. protein family. Nineteen of these top twenty patterns contain the binding residues that is crucial for the binding functionality of the Cytochrome C. family protein. However, each pattern, on its own, has a small fraction of supporting sequences and hence a single pattern alone cannot represent variety of the functional binding segments in the protein sequences. However,the APC, containing a set of similar patterns with variations, provides a much more detailed description of the binding segments and are able to capture their variability.

Table 1. Top 20 Patterns in the Full Sequences of the Cytochrome C. Family

Rank	Pattern	Frequency	Score	Binding Residues
1	ADRGEKLYQKVGCV	8	1179941.62	
2	CSMCHAREPVW	6	55750.35	H18
3	GRCSMCHAREP	6	23786.79	H18
4	RCSMCHAREP	8	12410.76	H18
5	IMPLGNITQMT	5	11628.94	M62
6	CSMCHAREP	11	5021.18	H18
7	MPLGNITQMT	6	3763.97	M62
8	GRCSMCHA	11	928.88	H18
9	RCSMCHA	16	576.93	H18
10	MCHAREP	13	250.46	H18
11	MPLGNITQ	7	202.92	M62
12	CSMCHA	19	174.14	H18
13	SHAMPPAN	6	117.56	M62
14	GVSHAMPP	6	117.14	M62
15	HAMPPANV	5	79.82	M62
16	MPLGNIT	8	57.37	M62
17	HAMPPAN	8	47.54	M62
18	MCHAAEP	6	33.14	H18
19	SHAMPP	12	32.01	M62
20	CAACH	22	27.97	H18

Table 2. Top 20 Patterns in the Full Sequences of the Ubiquitin Protein Family

Rank	Pattern	Frequency	Score	Binding Residues
1	TLHLVLRL	5	161.28	
2	DYNIQKE	5	104.63	Lys63
3	DYNIQK	7	55.28	Lys63
4	AGKQLED	5	53.62	Lys48
5	QQRLIF	7	39.87	
6	LIFAGK	7	39.25	Lys48
7	YNIQK	9	23	Lys63
8	DQQRLI	6	19.96	
9	LIYSGK	5	17.47	Lys48
10	QQRLI	11	16.88	
11	IFAGK	8	16.81	
12	QRLIF	9	16.61	
13	KEGIP	9	15.66	Lys33
14	KTLTGK	6	13.49	Lys6, Lys11
15	VKAKIQ	5	13.26	Lys27,Lys29
16	LHLVL	10	11.85	
17	QRLIY	7	10.95	
18	LIYSG	7	10.36	
19	LIYAG	6	9.51	
20	ESTLH	6	7.1	

In our first biological study, we showed that protein functional segments can be represented by a set of patterns called an APC , built using our PA Process. The set of discovered APCs are displayed with pFam's alignment represented by HMM Logo (Fig. 5a). The APCs contained invariant sites in their columns and APs in its rows. For the rightmost proximal APC, the three top invariant sites, His18, Cys17, and Cys14 in their proper location, are essential to the functionality of the Cytochrome C. protein family for binding the heme ligand. More precisely, the His18 invariant site acts as the proximal binding residue to the heme iron, and the two Cysteines invariant sites, Cys14 and Cys17, link the two thioether bonds to the two vinyl groups on the heme. Similarly, the Met62 invariant site in the distal APC acts as the distal binding residue to the heme iron from the opposite distal side of the protein. These resulting APCs contain invariant sites corresponding to the binding residues, which are the main biological function of Cytochrome C. protein family. Also, the binding residues, represented by invariant sites, are surrounded by APs that form the functional binding segment.

Our discovered proximal APC for Cytochrome C. is consistent with the proximal binding motif [C]-x(2)-[CH] from PROSITE [10,11]and the strong emission probability from pFam [12,13]. Moreover, our method identified the distal binding APC , whereas PROSITE does not annotate this APC as a binding motif and pFam only identifies it as a weak emission probability.

4.2   Ubiquitin Results and Discussion

We applied our method to the Ubiquitin protein family. The input sequences from pFam are from the Ubiquitin protein family, which is uniquely identified in pFam by the family identification PF00240 and contains 78 essential sequences that have a maximum length of 83. Table 2 shows that many top ranked patterns correspond to the seven binding residues of the Ubiquitin protein. Other patterns correspond to the conserved elements around the binding residues. Though the discovered patterns do indicate some important functional signals in this family of Ubiquitin proteins, each pattern on its own has only a small fraction of supporting sequences and thus achieve a low sensitivity in representing the binding segments of this protein family. Proteins often exhibit great variability and thus APC would represent its functional sites more effectively and explicitly.

In our Ubiquitin experiment, we executed our PA Process on the multiple unaligned sequences of the Ubiquitin protein family. The Ubiquitin contains seven lysine residues, Lys6, Lys11, Lys27, Lys29, Lys33, Lys48, and Lys63 that can be linked to another Ubiquitin to form a poly-Ubiquitin chain [14–18]. The six APCs contain five out of the seven binding residues, however two remaining binding residues, Lys27 and Lys29, was not sufficient variants to be aligned and grouped into APCs in the pattern alignment process (Fig. 5b). For Ubiquitin, our results did not agree with the PROSITE consensus motif for the Ubiquitin domain signature, K-x(2)-[LIVM]-x-[DESAK]-x(3)-[LIVM]-[PAQ]-x(3)-Q-x-[LIVM]-[LIVMC]-[LIVMFY]-x-G-x(4)-[DE], which misses 172 Ubiquitin proteins. However, our results did agree with the profile HMM's emission probability in pFam.

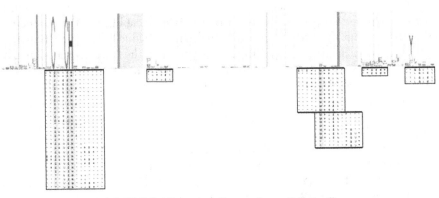

(a) HMM Alignment Comparison of Cyto C.

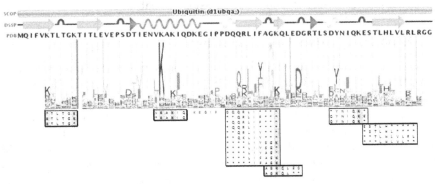

(b) HMM Alignment Comparison of Ubiquitin

Fig. 5. In figure (a) two of the largest resulting APCs represent the proximal and distal binding segments of the Cytochrome C. are compared to the HMM logo from pFam. In the largest APC Cys14, Cys17, and His18 are identified as the invariant sites. In the second and third largest APCs that overlap, Met62 is the invariant site of the distal binding segment where Met62 binds the heme iron. In Figure (b) the four resulting binding segments for the Ubiquitin protein family are compared to the HMM logo from pFam. The six discovered APCs contain five out of the seven binding residues: Lys6, Lys11, Lys33, Lys48, and Lys63.

5   Conclusion

In summary, our PA Process is able to identify APCs that correspond to protein binding segments for the Cytochrome C. and the Ubiquitin protein family. The APCs shows APs as its rows and residue variations in its aligned columns, which captures binding segment variations. In fact, for Cytochrome C., the invariant sites in the proximal APC are the binding residues as identified in PROSITE and pFam. However, the distal APC identifies an invariant site as the binding residue which is not identified in PROSITE. Hence, APCs can render much more effective protein analysis by automatically finding and grouping similar patterns from the sequences and narrowing down the important segments to be examined.

Acknowledgment. The authors thank the reviewers for their constructive feedback, Dr. K. Durston for insightful analysis of molecular biophysics, and D. Leong, C. M. Li, B. Chang, and D. Yuen for reading this manuscript. This research is supported by an NSERC Post Graduate Scholarship, Ontario Graduate Scholarship and an NSERC Discovery Grant.

References

1. Thompson, J.D., Higgins, D.G., Gibson, T.J.: Clustal w: improving the sensitivity of progressive multiple sequence alignment through sequence weighting, position-specific gap penalties and weight matrix choice. Nucleic Acids Res. 22(22), 4673–4680 (1994)
2. Notredame, C., Higgins, D.G., Heringa, J.: T-coffee: A novel method for fast and accurate multiple sequence alignment. J. Mol. Biol. 302(1), 205–217 (2000)
3. Subramanian, A.R., Kaufmann, A.M., Morgenstern, B.: Dialign-tx: greedy and progressive approaches for segment-based multiple sequence alignment. Algorithms Mol. Biol. 3, 6 (2008)
4. Durbin, R., Eddy, S.R., Krogh, A., Mitchison, G.: Biological Sequence Analysis: Probabilistic models of proteins and nucleic acids. Cambridge University Press (1998)
5. Wang, L., Jiang, T.: On the complexity of multiple sequence alignment. Journal of Computational Biology 1(4), 337–348 (1994)
6. Frith, M.C., Hansen, U., Spouge, J.L., Weng, Z.: Finding functional sequence elements by multiple local alignment. Nucleic Acids Res. 32(1), 189–200 (2004)
7. Bailey, T.L., Elkan, C.: Unsupervised learning of multiple motifs in biopolymers using expectation maximization. Machine Learning 21(1/2), 51–80 (1995)
8. Pisanti, N., Crochemore, M., Grossi, R., Sagot, M.F.: Bases of motifs for generating repeated patterns with wild cards. IEEE/ACM Transactions on Computational BIology and Bioinformatics 2(1), 40–50 (2005)
9. Lee, E.-S.A., Wong, A.K.C.: Synthesizing aligned random pattern digraphs from protein sequence patterns. In: Bioinformatics and Biomedicine Workshops (BIBMW), pp. 178–185 (2011)
10. Bairoch, A.: Prosite: a dictionary of sites and patterns in proteins. Nucleic Acids Research 19, 2241–2245 (1991)
11. Sigrist, C.J.A., Cerutti, L., de Castro, E., Langendijk-Genevaux, P.S., Bulliard, V., Bairoch, A., Hulo, N.: Prosite, a protein domain database for functional characterization and annotation. Nucleic Acids Res. 38(Database issue), 161–166 (2010)

12. Sonnhammer, E.L., Eddy, S.R., Durbin, R.: Pfam: A comprehensive database of protein domain families based on seed alignments. PROTEINS: Structure, Function, and Genetics 28, 405–420 (1997)
13. Finn, R.D., Mistry, J., Tate, J., Coggill, P., Heger, A., Pollington, J.E., Gavin, O.L., Gunasekaran, P., Ceric, G., Forslund, K., Holm, L., Sonnhammer, E.L., Eddy, S.R., Bateman, A.: The pfam protein families database. Nucleic Acids Research 211, D211–D222 (2010)
14. Peng, J., Schwartz, Elias, Thoreen, Cheng, Marsischky, Roelofs, et al.: A proteomics approach to understanding protein ubiquitination. Nature Biotechnology 21(8), 921–926 (2003)
15. Xu, P.P.: Characterization of polyubiquitin chain structure by middle-down mass spectrometry. Analytical Chemistry 80(9), 3438–3444 (2008)
16. Kirisako, T., Kamei, K., Kato, M., Fukumoto, Kanie, Sano, Tokunaga: A ubiquitin ligase complex assembles linear polyubiquitin chains. The EMBO Journal 25(20), 4877–4887 (2006)
17. Kim, H., Kim, Lledias, Kisselev, S., Skowyra, Gygi, Goldberg: Goldberg: Certain pairs of ubiquitin-conjugating enzymes (e2s) and ubiquitin-protein ligases (e3s) synthesize condegradable forked ubiquitin chains containing all possible isopeptide linkages. The Journal of Biological Chemistry 282(24), 17375–17386 (2007)
18. Ikeda, F.: Dikic: Atypical ubiquitin chains: new molecular signals. 'protein modifications: Beyond the usual suspects' review series. EMBO Reports 9 (6), 536–542 (2008)

Application of the Burrows-Wheeler Transform for Searching for Approximate Tandem Repeats

Agnieszka Danek[1], Rafał Pokrzywa[1],
Izabela Makałowska[2], and Andrzej Polański[1]

[1] Institute of Informatics, Silesian University of Technology,
Akademicka 16, 44-100 Gliwice, Poland
agnieszka.danek@polsl.pl
[2] Laboratory of Bioinformatics, Faculty of Biology, Adam Mickiewicz University,
Umultowska 89, 61-614 Poznań, Poland

Abstract. Tandem repeats (TRs) are contiguous copies of repeating patterns, which may be either exact or approximate. Approximate tandem repeats (ATRs) in a genomic sequences are adjacent copies of a repeating pattern of nucleotides, where similarity is defined by a suitable measure. Both TRs and ATRs are used in forensic analysis, DNA mapping, testing for inherited diseases and many evolutionary studies. All their functions and roles are not well defined and remains a subject of ongoing investigation. However, growing biological databases together with tools to look for such repeats may lead to better understanding of their behavior. This paper presents our method for searching for ATRs defined on the basis of the model of substitution mutations and its comparison to two other tools. The capabilities and limitations of methods are analyzed and results obtained with each tool are investigated.

Keywords: approximate tandem repeats, Burrows-Wheeler transform, suffix array, Hamming distance.

1 Introduction

Tandem repeats (TRs) are consecutive, repeating patterns in genomic sequences. TRs belong to the most important loci in genomes due to their abundance in DNA sequences and to their role both in evolution and in molecular mechanism of functioning of organisms. Evolution of tandem repeats loci is governed by a mechanism called slippage mutation, e.g. [1], which due to its high intensity belongs to the major factors of genomic dynamics. A very important issue is the dynamics of interaction between slippage and point mutation, which is still an area of an intensive research [2], [3], [4]. As for functional roles of TRs in cellular mechanisms there is a lot of evidence proving linkage of TRs to important molecular processes in cells. TRs play important roles in the gene expression and transcription regulations [5]. They are also widely used as markers for DNA mapping and DNA fingerprinting [7]. It is well known that when TRs are occurring in increased, abnormal number, they cause a series of inherited diseases [6] (i.e. trinucleotide repeat disorders).

T. Shibuya et al. (Eds.): PRIB 2012, LNBI 7632, pp. 255–266, 2012.

Critically important element in the research on TRs is development of tools for their efficient and accurate identification. Locations of TRs in genomes can be detected by using appropriately designed experimental techniques [8] based on DNA amplification methodologies. However, in the era of very high power of direct sequencing technology, methodologies of discovering TRs by using text mining techniques are coming to the first place. These methodologies allow for efficient and massive detection of both patterns and locations of TRs in genomic sequences. In the aspect of discovery by using text mining, TRs should be divided into two groups, exact tandem repeats (ETRs) and approximate tandem repeats (ATRs). For both groups of TRs many detection methods were published, some are overviewed in the recent survey papers [9] [10] [11] [12]. Detection tools based on text mining can be divided into two classes. The first class includes algorithms suitable for searching for ATRs based on introducing mathematical models or transformations to represent and measure repeatability of DNA sequences. These algorithms use models and methods such as autocorrelation distance between sequences, transforming nucleotide symbols to numerical values and then using frequency domain analyses, HMM models [13] [14]. The second class of methods involves combinatorial text searches through genomic databases [15] [16] [17] [18]. As result of intensive studies on TRs and developing methods of their efficient identification several growing-in-size biological databases of TRs have been developed [19] [20] [21]. These databases support many researches in molecular biology and evolution.

As observed in comparisons presented in the literature [9] [12], detection methods for TRs still need development and refinement due to their limitations and differences seen between various approaches. This observation is particularly important for ATRs, due to the fact that different approaches not only differ in algorithmic aspects but also often use different measures of similarity between motifs, which makes the results more difficult to compare. In this paper we present a new method for combinatorial, exhaustive detection of approximate tandem repeats based on application of the Burrows-Wheeler Transform, called BWatrs [15], [22]. We also show a study devoted to comparisons of different algorithms for discovery of ATRs. We compare three tools for discovery of ATRs, mreps published in [17], Tandem Repeat Finder published in [23] and our tool BWatrs. We compare tools for searching for ATRs by using quantitative performance measures suitable for evaluating combinatorial text search engines, number the detected pattern stratified with respect to motif length. Our research is based on developing an algorithm dedicated to analysis and comparison of lists of results returned by different ATRs search tools. To the best of our knowledge this study is the first one devoted exclusively to comparisons of combinatorial text mining algorithms for discovery of ATRs and presenting results of extensive browsing of returned lists of ATRs, including one-to-one identities of detected motifs and numbers of unique motives specific to each tool. The difficulty of task addressed in our study stems from inconsistent definitions of ATR among different papers and varying approaches to look for ATRs. In order to pursue scheduled research we had to set standards in parameter choices for different tools.

In the following sections, three evaluated tools are presented with the focus on our program. Next the methodology of comparison and performed experiment are described. Finally, the results of the comparisons are given together with some conclusions.

2 Evaluated Tools

In our research we evaluated three different tools designed to look for the approximate tandem repeats (ATRs) in the genomic sequences. In the succeeding subsections each of them is presented. The approaches are briefly described along with all parameters that can be tuned. Additionally, a more detailed explanation of our algorithm is provided.

2.1 Burrows-Wheeler Approximate Tandem Repeats Searcher

Burrows-Wheeler Approximate Tandem Repeats Searcher (BWatrs) is our approach to find ATRs. Its prior version was presented in [22]. It is a development of a method for searching for exact tandem repeats described in [24], [15]. To be aware of kind of ATR looked for by our method, it is necessary to present some related terminology. First, we define a measure of dissimilarity between two strings and the successive definitions help to understand a type of ATRs searched.

Definition 1. Given two strings A and B of equal length, $h(A, B)$ is a Hamming distance between A and B, that is a minimal number of substitution needed to be done in string A to transform it to string B.

Definition 2. A K-mismatch double ATR with period p is a string consisting of two consecutive strings S_1 and S_2, both of length p, that $h(S_1, S_2) \leq K$.

Definition 3. A K-mismatch ATR with period p is any string of length $n \geq 2p$ for which every substring of length $2p$ is a K-mismatch double ATR with period p. The ratio n/p is called an exponent.

Definition 4. A maximal K-mismatch ATR with period p is a string which is a K-mismatch ATR with period p and cannot be further extended to the left or to the right to still meet the definition of the K-mismatch ATR with period p.

Assuming we are given a string S over the alphabet Σ, range of acceptable periods $< p_1, p_2 >$ and maximal number of errors K, we are interested in finding all maximal K-mismatch approximate tandem repeats with period p, where $p \in < p_1, p_2 >$, within the string S. In our future consideration when we refer to ATR searched by the BWatrs we mean this kind of repeat.

Parameters. The BWatrs takes as an input a sequence S and several parameters determining the type of ATRs that are searched: minimum and maximum period (*minPeriod* and *maxPeriod*), minimum exponent *minExp*, minimum total length *minTotal*, maximum number of errors K and maximum percentage of errors *Kprc* between adjacent repeats, minimum total score *minScore* and a flag enabling/disabling marking *mark*.

The meaning of the *minPeriod*, *maxPeriod* and *K* parameters is directly connected with the Def. 4 of the searched ATR. The *minExp* and *minTotal* are straightforward and represent minimum acceptable exponent and total length of the ATR. The *Kprc* gives the possibility to define what percentage of the consecutive repeats (each *K-mismatch double ATR*) can be mismatched. In case of every period, the more restrictive of *K* and *Kprc* is taken into account. The *minScore* parameter defines the minimum total score (*totalScore*) of the ATR, which is calculated according to idea presented in [17], as:

$$totalScore = 1 - \frac{\#errors}{totalLength - period} \qquad (1)$$

where *#errors* is a sum of all errors (mismatches between the successive motifs) in the ATR. If after an error, the nucleotide return to its previous state, it is count as only one error. The *totalScore* was introduced to control the level of similarity between the whole ATR (not only between the adjacent motifs) and to determine the best period of the found ATR. Finally, the *mark* flag determine whether regions with ATRs already found should be marked to exclude them from future analysis.

Algorithm. At the beginning the input string S over the alphabet Σ is converted according to the Burrows-Wheeler transform (BWT) [25]. A special character # is appended to S, to indicate the end of the string. Then shift rotations of $S\#$ are made to obtain all suffixes of the input string. Next, all rotations are sorted alphabetically. Last column of such an array of suffixes is a BWT of S.

To find ATRs, three auxiliary arrays of length $|S\#|$ are also constructed. The mapping array *Map* determines at which position in the first column the character is located and, simultaneously, which character in the BWT precedes the current character. The *Pos* array determines the original positions of the suffixes in S. The *Prm* array, is an inverse of the *Pos* array ($Prm(Pos(i)) = i$).

First step of the presented algorithm is finding candidates for *K-mismatch double ATRs*. Converted input string, together with the auxiliary arrays, allows to make use of the alphabetically sorted array of input string suffixes, without the need of storing the whole suffix array structure. The *Map* array is used together with an auxiliary array C of length $|\Sigma|$ and the function *occ* to determine the number of occurrences of a certain pattern in the input string S according to the algorithm presented by Ferragina and Manzini [26]. The value $C\,[ch]$ is the total number of occurrences of all characters preceding character ch in the BWT string. The function $occ\,(ch, 1, x)$ reports the number of occurrences of character ch in the BWT string from 1 to x in a constant time. The procedure starts with an empty pattern P, *startPos = 0*, *endPos = |S|* and recursively appends each character ch from the considered alphabet in front of P. This approach uses the results from the previous iteration to calculate a range of positions for a longer pattern $(ch + P)$: *newStartPos = C[ch] + occ(ch, 1, prevStartPos)*, *newEndPos = C[ch] + occ(ch, 1, prevEndPos)*. If there is only one occurrence of the current pattern, the recursion is stopped. This way it is possible to go recursively through all groups of repeats found.

Observation. *Two strings of length p with the Hamming distance k between them have always a common, matching substring of length d at corresponding positions, such that:*

$$d \geq \left\lfloor \frac{p}{k+1} \right\rfloor \tag{2}$$

The above observation, previously made in a similar form by Kurtz et al. [18], allows to determine how far away from each other a certain pair of repeats should be positioned to be considered a candidate for a *K-mismatch double ATR*. Therefore, a group of repeats of length d will be used to find candidates for *K-mismatch double ATR* only for periods p satisfying the equation (2) for all $k \leq K$ (or some $K' < K$, if K is restricted by *Kprc*) and conforming to input parameters *minPeriod* and *maxPeriod*. A pair of repeats from the considered group of repeats, with indexes i and j in the suffix array structure, can be a part of a *k-mismatch double ATR* with calculated period p if a difference between their positions in the input string is equal to p, that is $Pos(i) - Pos(j) = p$. To find all such pairs of repeats it is enough to check for each repeat with index i from the group, if p positions to left another repeat from the same group of repeats exists, that is, if $Prm(Pos(i) - p)$ is in the range of indexes of the current group of repeats. If so, the pair of repeats of length d is reported as a candidate for a *double k-mismatch tandem repeat* with period p and number of mismatches k.

The next step is validating the reported candidates. It is checked if a pair of repeats is indeed a part of one (or more) *k-mismatch double ATR with period p*. If the index of the left repeat is j, a *k-mismatch double ATR* of length $2p$ can begin anywhere in the input string between positions $Pos(j) - (p-d)$ and $Pos(j)$. The Hamming distance is calculated between consecutive strings of length p beginning at all possible positions. If for any of these positions it is equal exactly to k and the *totalScore* of the repeat is acceptable, the *k-mismatch double ATR* is reported at this position. Otherwise, the candidate is rejected.

In the following stage, the *k-mismatch double ATR* found is extended to the left and to the right to obtain a *maximal K-mismatch ATR*. It is done one character at a time, by checking if a string of the length $2p$ that begins one character to the left (or right) from the current *double ATR* is a *K-mismatch double ATR*. For each found *K-mismatch maximal ATR with period p*, the *totalScore* is calculated for all periods from 1 to p and the final period is changed to the one that corresponds to the maximum *totalScore*. This way every repeat is reported with the best fitting period. Also, all erroneous edges are cut off from the final ATR. Lastly, it is checked if the found repeat fulfills all the conditions determined by the input parameters. If yes, it is eventually accepted.

Algorithm described will find the same *K-mismatch maximal ATR* many times, extending different *k-mismatch double ATRs* found within that maximal repeat. Thus, all ATRs found, contained entirely in other repeats (or being the same repeat) are filtered out. To meaningfully decrease the amount of computations performed, the marking can be enabled with the *mark* parameter. An additional array of bits of length $|S|$ is used. Initially all bits are set to 0. After an ATR is found, all bits corresponding to its complete position are set to 1. This array of bits is a simple way to mark regions with found ATRs. If a candidate is entirely placed within a region already occupied by other, previously found, repeat, it is immediately rejected. When marking is on the results can be slightly different, because of change in order of searching for ATRs. Nevertheless, all regions with ATRs will always be reported, while the reduction in the amount of computation is significant.

2.2 Mreps

Mreps is a software for identifying tandem repeats in DNA sequences, presented in [17] and available on-line [27]. It exhaustivity looks for all tandem repeats with substitutions that satisfy some assumed criteria (these repeats meet the Definition 4). Then, the repeats go through a heuristic treatment in order to obtain more biologically relevant repetitions.

Parameters. Mreps takes as an input a sequence S and several parameters defining ATRs that are searched: minimum period and maximum period ($minPeriod$ and $maxPeriod$), minimum exponent $minExp$, minimum total length $minSize$, maximum total length $maxSize$, resolution res, positions in the string S to start and end search for ATRs ($from$ and to) and a flag enabling outputting small repeats $allowsmall$.

Algorithm. In the exhaustive search, the algorithm looks for the longest common extensions with res mismatches at each position of the input string S. These are used to find res-mismatches double ATRs. Then found double repeats are joined to obtain res-mismatches maximal ATRs. In the following heuristic approach, the period of each found repeat is changed according to the internal total score of that repeat and basing on a statistical analysis made by the authors on some artificial genomic sequences.

The main drawback of $mreps$ is its inability to handle the N-regions of the input sequence. The program randomly changes every N to one of the characters it can process: A, C, T or G. Additionally, If there are too many N characters, the program does not operate at all, outputting an error. As a consequence, $mreps$ can report an artificial repeats in regions where originally only N characters were present. Thus it is difficult to apply $mreps$ to look for tandem repeats in e.g. human genome, as it contains large number of regions of N characters.

2.3 Tandem Repeat Finder

The Tandem Repeat Finder (TRF) it is a widely used statistically based method for searching for ATRs that can contain mismatches and indels, described in [23] and available on-line [28]. It collects exact matches as seeds and then processes and extends them, to find ATRs that satisfy the assumed statistical criteria.

Parameters. The TRF takes as an input a sequence S and parameters defining ATRs that are searched: alignment weights for match, mismatch and indels for Smith-Waterman style local alignment using wraparound dynamic programming (*match*, *mismatch* and *delta*), matching and indel probability (*PM* and *PI*), minimum alignment score to report the repeat *minScore* and maximum period size *maxPeriod*.

Algorithm. The algorithm has two main phases: detection and analysis. Alignment of two copies of a pattern of length n is modeled as n independent Bernoulli trials (coin-tosses), where head is interpreted as a match between aligned nucleotides and tail is a mismatch, insertion or deletion. During the detection component, for some small integer k, all possible k-length strings are listed and the input sequence is scanned to look for all positions of them. Next, the distance lists are created, collecting all pairs of matching strings placed at the same distance from each other. Then, the statistical criteria based on four distributions (depending on period, *PM*, *PI* and k) are applied, to find candidate tandem repeats. In the analysis component, found candidates are aligned with the surrounding sequence using wraparound dynamic programming. If at least two copies of a pattern are aligned and satisfy the *minScore* parameter, the ATR is reported.

The main disadvantage is that not all ATRs will be find. If consecutive copies of the repeat do not contain a series of k matching nucleotides at corresponding positions, such ATR will not be discovered. Also, the input parameters do not give much freedom in specifying kind of repeats searched.

3   Methodology

In this section the methodology used to compare results obtained with three evaluated tools is described. We present our program, the ATR-compare, which takes as an input two lists of ATRs and makes a summary and comparison between them. Next, the experiment performed is specified.

3.1   ATR-Compare

The ATR-compare is a program that we designed to make a comparison of the outputs of the evaluated tools for searching for ATRs. It takes as an input two sets of ATRs, each being a text file formatted in such a way, that every line represents one tandem repeat. Each repeat/line is composed of five features of the repeat (start position, end position, period, exponent, sequence), separated by a whitespace. The ATR-compare perform an extensive browsing through the two input lists of ATRs, to give a detailed comparison of them.

The output consists of number of text files. First is a quantitive comparison of the two input sets of ATRs, with division to different periods. Next, subsets of the input sets, satisfying a certain criteria, are reported in separate text files. Each such subset has a corresponding text file with statistics showing number

of ATRs in the subset, classified according to period. The subsets are: set of
ATRs identical in both input sets, set of ATRs almost identical in both input
sets, differing only by period, set of ATRs from the first list contained entirely
in one of ATR of the second list (but not equal in length) and vice versa, set
of overlapping ATRs from both input lists, not belonging to any of the previous
sets and lastly, two sets of ATRs unique to each of the input sets.

3.2 Experiment

The aim of the experiment was to compare ATRs that can be found with the evaluated tools. For that purpose all three programs were run for the same genomic
sequence: Homo sapiens chromosome 22, GRCh37.p5 Primary Assembly (Acces-
sion.Version: NC_000022.10). In case of the mreps tool, because of its limitations,
it was necessary to first find large N-regions of the input sequence and exclude
them from the search. The *from* and *to* parameters were used to run the mreps
for all other regions and then the obtained results were joined.

The input parameters of each tool were chosen to define searched ATRs in as
similar way to other tools as it is possible. The parameters selected for BWatrs
were: $minPeriod = 1$, $maxPeriod = 100$, $minExp = 3$, $minTotal = 4$, $K = 7$,
$Kprc = 35$, $minScore = 75$, $mark = 1$. The parameters chosen for mreps were:
$minPeriod = 1$, $maxPeriod = 100$, $minExp = 3$, $minSize = 4$, $res = 7$ and *allows-
mall* enabled. For the TRF set of standard parameters was chosen: $match = 2$,
$mismatch = 3$, $delta = 5$, $PM = 80$, $PI = 10$, $minscore = 14$, $maxperiod = 100$
and obtained results were filtered to contain only repeats with period larger or
equal to 1, exponent larger or equal to 3 and total size larger or equal to 4.
The output of each program was converted to satisfy the convention for input
sets used by the ATR-compare. Then the ATR-compare was run for each pair
of results sets and to set of ATRs found by one tool and the set of appropriate
results of comparison of the two other tools.

4 Results

The experiment described in the previous section reveal that BWatrs found
the highest total amount of ATRs: 1281009, while mreps and TRF found 1018136
and 158636 ATRs, respectively. In the Fig. 1 there is a quantitive comparison
of the number of ATRs found by different tools stratified with respect to period.
It can be observed that application of both tools, BWatrs and mreps has led to
detection of much more ATRs with small periods than application of TRF and,
conversely, to detection of less ATRs than TRF, for larger periods. The latter
observation is caused by the fact that for TRs with larger period the restriction
of allowable number of mismatches to 7, caused by assuming values of the parameters ($K = 7$ for BWatrs and $res = 7$ for mreps) can be violated when TRF
is applied.

Further results of application of our program ATR-compare demonstrate that
BWatrs has found 619250 ATRs identical to mreps and 38890 repeats identical

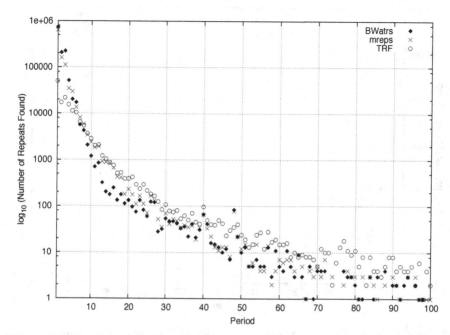

Fig. 1. Number of all ATRs (with distinction to different periods) found by each of the evaluated tools

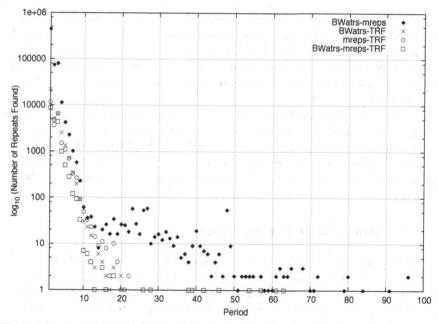

Fig. 2. Number of identical ATRs (with distinction to different periods) found by pairs of the evaluated tools and by all three tools together

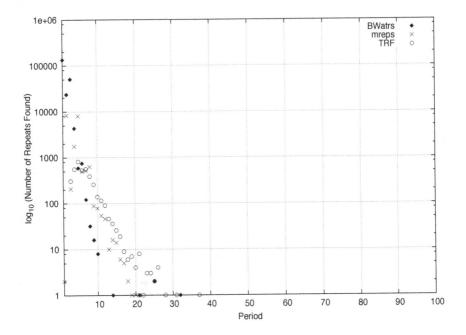

Fig. 3. Number of unique ATRs (with distinction to different periods) found by each of the evaluated tools

to TRF. Also, 27103 identical repeats were found between mreps and TRF. Lastly, 18759 exactly the same repeats were found by all evaluated tools. In the Fig. 2, the amount of found ATRs identical to each pair of tools and to all three tools is presented, with distinction with respect to different periods. Outcomes of BWatrs and mreps are quite similar one to another, but still there are many ATRs that differ.

In the Fig. 3, unique ATRs found by each tool are shown (also classified according to the period). BWatrs located the largest number of unique repeats: 214708, while mreps and TRF discovered 20311 and 3951 of such repeats, respectively. The subsets representing unique ATRs of BWatrs consist mainly of short repeats with small period, but there are also some, potentially interesting, longer repeats.

5   Conclusions and Future Work

It should be noted that both algorithms mreps and TRF are already frequently used in many genomic studies and are considered by their users as reliable tools for ATRs detection. Our study confirms efficiencies of existing tools for detection of ATRs. Nevertheless, application of two different tools mreps and TRF for exemplary chromosomal data and comparison to our new tool BWatrs show considerable variations between outcomes of different algorithms. Results of comparisons are presented in Figures 1, 2 and 3. Differences seen in figures 1, 2 and

3 stem from different constructions, different parameters, different definitions of repeatability and different search strategies between different algorithms. Some effects of using varying approaches for ATRs of different lengths were explained in the previous section. Our tool BWatrs is most sensitive for short ATRs. It also leads to detection of largest numbers of unique ATRs, as seen in figure 3. These properties can be advantageous in some applications, like inter species comparisons, homology detection between genomic sequences or more specialized, DNA assembly quality control.

As a conclusion, using effective text searching algorithms based on Burrows-Wheeler transform of entire chromosomal sequences allows us to obtain a new tool competitive to existing methodologies.

Our further research will involve further comparisons of results of ATR searchers focusing on biological significance of the ATRs found by different tools. One of the main features of the tandem repeat is its high probability to change in number of copies. It is caused by its structure that increases the likelihood of slippage mutation to occur [1], [4], [29]. Thus, as a meaningful ATR, we can understand a tandem repeat that differs in number of copies between two individuals. We plan to look for such repeats in the reference and the alternative human genomes, taking into account the subsets generated by the ATR-compare tool.

Acknowledgments. This work was supported by the European Union from the European Social Fund (grant agreement number: UDA-POKL.04.01.01-00-106/09) A.D., R.P. and by Polish National Science Center under project DEC-2011/01/B/ST6/06868 I.M, A.P.

References

1. Chakraborty, R., Kimmel, M., Stivers, D.N., Davison, L.J., Deka, R.: Relative mutation rates at di-, tri-, and tetranucleotide microsatellite loci. PNAS 94, 1041–1046 (1997)
2. Kruglyak, S., Durrett, R.T., Schug, M.D., Aquadro, C.F.: Equilibrium distributions of microsatellite repeat length resulting from a balance between slippage events and point mutations. PNAS 95, 10774–10778 (1998)
3. Pumpernik, D., Oblak, B., Borštnik, B.: Replication slippage versus point mutation rates in short tandem repeats of the human genome, Mol. Genet. Genomics 279(1), 53–61 (2008)
4. Leclercq, S., Rivals, E., Jarne, P.: DNA slippage occurs at microsatellite loci without minimal threshold length in humans: a comparative genomic approach. Genome Biol. Evol. 2, 325–335 (2010)
5. Vinces, M.D., Legendre, M., Caldara, M., Hagihara, M., Verstrepen, K.J.: Unstable Tandem Repeats in Promoters Confer Transcriptional Evolvability. Science 324, 1213 (2009)
6. McMurray, C.T.: Mechanisms of trinucleotide repeat instability during human development. Nat. Rev. Genet. 11(11), 786–799 (2010)
7. Jeffreys, A.J., Wilson, V., Thein, S.L.: Individual-specific 'fingerprints' of human DNA. Nature 316, 76–79 (1985)

8. Weber, J.L., Wong, C.: Mutation of human short tandem repeats. Hum. Mol. Genet. 2, 1123–1128 (1993)
9. Merkel, A., Gemmell, N.: Detecting short tandem repeats from genome data: opening the software black box. Brief. Bioinform. 9(5), 355–366 (2008)
10. Saha, S., Bridges, S., Magbanua, Z.V., Peterson, D.G.: Empirical comparison of ab initio repeat finding programs. Nucleic Acids Res. 36(7), 2284–2294 (2008)
11. Lerat, E.: Identifying repeats and transposable elements in sequenced genomes: how to find your way through the dense forest of programs. Heredity 104(6), 520–533 (2009)
12. Leclercq, S., Rivals, E., Jarne, P.: Detecting microsatellites within genomes: significant variation among algorithms. BMC Bioinformatics 8, 125 (2007)
13. Smit, A.F.A., Hubley, R., Green, P.: RepeatMasker, http://repeatmasker.org
14. Frith, M.C.: A new repeat-masking method enables specific detection of homologous sequences. Nucleic Acids Res. 39(4), e23 (2011)
15. Pokrzywa, R., Polanski, A.: BWtrs: A tool for searching for tandem repeats in DNA sequences based on the Burrows-Wheeler transform. Genomics 96, 316–321 (2010)
16. Pellegrini, M., Renda, M.E., Vecchio, A.: TRStalker: an efficient heuristic for finding fuzzy tandem repeats. Bioinformatics 26(12), 358–366 (2010)
17. Kolpakov, R., Bana, G., Kucherov, G.: mreps: efficient and flexible detection of tandem repeats in DNA. Nucleid Acids Research 31, 3672–3678 (2003)
18. Kurtz, S., Choudhuri, J.V., Ohlebusch, E., Schleiermacher, C., Stoye, J., Giegerich, R.: REPuter: The Manifold Applications of Repeat Analysis on a Genomic Scale. Nucleic Acids Res. 29(22), 4633–4642 (2001)
19. Ruitberg, C.M., Reeder, D.J., Butler, J.M.: STRBase: a short tandem repeat DNA database for the human identity testing community. Nucleic Acids Res. 29(1), 320–322 (2001)
20. Gelfand, Y., Rodriguez, A., Benson, G.: TRDB—The Tandem Repeats Database. Nucleic Acids Res. 35 (suppl. 1), D80–D87 (2007)
21. Sokol, D, Atagun, F.: TRedD—a database for tandem repeats over the edit distance. Database 2010, article ID baq003, 10.1093/database/baq003 (2010)
22. Danek, A., Pokrzywa, R.: Finding Approximate Tandem Repeats with the Burrows-Wheeler Transform. International Journal of Medical and Biological Sciences 6, 8–12 (2012)
23. Benson, G.: Tandem Repeats Finder: a program to analyze DNA sequences. Nucleic Acids Research 27, 573–580 (1999)
24. Pokrzywa, R.: Application of the Burrows-Wheeler Transform for searching for tandem repeats in DNA sequences. Int. J. Bioinf. Res. Appl. 5, 432–446 (2009)
25. Burrows, M., Wheeler, D.J.: A block-sorting lossless data compression algorithm, SRC Research Report 124, Digital Equipment Corporation, California (1994)
26. Ferragina, P., Manzini, G.: Opportunistic data structures with applications. In: Proceedings of the 41st Annual Symposium on Foundations of Computer Science, pp. 390–398. IEEE Computer Society, Washington, DC (2000)
27. mreps, http://bioinfo.lifl.fr/mreps
28. Tandem Repeat Finder, http://tandem.bu.edu/trf/trf.html
29. Bhargava, A., Fuentes, F.F.: Mutational Dynamics of Microsatellites. Molecular Biotechnology 44(3), 250–266 (2010)

Pattern Recognition for Subfamily Level Classification of GPCRs Using Motif Distillation and Distinguishing Power Evaluation

Ahmet Sinan Yavuz, Bugra Ozer, and Osman Ugur Sezerman

Faculty of Engineering and Natural Sciences
Sabanci University
Istanbul, Turkey
{asinanyavuz,bozer,ugur}@sabanciuniv.edu

Abstract. G protein coupled receptors (GPCRs) are one of the most prominent and abundant family of membrane proteins in the human genome. Since they are main targets of many drugs, GPCR research has grown significantly in recent years. However the fact that only few structures of GPCRs are known still remains as an important challenge. Therefore, the classification of GPCRs is a significant problem provoked from increasing gap between orphan GPCR sequences and a small amount of annotated ones. This work employs motif distillation using defined parameters, distinguishing power evaluation method and general weighted set cover problem in order to determine the minimum set of motifs which can cover a particular GPCR subfamily. Our results indicate that in Family A Peptide subfamily, 91% of all proteins listed in GPCRdb can be covered by using only 691 different motifs, which can be employed later as an invaluable source for developing a third level GPCR classification tool.

Keywords: g-protein coupled receptors, data mining, pattern recognition.

1 Introduction

G protein coupled receptors (GPCRs) represent the largest family of membrane proteins in the human genome. As their dysfunction contributes to some of the most prevalent human diseases, they are of exceptionally high interest in various areas including the drug industry as more than 50% of modern drugs have GPCRs as their main targets [1]. An important property of the GPCRs is that certain aminoacid residues are well conserved across specific families [2]. This property has been utilized in numerous studies, such as synthesizing new GPCRs [3-6], and developing family classifiers. In addition, all GPCRs share a particular structural framework. Structure of a G-protein-coupled receptor comprises seven α-helical transmembrane domains, an extracellular N-terminus, and an intracellular C-terminus [7].

GPCRs are activated by a diverse range of ligands such as small peptides, amino acid derivatives, taste, light or smell [8]. The general classification for GPCRs in vertebrates is as follows: rhodopsin-like (Family A), secretin-like (Family B), glutamate-like (Family C), Adhesion and Frizzled/Taste2 [9, 10]. In addition to this

T. Shibuya et al. (Eds.): PRIB 2012, LNBI 7632, pp. 267–276, 2012.
© Springer-Verlag Berlin Heidelberg 2012

classification, there are 4 levels of classification down in classification tree. Family A is the family of highest interest from a pharmaceutical research perspective as besides being more than 80% of all human GPCRs are in this family alone [11], the number of sequences in this family is significantly higher than the others. Therefore, we will also emphasize our efforts on peptides subfamily, which belong to Family A.

Due to their significant roles and their importance in drug design, it is highly crucial to be able to distinguish which ligands a specific GPCR interacts with and which regions of the sequence have a particularly crucial role in ligand binding. However, this process is complex, and it is not easy to identify corresponding regions. Despite the significant amount of pharmaceutical research done in this field, 3D structures of only few GPCR structures are known [9], whereas there are large numbers of GPCR primary sequences have been identified [12]. Therefore, in order to identify and characterize the novel receptors, it is crucial to develop *in silico* methods that only work with primary sequences to determine the ligand binding sites and motifs of these novel receptors.

Additional serious challenge is the classification of orphan GPCR sequences. An orphan GPCR is a sequence that has high similarity to known and annotated GPCR sequences but nothing is known about its structure, physiologic function or the activating ligand. As the difference between the number of annotated sequences and the number of identified sequences raises, so does the number of orphan GPCRs. Besides, considering the contribution of GPCRs to cancer initiation, growth and metastatic spread, identification of orphan GPCRs and revealing the pathways related with these GPCRs is placed in the spotlight as prime candidates for cancer prevention and treatment and the orphan GPCRs are of very high interest as they are not yet identified. Therefore, it is essential to find the rules that cover most of the GPCR sequences especially those in the Family A, which is the family most relevant to human drug design. In this work, we will focus utterly on motif coverage within the Peptides subfamily, which belongs to Family A.

As a quick summary, there are two dominant goals for *in silico* GPCR researches: first is to classify GPCR sequences according to their subfamilies, and second is to identify the key ligand-receptor binding sites and family specific motifs using only the protein sequence information. Unlike many of the previous efforts, major concern of this work is only 3rd level classification of GPCR sequences and exploration and analysis of the presence of any layered motifs that are effective in the determination of sub-subfamily classes. Hence, this work is concerned with not only aiming for an *in silico* motif mining for GPCR classification but also providing a valuable source of conserved motifs for experimentalists and other groups working in 3rd level GPCR classification.

1.1   Related Work

There are many current GPCR classification methods involving various machine learning techniques. One of the most common methods employed in GPCR classification is support vector machines (SVM). In this sense, GPCRpred server [13] is based on 20 different SVMs for different levels of classification where the feature vectors are derived from the dipeptide arrangement of each protein. As reported in [14], SVM classification gives better results compared to BLAST and profile HMMs with around

90% valid classification level. However, there are several failures attached with this approach as it misses the physiochemical properties of the receptors which are vital in determining the matching ligand, leading to inaccurate results.

Another common approach to GPCR classification is usage of Hidden Markov Models (HMM). PRED-GPCR [15] server uses this approach with employing 265 signature profile HMMs in the classification of GPCR sequences. However, HMM based prediction methods are not optimal in predicting subfamilies. In addition, likewise SVM based methods, HMM-based methods also bears the problem of opaqueness, yet they are not straightforward to discover key ligand interacting sites of the receptors.

In addition to these techniques, a HMM/SVM hybrid method is utilized for GPCR classification. Named the GRIFFIN Project [16], this project combines the efficiency of HMM-based prediction with predictive power of SVM.

In addition to these widely used methods, there were also some other methods [17,18] proposed which use a number of metrics to make classification efforts more successful and summarize the amino acids of a sequence in a number of continuous parameters. Additionally, Davies et al. [19] proposed a method using 10 different classification algorithms, which employs the structural and physiochemical properties of amino acids, to perform a hierarchical GPCR sequence classification. In this method, best resulting classification method at each level is employed in progressing down the classification tree. Even though, they have various superiorities, all these methods lack the necessary transparency to determine the key ligand receptor interaction sites and identify specific residues.

In overall, current methods in GPCR prediction are suffering mainly from opaqueness of models and impossibility of extracting information out of models in addition to classification. Identifying key interaction sites conserved in families, sub-families or sub-sub families will be beneficial in classification of orphan GPCR sequences. Hence, this work mainly aims to extract possible ligand-receptor interaction sites for each sub-subfamily via identifying the key motifs that cover protein families.

2   Methods

In order to form our training set, we have obtained 304 peptide subfamily sequences from GPCRdb, which includes 32 different sub-subfamilies, such as angiotensin, bombesin, and bradykinin. We aimed to find covering motifs for each of these sub-subfamilies via our pattern recognition method.

Our proposed method in this work can be summarized as follows:

1. Motif distillation by Motif Specificity Measure
2. Distinguishing Power Evaluation of distilled motifs
3. Motif selection with general weighted set cover problem

Briefly, motif distillation step is used to discriminate family specific motifs from randomly generated pool of motifs. Subsequently, distinguishing power evaluation (DPE) of the distilled motifs is used to determine the efficiency of the motifs in sub-subfamily classification with assigned DP score value to enable comparison between each other. Lastly, DP score assigned top selected motifs are used in general weight set cover problem to find out the smallest set of motifs that can cover the maximum amount of proteins located in peptide subfamily.

2.1 Amino Acid Grouping

It is commonly known that there 20 amino acids present considering the proteins. It is challenging to determine fixed length conserved motifs within a protein family using 20-letter amino acid alphabet. Through the evolution, families binding to the same ligand change their sequence while preserving the physicochemical properties of the binding site the same. Therefore, it is very difficult to find identical binding signals within a family. To be able to capture similar motifs, which are different in their sequence, a common approach is to reduce this 20-letter alphabet to a smaller number by grouping the similar amino acids together. At this stage, there are several basic physicochemical properties such as hydrophobicity, charge, and mass, which can be used as an origin of grouping. For our approach, Sezerman grouping of amino acids is used, which is proposed at Cobanoglu et al. [20], and its efficiency tested over other amino acid grouping techniques [21].

Table 1. The amino acid grouping scheme in Sezerman's grouping

Groups	A	B	C	D	E	F	G	H	I	J	K
Amino Acids	IVLM	RKH	DE	QN	ST	A	G	W	C	YF	P

2.2 Motif Definition and Motif Specificity Measure

The sequence information without any feature selection cannot be used to perform any rule extraction. The main idea behind motif specificity measure is within a sub-subfamily, certain length aminoacid sequences at specific positions of the same exocellular region would be preserved in comparison to sequences of other sub-subfamilies. The main idea behind this project is that amino acids might be fundamental to the binding process since otherwise they would not have been conserved. The motifs are essential to represent some location specific properties of the sequences, as the objective of this study is to determine key interaction sites as well as extracting set of rules for classification. For this purpose, motifs used in this work are defined similar to the motifs proposed by Cobanoglu et al. [20], which includes information of triplet of residues, the exocellular region of occurrence (n-terminus, exoloop1, exoloop2 or exoloop3) and lastly the position of first residue of triplet relative to the length of the amino acid sequence. In order to determine the transmembrane regions reported in the motif definition, we used TMHMM tool [22].

In general, total number of possible motifs is over hundreds of thousands; nevertheless, most of them occur very infrequently. The ideal motif would be the one that occurs in all the sequences that belongs to a particular sub-subfamily but never in a sequence from another sub-subfamily. In other words, motifs that are unique to a sub-subfamily would be rewarded, whereas motifs that occur either in few sequences or numerous sub-subfamilies, would be penalized. The Term Frequency Inverse Document Frequency (TF-IDF) [23] weight is a metric that measures the occurrences of a word in a family in relation to the overall number of the family members, thus enabling determination of highly family specific motifs. TF-IDF is designed for a parallel purpose and considered as a valid tool at text mining applications, and in this work, the pre-defined TF-IDF weights are used in defining the Motif Specificity Measure, originally with detailed definitions given in Cobanoglu et al. [20]. In short, as its name suggests, motif specificity measure quantifies the specificity of a motif to a family; hence, indicates that the motif is a random motif or a possibly useful one.

2.3   Distinguishing Power Evaluation

Distinguishing power evaluation (DPE) method aims to determine the best motifs for classification in the training set. Main notion of DPE is repeatedly building decision trees from randomly partitioned test and training data, and looking for those motifs that occur very frequently in each of these decision trees [20]. During this process, DPE picks the motifs initially determined by TFIDF with the highest sub-subfamily specificity using the motif specificity measure.

Apart from its specificity to a certain sub-subfamily, there is a need for an independent comparison criterion between motifs in their distinguishing power. To create such a comparable criterion for assessing a motif's importance in classification, DPE method calculates a distinguishing power (DP) score, which is simply the sum of the accuracies of the decision trees in which that motif occurs [20]. By this score, it is possible to identify the motifs with high information gain and which may be vital in classification. More detailed instructions on DPE can be found in Cobanoglu et al. [20].

In the use of DPE method, three different total number of motifs are tested,   250, 500 and 1000 motifs, from the top of the list of distilled motifs with descending DP scores and assessed their power to cover the whole dataset, individually.

2.4   General Weighted Set Covering Model

DPE selected motif set contains various weak motifs that have a limited contribution in covering the subfamily dataset. These weak motifs may cause overlearning of training data. Therefore, frequently occurring and a complete sub-subfamily covering minimum set of motifs would be more reliable in correct classification of unseen data. Otherwise, motifs that have smaller coverage will optimize the performance on training data and decrease the accuracy of classification algorithm in unseen data.

In order to achieve a minimum number of motifs that can explain maximum portion of each given sub-subfamily datasets, it has been implemented a general weighted set covering model on DPE selected motifs. For each motif, a set of proteins which have that motif is defined separately. Additionally, considering these sets, we applied general weight set covering model to determine the minimum number of motifs which cover all of the proteins in sub-subfamilies, but not all the sub-subfamilies. In detail, this model initially calculates occurrences of each motif in all dataset proteins and each sub-subfamily proteins separately. Afterwards, calculated presence counts were used for calculating ratio1 and assessing the weight, or importance, of that motif in sub-subfamily coverage. Motifs were then sorted according to their weights, and for each sub-subfamily, highest ranked motifs were selected until no further improvement in coverage occurs.

Motif weights have been calculated via different weighting schemes including but not limited to equal weighting of motifs and maximum cardinality of motifs [26]; however, both of these criterions lack the information on specificity of a motif to a subfamily. In other words, these criterions do not provide sufficient information to distinguish subfamilies from each other, but they merely provide information on their presence in whole dataset. In order to overcome this problem, we used a maximum ratio1 criterion, which represents motif coverage in a particular family of proteins in comparison with its existence in all other families [24].

A weight of a motif for all sub-subfamilies based on maximum ratio1 criterion is calculated as:

$$W_i = \frac{|Proteins\ covered\ in\ sub-subfamily\ i|}{|Protein\ covered\ in\ whole\ subfamily|}, \forall i \in Selected\ Subfamily \qquad (1)$$

where, i is a sub-subfamily in given subfamily dataset. According to this criterion, if a motif only appears in one subfamily with high coverage, which is a desirable result for classification purposes, its ratio1 value will be become 1, and it will be regarded as an important motif in the model.

3   Experimental Results

Associated DPE runs with motif count 250, 500 and 1000 resulted in 43%, 51%, and 91% coverage of sub-subfamily proteins, respectively. In DPE count = 250 case, the set covering model only selected 174 motifs for maximum coverage, while for DPE count = 500, there were 329 selected motifs selected. For DPE count = 1000 experiment, our model picked 691 motifs for maximum coverage, which as a result came out to be the most efficient DPE count. Full list of the motifs is available as supplementary material (Supplementary Table 1). Although increasing DPE motif count shows parallel behavior to the sub-subfamily coverage trend, each step increases in DPE motif count results in a significant increase in computational time. Besides, in order to avoid overlearning, we decided to keep motif count in a limit. Therefore, we chose DPE count = 1000 as our best result for further studies. The detailed results of the selected DPE experiment and applied general weighted set covering model is included in Table 2.

Several sub-subfamilies, namely Duffy-antigen and GPR37 endothelin B-like, shows a consistent low coverage between different DPE counts, indicating motifs that are effective on classification of these sub-subfamilies have low distinguishing power (DP score) and therefore are not selected within given DP motif counts. Hence, classification to these sub-subfamilies can be more difficult, since covering motifs are not specific enough. On the other hand, adrenomedullin family indicates a high coverage for each count set (100%, 76%, 73% respectively for DP count = 1000, 500, and 250). Consistent high coverage for adrenomedullin sub-subfamily indicates that family-specific motifs have a high DP scores showing family's distinct nature, and these motifs ranked mostly in top 250 motifs. Similar kind of behavior is also seen in anaphylatoxin sub-sub family with 96%, 74%, 43% protein coverage for given DPE counts respectively. The significant reduction in DPE count = 250 indicates that these sub-subfamily specific motifs are mostly ranked in 250-500 range. Another significant sub-subfamily result has been obtained at the case of allostatin C. Within the same trend of adrenomedullin and anaphylatoxin, 95%, 66%, 46% protein coverages were obtained for tested DPE counts. As a summary, these results validate our findings on the DPE motif selection thresholds and have an important effect on the scope of possible protein coverage. The most important 5 motifs and their locations for adrenomedullin, anaphylotoxin, and allostatin C sub-subfamily are given in the Table 3. Also, location distributions of selected motifs for these three sub-subfamilies are summarized in a histogram, Figure 1. Via analyzing the difference between high and low DP scored motifs for different sub-subfamilies, the complex sub-subfamilies can be identified and used as an additional insight in developing 3rd level GPCR classification methods.

Table 2. General weight set covering model on DPE evaluated motifs resulted in listed coverage for each sub-subfamily in GPCRdb

Family	Total Protein	Covered Protein	Percentage
Adrenomedullin	33	33	1.00
Allatostatin C	41	39	0.95
Anaphylatoxin	98	94	0.96
Angiotensin	180	147	0.82
APJ like	90	59	0.66
Bombesin	163	149	0.91
Bradykinin	223	222	1.00
Chemokine	1286	1080	0.84
Chemokine receptor-like	77	73	0.95
Cholecystokinin	170	163	0.96
Duffy antigen	51	25	0.49
Endothelin	144	130	0.90
Fmet-leu-phe	305	279	0.91
Galanin-like	405	371	0.92
GPR37 endothelin B-like	78	44	0.56
Interleukin-8	118	112	0.95
Melanin-CHormone Recep family	228	219	0.96
Melanocortin	789	749	0.95
Neuromedin U-like	215	177	0.82
Neuropeptide Y	1329	1262	0.95
Neurotensin	59	57	0.97
Opioid	288	256	0,89
QRFP family	86	81	0.94
Prokineticin receptors	98	93	0.95
Prolactin-releasing peptide	113	89	0.79
Proteinase-activated like	250	243	0.97
Somatostatin- and angiogenin-like	96	95	0.99
Somatostatin	376	361	0.96
Sulfakinin CCKLR	9	9	1.00
Tachykinin	270	265	0.98
Urotensin II	47	31	0.66
Vasopressin-like	436	412	0.94
TOTAL:	8151	7419	0.91

Table 3. Highest ranked 5 reduced alphabet motifs and their location for adrenomedullin, allastostatin C and anaphylotoxin. Position within a loop is defined as being the sequential position of the first letter of triplet, normalized by length of the corresponding loop. 0,1,2,3 correspond to the first, second, third and fourth quarter of the exoloops and n-terminus respectively.

Sub-subfamily	Motif	Location	Position within Loop
Adrenomedullin	CAA	exoloop 2	0+1
Adrenomedullin	JEA	exoloop 1	0
Adrenomedullin	KCA	exoloop 2	0
Adrenomedullin	JBE	exoloop 3	0
Adrenomedullin	GFA	n-terminus	2
Allatostatin C	JAA	exoloop 1	0+1
Allatostatin C	AAA	exoloop 3	0
Allatostatin C	EEJ	n-terminus	1+2
Allatostatin C	ACD	n-terminus	1+2
Anaphylatoxin	AEA	exoloop 2	0+1
Anaphylatoxin	EJA	exoloop 2	1
Anaphylatoxin	AAC	exoloop 3	2
Anaphylatoxin	BBA	exoloop 2	2
Anaphylatoxin	CJC	n-terminus	2

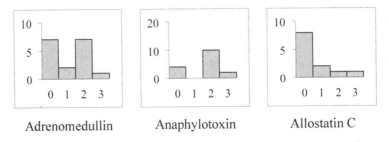

Fig. 1. Histograms of motif locations present in selected motifs for adrenomedullin, anaphlotoxin and allostatin C sub-subfamilies. In x-axis locations are denoted as 0 for n-terminus, 1 for exoloop 1, 2 for exoloop 2 and 3 for exoloop 3.

As DPE counts and selected motif counts differ notably, it can be concluded that our motif selection step helps to eliminate the motifs with high DP score and limited in the information they bring to coverage of sub-subfamily. Besides, large numbers in selected motif sets with large coverage rates indicate that these motifs can be used rule out complex patterns in transmembrane regions of GPCR receptors determining the sub-subfamily of the protein.

4   Conclusions and Future Work

In the light of recent findings, DPE method and combined applied general weight set model can be used for determining the motif set that can be used for developing classifiers for 3rd level GPCR classification problem. As 3rd level GPCR motif identification has not explored extensively in literature before, we hope that our method of obtaining minimal set of important motifs with high specificity will be a stepping stone for further developments in sub-subfamily GPCR classification. Our example case of Peptide sub-subfamily showed that our method can find important motifs for obtaining significantly large family coverage.

As can be seen from Figure 1, adrenomedullin family mostly binds from the motifs in the n-terminus and exoloop 2. These motifs mostly include negatively charged residues followed by aliphatic hydrophobic residues or ring structures and positively charged residues and/or Serine or Threonine (Table 3). Whereas anaphylotoxin mostly binds from the motifs occurring at exoloop 2 and allostatin C mostly binds from the n-terminus.   Our method provides location and binding motif information each of the peptide sub-subfamilies, which are very valuable for drug development.

The future work lies in quantifying the actual predictive performance of selected motifs and developing a classification server via generalizing motif sets for each and every sub-subfamily present in the GPCRdb.

Acknowledgments. Authors would like to express their gratitude to Cem Meydan (Sabanci University) and   Ceyda Sol (Sabanci University) for their valuable discussions.

Supplementary Material

Supplementary Table 1 – A complete list of motifs for all sub-subfamilies can be accessed online via http://bit.ly/O2Kk6N.

References

1. Filmore, D.: It's a GPCR World. Modern Drug Discovery 7(11), 24–28 (2004)
2. Joost, P., Methner, A.: Phylogenetic analysis of 277 human G-protein- coupled receptors as a tool for the prediction of orphan receptor ligands. Genome Biology 3(11), research0063.1–research0063.16 (October 2002)
3. Davey, J., Ladds, G.: Heterologous Expression of GPCRs in Fission Yeast. Methods in Molecular Biology 746, 113–131 (2011)
4. Gerber, S., Krasky, A., Rohwer, A., Lindauer, S., Closs, E., Rognan, D., Gunkel, N., Selzer, P.M., Wolf, C.: Identification and characterisation of the dopamine receptor II from the cat flea Ctenocephalides felis (CfDo- pRII). Insect Biochemistry and Molecular Biology 36(10), 749–758 (2006)
5. Libert, F., Parmentier, M., Lefort, A., Dinsart, C., Van Sande, J., Maenhaut, C., Simons, M.J., Dumont, J.E., Vassart, G.: Selective amplification and cloning of four new members of the G protein-coupled receptor family. Science 244(4904), 569–572 (1989)

6. Methner, A., Hermey, G., Schinke, B., Hermans-Borgmeyer, I.: A novel G protein-coupled receptor with homology to neuropeptide and chemoattractant receptors expressed during bone development. Biochemical and Biophysical Research Communications 233(2), 336–342 (1997)
7. Horn, F., Bettler, E., Oliveira, L., Campagne, F., Cohen, F.E., Vriend, G.: GPCRDB information system for G protein-coupled receptors. Nucleic Acids Research 31(1), 294–297 (2003)
8. Gether, U.: Uncovering molecular mechanisms involved in activation of G protein-coupled receptors. Endocrine Reviews 21(1), 90–113 (2000)
9. Rosenbaum, D.M., Rasmussen, S.R.G.F., Kobilka, B.K.: The structure and function of G-protein-coupled receptors. Nature 459(7245), 356–363 (2009)
10. Foord, S.M., Bonner, T.O.M.I., Neubig, R.R., Rosser, E.M., Pin, J.P., Davenport, A.P., Spedding, M., Harmar, A.J.: International Union of Pharmacology. XLVI. G Protein-Coupled Receptor List. Pharmacological Reviews 57(2), 279–288 (2005)
11. Davies, M.N., Secker, A., Halling-Brown, M., Moss, D.S., Freitas, A.A., Timmis, J., Clark, E., Flower, D.R.: GPCRTree: online hierarchical classification of GPCR function. BMC Research Notes 1, 67 (2008)
12. Gaulton, A., Attwood, T.K.: Bioinformatics approaches for the classification of G-protein-coupled receptors. Current Opinion in Pharmacology 3(2), 114–120 (2003)
13. Bhasin, M., Raghava, G.P.S.: GPCRpred: an SVM-based method for prediction of families and subfamilies of G-protein coupled receptors. Nucleic Acids Research 32(Web Server Issue), W383–W389 (2004)
14. Karchin, R., Karplus, K., Haussler, D.: Classifying G-protein coupled receptors with support vector machines. Bioinformatics 18(1), 147–159 (2002)
15. Papasaikas, P.K., Bagos, P.G., Litou, Z.I., Promponas, V.J., Hamod- Rakas, S.J.: PRED-GPCR: GPCR recognition and family classification server. Nucleic Acids Research 32(Web Server Issue), W380–W382 (2004)
16. Yabuki Y., Muramatsu T., Hirokawa T., Mukai H., Suwa M.: GRIFFIN: a system for predicting GPCR–G-protein coupling selectivity using a support vector machine and a hidden Markov model. Nucleic Acids Research, 33(Web server issue), W148–W153 (2005)
17. Cui, J., Han, L.Y., Li, H., Ung, C.Y., Tang, Z.Q., Zheng, C.J., Cao, Z.W., Chen, Y.Z.: Computer prediction of allergen proteins from sequence-derived protein structural and physicochemical properties. Molecular Immunology 44(4), 514–520 (2007)
18. Atchley, W.R., Zhao, J., Fernandes, A.D., Druke, T.: Solving the protein sequence metric problem. PNAS 102(18), 6395–6400 (2005)
19. Davies, M.N., Secker, A., Freitas, A.A., Mendao, M., Timmis, J., Flower, D.R.: On the hierarchical classification of G protein-coupled receptors. Bioinformatics 23(23), 3113–3118 (2007)
20. Cobanoglu, M.C., Saygin, Y., Sezerman, U.: Classification of GPCRs using family specific motifs. IEEE Transactions on Computational Biology 8(6), 1495–1508 (2011)
21. Davies, M.N., Secker, A., Freitas, A.A., Clark, E., Timmis, J., Flower, D.R.: Optimizing amino acid groupings for GPCR classification. Bioinformatics 24(18), 1980–1986 (2008)
22. Krogh, A., Larsson, B., von Heijne, G., Sonnhammer, E.L.L.: Predicting trans- membrane protein topology with a hidden Markov model: application to complete genomes. Journal of Molecular Biology 305(3), 567–580 (2001)
23. Salton, G.: Developments in automatic text retrieval. Science 253(5023), 974–980 (1991)
24. Sol, C.: Identification of disease related significant SNPs. M.Sc. Thesis. Faculty of Engineering and Natural Sciences. Sabanci University (2010)

Author Index